Miniaturized Transistors

Miniaturized Transistors

Special Issue Editors

Lado Filipovic
Tibor Grasser

MDPI • Basel • Beijing • Wuhan • Barcelona • Belgrade

MDPI

Special Issue Editors

Lado Filipovic
Institute for Microelectronics
Austria

Tibor Grasser
Institute for Microelectronics
Austria

Editorial Office
MDPI
St. Alban-Anlage 66
4052 Basel, Switzerland

This is a reprint of articles from the Special Issue published online in the open access journal *Micromachines* (ISSN 2072-666X) from 2018 to 2019 (available at: https://www.mdpi.com/journal/micromachines/special_issues/Miniaturized_Transistors)

For citation purposes, cite each article independently as indicated on the article page online and as indicated below:

LastName, A.A.; LastName, B.B.; LastName, C.C. Article Title. *Journal Name* **Year**, *Article Number*, Page Range.

ISBN 978-3-03921-010-7 (Pbk)
ISBN 978-3-03921-011-4 (PDF)

Contents

About the Special Issue Editors

Lado Filipovic studied electrical engineering at Carleton University, in Ottawa, Canada, where he received the degree of Bachelor in electrical engineering (B.Eng.) in 2006 and Master in applied science (M.A.Sc) in 2009. He joined the Institute for Microelectronics in January 2010, where he completed his doctoral degree in December 2012, which focused on topography simulations of novel semiconductor processes. Since then his scientific interests have broadened to studying and modeling the effects of fabrication-induced variability on semiconductor geometries and their application in the fabrication of quantum metrology devices. His added research interests involve the design and simulation of environmental sensors and, more specifically, using metal oxide semiconductors and two-dimensional semiconductors for the detection of toxic gases and pollutants.

Tibor Grasser is a professor of microelectronics reliability and an IEEE Fellow. He has been the head of the Institute for Microelectronics since 2016. He has edited various books, e.g. on advanced device modeling (World Scientific), the bias temperature instability (Springer) and hot carrier degradation (Springer), is a distinguished lecturer of the IEEE EDS, is a recipient of the Best and Outstanding Paper Awards at IRPS (2008, 2010, 2012, and 2014), IPFA (2013 and 2014), ESREF (2008) and the IEEE EDS Paul Rappaport Award (2011). He currently serves as an Associate Editor for the *IEEE Transactions on Electron Devices* following his assignment as Associate Editor for *Microelectronics Reliability* (Elsevier) and has been involved in various outstanding conferences such as IEDM, IRPS, SISPAD, ESSDERC, and IIRW. Prof. Grasser's current research interests include theoretical modeling of performance aspects of 2D and 3D devices (charge trapping, reliability), starting from the ab initio level over more efficient quantum-mechanical descriptions up to TCAD modeling. The models developed in his group have been made available in the most important commercial TCAD environments.

micromachines

MDPI

Editorial

Editorial for the Special Issue on Miniaturized Transistors

Lado Filipovic * and Tibor Grasser *

Institute for Microelectronics, TU Wien, Gußhausstraße 27-29/E360, 1040 Vienna, Austria
* Correspondence: filipovic@iue.tuwien.ac.at (L.F.); grasser@iue.tuwien.ac.at (T.G.)

Received: 25 April 2019; Accepted: 25 April 2019; Published: 2 May 2019

Complementary Metal Oxide Semiconductor (CMOS) devices and fabrication techniques have enabled tremendous technological advancements in a short period of time. In recent decades transistor scaling has enabled us to fit into our pockets what would be considered a supercomputer a few decades ago. However, as we approach the physical limits of scaling, the question frequently asked is: What is the future of CMOS? Sustaining increased transistor densities along the path of Moore's Law has become increasingly challenging with limited power budgets, interconnect bandwidths, and fabrication capabilities. In the last decade alone, transistors have undergone significant design makeovers; from planar transistors of 10 years ago, technological advancements have accelerated to today's FinFETs, which hardly resemble their bulky ancestors. FinFETs could potentially take us to the 5-nm node, but what comes afterwards? From gate-all-around devices to single electron transistors and two-dimensional semiconductors, a torrent of research is being carried out in order to design the next transistor generation, engineer the optimal materials, improve the fabrication technology, and properly simulate future devices.

There are 13 papers published in this Special Issue, covering recent advances in research aspects related to transistor miniaturization, including process and device simulation as well as novel transistor designs and innovative working principles for future transistor technologies. Two reviews are included in this Special Issue, covering technology computer aided design (TCAD) process and device simulations. To enable high performance TCAD and accelerated simulations, alternative meshing strategies are sought after, another topic addressed herein. High performance TCAD is indispensable for the design of future transistor structures. The sophistication and physical accuracy of the semiconductor models which are used today have reached unprecedented levels, allowing researchers to predict the best candidates for next generation devices without ever stepping foot into a fabrication facility. Using semiconductor TCAD, several authors in this Special Issue have proposed and analyzed different transistor materials and geometries. These include transistors based on ferroelectric materials, those based on 2D semiconductors, semi-floating-gate synaptic transistors (SFGSTs), and drain-engineered InGaN heterostructure tunnel FETs (TFETs). Furthermore, several groups have successfully used TCAD to optimize existing structures including SOI MOSFETs, SiGe tunnel FETs, n-channel MOSFETs, and silicon nanowire (SiNW) transistors, These novel transistor geometries and materials require a complex combination of processing steps, which can lead to significant variation in the geometry and ultimately operation of real-world devices. The variability and variation in advanced three-dimensional devices such as FinFETs and stacked nanowires, as well as future gate-all-around (GAA) structures has been addressed by researchers in this Special Issue. In the following, we summarize the individual contributions in this Special Issue, starting with the two reviews followed by 11 scientific research manuscripts.

Klemenschits et al. [1], in their review, summarize methods used to pattern modern gate stacks, which are no longer a single metal or polysilicon layer, but rather a complex stack of materials which must be carefully patterned to create the gate contact. A review of methods used for topography simulations is given therein, including a discussion on the use of explicit and implicit methods to define surfaces during process simulations. Ultimately, the authors describe the methods which made possible the recent advances in the modeling of gate stack patterning using advanced geometries. Enhanced capabilities of today's simulators and algorithms which accelerate simulation times have been the backbone for allowing modern TCAD to reach such a sophistication to enable the types of research presented in several papers in this Special Issue. In [2], Gnam et al. describe an algorithm they developed in order to accelerate flux calculations when performing process simulations, an essential component of modern process TCAD. With their method they obtained speedups of up to eight times while keeping surface deviations below 3%, ensuring that the simulations retain the high quality and accuracy expected from TCAD models. In the second review in this Special Issue, thin film transistors are addressed. TFTs have recently shown broad potential in applications from RFID tags, logical calculations, and many more. In order to enable circuit simulations with TFTs, fully physical models are not convenient, which is why compact models are indispensable. Lu et al. [3] provide a review of existing compact models for TFTs with different active layers while paying special attention to surface-potential-based compact models of silicon-based TFTs. Ultimately, the review authors propose models which provide accurate circuit-level performance predictions and RFID circuit designs.

Furthermore, Hueting [4], who analyzed current research into ferroelectric transistors. This research field looks at employing ferroelectric materials to obtain positive feedback in the gate control of a switch. The two device architectures analyzed are the NC-FET and the π-FET. The author showed that while the NC-FET shows better performance in terms of subthreshold swing and on current, the π-FET offers a much higher speed of operation. Ultimately a hybrid solution is proposed using a ferroelectric material with a high piezocoefficient. Chang et al. [5] have studied two-dimensional (2D) field effect transistors (FETs) based on indium selenide (InSe), noting that remote phonon and Fröhlich interaction plays a comparatively major role in determining electron transport in InSe. Cho et al. [6] demonstrate a semi-floating-gate synaptic transistor (SFGST) for energy-efficient hardware-driven neuromorphic systems. The authors utilize a poly-Si semi-floating gate and a SiN charge-trap layer which is charged by a tunneling FET which is embedded between the channel and the drain junction. The design is intended to operate as fast as the human brain with low power consumption and high integration density. Duan et al. [7] propose a drain engineered InGaN heterostructure field effect transistor (DE-HTFET) which uses an additional metal on the drain region to modulate the energy band near the drain/channel interface. Their design showed a reduction in the subthreshold swing by 53.3% and a doubling of I_{ON} compared to nonpolar DE-HTFETs.

In order to improve the performance of inversion-channel and buried-channel SOI MOSFETs, Omura [8] looks at their low-frequency noise behavior at sub-100 nm channel widths. The author proposes models which explain why the low-frequency noise in the buried channel MOSFET is primarily influenced by interface traps near the top of the surface of the SOI layer and not the traps near the bottom surface of the SOI. Yang et al. [9] proposed a TFET using SiGe source and drain regions which increase the ESD failure current by 17% compared to conventional Si source/drain TFETs. Simulation studies such as this one are essential in optimizing devices without costly fabrication and laboratory measurements. Wang et al. [10] proposed a novel Z-gate n-channel MOSFET layout to improve its radiation tolerance. The novel layout can be radiation-hardened with a fixed charge density at a shallow trench isolation of 3.5×10^{12} cm^{-2} while offering a small footprint and small gate capacitance when compared to the enclosed gate layout. In [11] Jiang et al. propose a method for phosphorus doping in SiNW using plasma in order to improve the electrical characteristics of the nanowire. The method showed a positive effect on wires with diameters down to 5 nm and improves the I_{ON}/I_{OFF} ratio.

Variability in device performance is another aspect of novel and miniaturized designs which must be addressed if the design is ever to make the leap from theoretical feasibility to industrial relevance. Lorenz et al. [12] examine the statistical and systematic process variations in three-dimensional nanoscale devices such as FinFETs and stacked nanowire transistors. The authors demonstrate the achievements and feasibility of a full simulation of the impact of relevant systematic and stochastic variations on advanced devices and circuits. In [13], Lee et al. used TCAD in order to study the

Micromachines **2019**, *10*, 300

impact of variability on the next generation Si_xGe_{1-x} channel gate-all-around (GAA) nanowire metal MOSFETs by looking at the effects of random discrete dopants, line edge roughness, and metal gate granularity. After generating 7200 transistor samples and performing 10,000 quantum transport simulations, a statistical analysis is performed, revealing metal gate granularity as the dominant variability source which should be considered.

We would like to take this opportunity to thank all the authors for submitting exceptional and highly relevant research papers to this Special Issue. We would also like to sincerely thank all the reviewers who took precious time to carefully examine and help improve the quality of all submitted papers. Peer review is an essential component of good science and they deserve recognition for the success of this Special Issue. It is our sincere hope that the results provided in this Special Issue prove useful to scientists and engineers who find themselves at the forefront of this rapidly evolving field. Now, more than ever, it is essential to look for solutions to find the next disrupting technologies which will allow for transistor miniaturization well beyond Silicon's physical limits and the current state-of-the-art.

Conflicts of Interest: The authors declare no conflict of interest.

References

1. Klemenschits, X.; Selberherr, S.; Filipovic, L. Modeling of Gate Stack Patterning for Advanced Technology Nodes: A Review. *Micromachines* **2018**, *9*, 631. [CrossRef] [PubMed]
2. Gnam, L.; Manstetten, P.; Hössinger, A.; Selberherr, S.; Weinbub, J. Accelerating Flux Calculations Using Sparse Sampling. *Micromachines* **2018**, *9*, 550. [CrossRef] [PubMed]
3. Lu, N.; Jiang, W.; Wu, Q.; Geng, D.; Li, L.; Liu, M. A Review for Compact Model of Thin-Film Transistors (TFTs). *Micromachines* **2018**, *9*, 599. [CrossRef] [PubMed]
4. Hueting, R.J. The Balancing Act in Ferroelectric Transistors: How Hard Can It Be? *Micromachines* **2018**, *9*, 582. [CrossRef] [PubMed]
5. Chang, P.; Liu, X.; Liu, F.; Du, G. Remote Phonon Scattering in Two-Dimensional InSe FETs with High-κ Gate Stack. *Micromachines* **2018**, *9*, 674. [CrossRef] [PubMed]
6. Cho, Y.; Lee, J.; Yu, E.; Han, J.H.; Baek, M.H.; Cho, S.; Park, B.G. Design and Characterization of Semi-Floating-Gate Synaptic Transistor. *Micromachines* **2019**, *10*, 32. [CrossRef] [PubMed]
7. Duan, X.; Zhang, J.; Chen, J.; Zhang, T.; Zhu, J.; Lin, Z.; Hao, Y. High Performance Drain Engineered InGaN Heterostructure Tunnel Field Effect Transistor. *Micromachines* **2019**, *10*, 75. [CrossRef] [PubMed]
8. Omura, Y. Empirical and Theoretical Modeling of Low-Frequency Noise Behavior of Ultrathin Silicon-on-Insulator MOSFETs Aiming at Low-Voltage and Low-Energy Regime. *Micromachines* **2019**, *10*, 5. [CrossRef] [PubMed]
9. Yang, Z.; Yang, Y.; Yu, N.; Liou, J.J. Improving ESD Protection Robustness Using SiGe Source/Drain Regions in Tunnel FET. *Micromachines* **2018**, *9*, 657. [CrossRef] [PubMed]
10. Wang, Y.; Shan, C.; Piao, W.; Li, X.; Yang, J.; Cao, F.; Yu, C. 3D Numerical Simulation of a Z Gate Layout MOSFET for Radiation Tolerance. *Micromachines* **2018**, *9*, 659. [CrossRef] [PubMed]
11. Jiang, Y.; Wang, W.; Wang, Z.; Wang, J.P. Incorporation of Phosphorus Impurities in a Silicon Nanowire Transistor with a Diameter of 5 nm. *Micromachines* **2019**, *10*, 127. [CrossRef] [PubMed]
12. Lorenz, J.; Bär, E.; Barraud, S.; Brown, A.; Evanschitzky, P.; Klüpfel, F.; Wang, L. Process Variability—Technological Challenge and Design Issue for Nanoscale Devices. *Micromachines* **2019**, *10*, 6. [CrossRef] [PubMed]
13. Lee, J.; Badami, O.; Carrillo-Nuñez, H.; Berrada, S.; Medina-Bailon, C.; Dutta, T.; Adamu-Lema, F.; Georgiev, V.P.; Asenov, A. Variability Predictions for the Next Technology Generations of n-type Si_xGe_{1-x} Nanowire MOSFETs. *Micromachines* **2018**, *9*, 643. [CrossRef] [PubMed]

![micromachines logo] *micromachines*

MDPI

Article

Incorporation of Phosphorus Impurities in a Silicon Nanowire Transistor with a Diameter of 5 nm

Yanfeng Jiang [1,2,*], Wenjie Wang [1], Zirui Wang [1] and Jian-Ping Wang [2]

[1] Department of Electrical Engineering, IoT College, Jiangnan University, Wuxi 214122, China; wdz@ncut.edu.cn (W.W.); hightoppub@126.com (Z.W.)

[2] Department of Electrical & Computer Engineering, University of Minnesota, Minneapolis, MN 55414, USA; jiangy@umn.edu

* Correspondence: jiangyf@jiangnan.edu.cn; Tel.: +86-510-8591-0633

Received: 15 November 2018; Accepted: 3 February 2019; Published: 15 February 2019

Abstract: Silicon nanowire (SiNW) is always accompanied by severe impurity segregation and inhomogeneous distribution, which deteriorates the SiNWs electrical characteristics. In this paper, a method for phosphorus doping incorporation in SiNW was proposed using plasma. It showed that this method had a positive effect on the doping concentration of the wires with a diameter ranging from 5 nm to 20 nm. Moreover, an SiNW transistor was assembled based on the nanowire with a 5 nm diameter. The device's I_{ON}/I_{OFF} ratio reached 10^4. The proposed incorporation method could be helpful to improve the effect of the dopants in the silicon nanowire at a nanometer scale.

Keywords: doping incorporation; plasma-aided molecular beam epitaxy (MBE); segregation; silicon nanowire

1. Introduction

Semiconductor nanowire (NW) shows potential for its application as a fundamental building block for nano-electronic and nanophotonic devices. It also offers substantial promise for integrated nanosystems [1–3]. The transistors based on the nanowires are attracting increasing attention due to their potential applications in electronics and biomolecule detection [4]. Until now, many works on the transistors assembled using semiconductor nanowires [3–6] have been reported. Promising device characteristics, such as high hole/electron mobilities, large ON-currents, large I_{ON}/I_{OFF} ratio, and good subthreshold swings, are shown on the fabricated devices [1–3].

Among the proposed various nanowires, silicon nanowire (SiNW) is considered as the most promising candidate as it shows excellent characterizations, and it is compatible with the main-stream technology of the integrated circuit. For a transistor based on silicon nanowire to be used in a modern integrated circuit, its size should be scaled down continuously while keeping its electrical characteristics. In this way, the silicon nanowire transistor could be a strong candidate as the building block for an advanced integrated circuit to follow the Moore's Law in the Post-Moore Era [3,7]. A silicon nanowire transistor with a smaller diameter can have a higher density, smaller capacitance, more transconductance, and controllability of the gate voltage on drain current. Thus, it is meaningful to explore the SiNW transistor with a smaller diameter.

However, there are some unsolved problems encountered during the miniaturization process. One of the key issues is the doping incorporation in the nanowire. Based on the experimental results in Reference [8], the conductivity of the doped SiNW with diameter of less than 20 nm is much lower than the predicted conductivity based on the actual doping concentration. This means part of the impurities in the wire are not incorporated when the diameter of the SiNW is smaller than 20 nm. Some theories were proposed to explain the discrepancy [9–12], where the basic ideas included the influences of the doping profiles [9], surface states [10,11], and diameter variation [12] of the thin wire. For example, it was demonstrated that the surface state of the nanowire was very sensitive to the variation of the diameter along the length direction of the nanowire. Thus, the surface state was influenced by the diameter. Then, the carrier concentration and carrier's lifetime were changed accordingly [11]. Mikael et al. [13] reported the direct observation of the influence of the segregation of the impurities in the silicon nanowire. An abrupt decrease of the free charge carrier concentration was observed on the nanowire with a diameter smaller than 15 nm. Therefore, the deterioration in the effective carrier concentration of the nanowire with a diameter in tens nanometers was a severe problem, which could influence the characteristics of the SiNW transistor at a nanometer scale.

Until now, only a few works on how to incorporate the impurity atoms in silicon nanowires were reported. Some researchers used the annealing method to incorporate the impurities into the nanowires by tuning the interface states [14,15]. In this paper, a plasma-assisted molecular beam epitaxial (MBE) method was proposed for the preparation of the nanowires with a diameter varied from 5 nm to 20 nm. It was demonstrated that the plasma could help to increase the conductivities of the doped nanowires. The SiNW transistor with 5 nm diameter was fabricated based on the prepared nanowire and it showed good characteristic results.

2. Materials and Methods

The growth system was a molecular beam epitaxy (MBE) apparatus (Riber 32), with a base pressure below 5×10^{-11} Torr, including electron-beam guns for the evaporation of Au and Si, as well as a substrate heater. A radio frequency plasma source was equipped to assist the growth of the nanowires. The growth substrate was boron-doped Si (111) wafer. It was cleaned in a hot acetone bath for 10 min and subsequently etched in a NH_4OH: H_2O_2:H_2O (1:1:5) solution for 10 min at 70 °C, and then dipped in a HF:H_2O (1:5) solution for 10 s. After the cleaning process, the wafer was loaded onto a sputtering system. Silicon wafer was ion bombarded in the system. The substrate was irradiated with 120 keV Ar ions. Energy of 120 keV was chosen because the projected range of the ions with such energy created a 2-nm roughness on the surface. The generated roughness was helpful for the formation of the Au droplets. The sizes of the Au droplets would determine the diameters of the grown silicon nanowires afterwards. The periodic rough surface can lead to a good templating of Au droplets.

After the sputtering, the substrate was loaded onto the MBE system. The silicon substrate was annealed in an ultra-high vacuum at 925 °C for 15 min, to obtain an oxygen-free surface. The 7×7 reconstruction was obtained after annealing of the substrate, as checked by low-energy electron diffraction (LEED). The temperature of the substrate was controlled by a thermocouple, as well as a pyrometer.

Au film was deposited at a substrate at room temperature, with a thicknesses between 1 nm and 1.2 nm as measured by a quartz monitor. After that, the temperature was increased to 525 °C. The technique utilizing spontaneous dewetting can engineer patterns in soft materials on the nanometer scale, without the conventional lithographic process [14]. The sizes of the formed Au droplets were not uniform because of the existence of the roughness of the silicon substrate. Given the diameters of the grown nanowires depended on the sizes of the Si–Au alloy droplets, the silicon nanowires with different diameters were grown simultaneously on the same substrate.

During the SiNW growth, constant Si flux amounted to 4.7 Å/s. A solid phosphorus source with 99.9995% purity was used, where the concentration was related to the source's temperature. Based on the calibrated method by Pratyush, et al. [15], the source temperatures were monitored between 900–1200 °C to obtain concentrations varying from 1×10^{16} cm^{-3} to 1×10^{19} cm^{-3}. The evaporation time lasted for 1.8 h.

For the convenience of comparison, two groups of samples with diameters from 5 nm to 20 nm were prepared. Group A was grown with plasma-assistance while group B was without plasma. Except the plasma, the other parameters during the growth of the two sets of the SiNWs were all the same. The plasma source was 250 W radio frequency (RF) power, with a frequency of 13.56 MHz.

In this way, the two groups of samples were prepared using the same conditions, except that group A was with plasma assistance while the group B was without plasma. After the substrates were loaded out, the gold caps on the top of the grown SiNWs were removed using an aqueous solution of KI and I_2.

3. Results

The scanning electron microscopy (SEM) image and high resolution transmission electron microscopy (HRTEM) are shown in Figure 1. A bunch of the grown SiNWs can be seen clearly, with diameters from 5 nm to 20 nm, as shown in Figure 1a. The TEM image in Figure 1b shows the grown nanowire with a single-crystalline structure and vertically along the <111> crystalline direction.

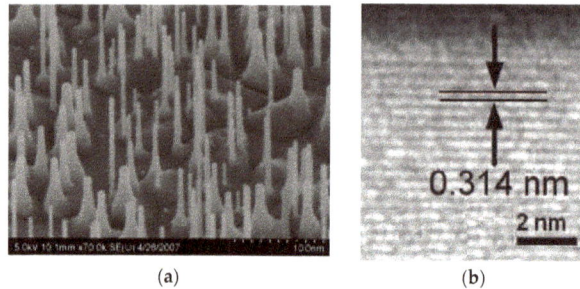

(a)   (b)

Figure 1. (**a**) Scanning electron microscopy (SEM) photograph of the grown silicon nanowires (SiNWs). Diameters are varied from 5 nm to 20 nm. The scale bar is 100 nm. (**b**) Transmission electron microscopy (TEM) photograph of the silicon nanowire, showing its crystal structure in the <111> direction. The scale bar is 2 nm.

The prepared SiNWs were put into ethanol for ultrasonic dispersion. Afterwards, the wires were transferred onto the SiO_2/Si substrate. Multiple electrical contacts as shown in Figure 2 were used for characterization of the prepared wires.

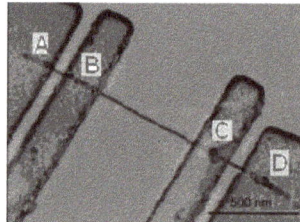

Figure 2. Image of the electrical contacts of the prepared SiNW. Scale bar is 500 nm.

Characterizations were conducted to obtain the resistivities with diameters from 5 nm to 20 nm for both groups (A and B), corresponding to the different concentrations, as shown in Figure 3. It could be clearly observed that the resistivities of the SiNWs in group A, with plasma assistance during growth, were much lower than those in the group B without plasma assistance. For group B, the resistivities remained at almost the same level from 20 nm to 13 nm with different doping concentrations. However, for the wires smaller than 12 nm, the resistivities increased dramatically, corresponding to the decrease of the number of the incorporated carriers. When the diameters were smaller than 8 nm, the resistivities of group B were unreadable, as is not shown in Figure 3. The results coincide with other reported results [13,16–18]. Theoretical explanations on this phenomenon could be attributed to quantum confinement [16], surface segregation of dopants [17], or ionization energy due to a dielectric mismatch at the wire surface [18]. With the diameter decreasing, there will be an increased possibility of the occurrence of surface depletion of charge carriers due to interface states, trapped charges, changeable mobilities, and a size-dependent incorporation of dopants during growth. Therefore, the amount of the incorporated dopants in the SiNW could be decreased [13].

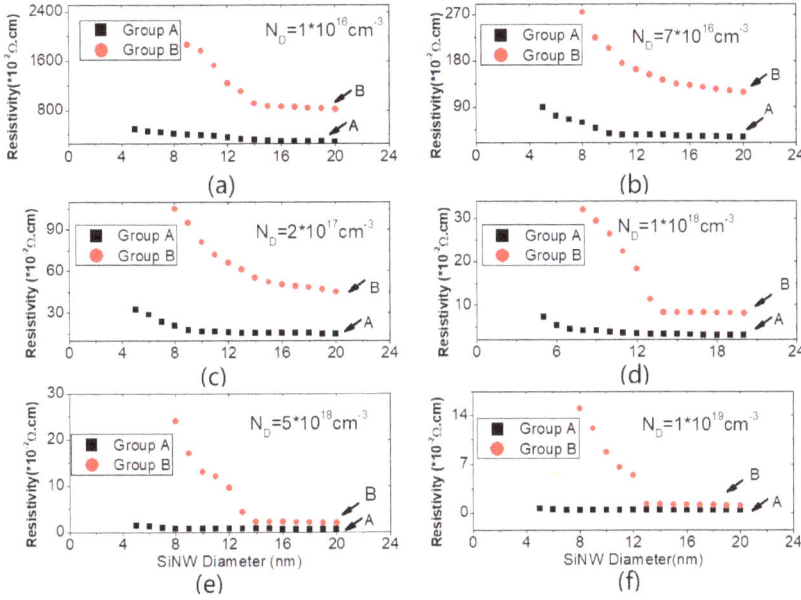

Figure 3. The measured resistivities of the SiNWs with different diameters for groups A and B. (a) $N_D = 1 \times 10^{16}$ cm^{-3}; (b) $N_D = 7 \times 10^{16}$ cm^{-3}; (c) $N_D = 2 \times 10^{17}$ cm^{-3}; (d) $N_D = 1 \times 10^{18}$ cm^{-3}; (e) $N_D = 5 \times 10^{18}$ cm^{-3}; (f) $N_D = 1 \times 10^{19}$ cm^{-3}.

Compared with the results of group B, group A showed much better results, as shown in Figure 3. For all the six different concentrations ranging from 1×10^{16} cm^{-3} to 1×10^{19} cm^{-3}, the resistivities always remained the same with the diameters changing from 20 nm to 10 nm. Slight increments occured on the samples with diameters from 10 nm to 5 nm. This demonstrated that the segregation effect was alleviated by using the plasma during growth. It was directly observed that the plasma assistance could be helpful in incorporating the impurities in the SiNWs with diameters down to 5 nm.

The fabricated 5-nm SiNW was used to assemble the transistor. The source and drain electrodes were defined by the lithography process. The grown SiNWs were transferred onto the electrodes using Cui's approach [1]. The metal contacts were evaporated on the electrodes. The naturally grown oxide layer on the SiNWs can be used as the gate oxide.

Figure 4 shows the SEM photograph of the transistor, where the drain, source, and gate are annotated. N.Singh et al. [7] showed the transistor with a diameter $\leq$5 nm using the top-down method. In this study, the 5-nm SiNW transistor was fabricated using the "bottom-up" approach. The following characterizations on the SiNW transistor were based on the prepared 5-nm-wire with 2×10^{17} cm^{-3} doping concentration.

Figure 5 shows the electrical properties of the I_{DS}-V_{DS} with different gate voltages ranging from -0.5 V to 0.5 V. It shows a linear response characteristic of the I_{DS}-V_{DS}, indicating ohmic terminal contacts. When $V_G = 0$ V, the I_{DS}-V_{DS} curve is linear with a resistance of 2 MΩ. When $V_G > 0$ V, the I_{DS}-V_{DS} curves still remain linear, whereas the resistance decreases with the gate voltage rises. When $V_G = 0.5$ V, the resistance is 0.42 MΩ. Nonlinear relationships of the I_{DS}-V_{DS} exhibit at $V_G < 0$. Thus, it can be seen that the controllability of V_G on the transistor is sensitive.

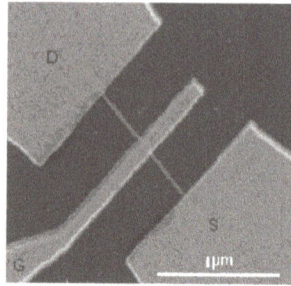

Figure 4. SEM image of the 5 nm SiNW transistor. Scale bar is 1 μm.

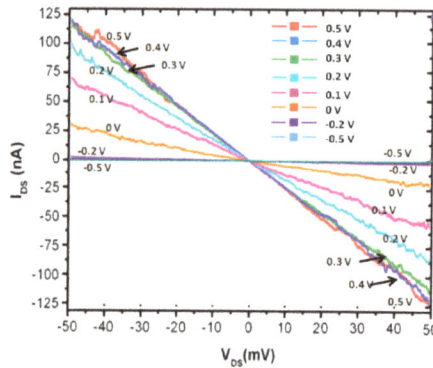

Figure 5. I_{DS}-V_{DS} relationships of the 5 nm SiNW transistor with different V_G voltages, ranging from -0.5 V to 0.5 V.

Figure 6 shows the relationship between the drain current versus the gate voltage (I_{SD}-V_G), with V_{SD} ranging from 5 mV to 50 mV. The inset shows the I_{SD}-V_G characteristic in an algorithm scale with V_{SD} = 50 mV. It can be clearly seen that the I_{ON}/I_{OFF} ratio reaches 10^4.

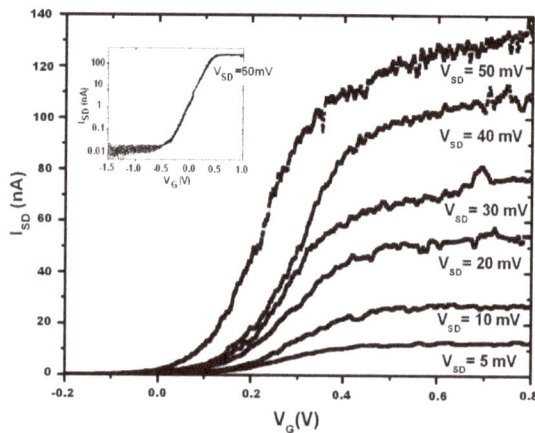

Figure 6. I_{SD}-V_G relationship of the 5-nm SiNW transistor, with V_{SD} sweeping from 5 mV to 50 mV. Inset shows the algorithm relationship of I_{SD}-V_G when V_{SD} = 50 mV, where the I_{ON}/I_{OFF} ratio is 10^4.

4. Discussion

The growing process is based on the ion bombardment on the silicon substrate. Roughness is generated on the surface. Thus, the tension remains on the deposited Au layer. A post-annealing step can help to generate nanosized Au-Si droplets. The following growth of the SiNWs is directly influenced by the droplets. The diameters of the grown SiNWs are determined by the dimensions of the droplets. As such precise control on the size of the droplets is critical for the grown SiNWs. Figure 1a shows that the diameters of the grown NWs are randomly varied from 5 nm to 20 nm. This is helpful for the experiment in this paper, since a several samples with different diameters can be prepared on the same substrate at the same time, with the same conditions. For future application in integrated circuits, it is important to uniformly prepare the SiNWs. To satisfy the mass production requirement, an advanced ultrafine lithographic process can be adopted instead of the ion bombardment approach.

For the fabricated SiNW transistor, its gate oxide layer was grown naturally after the SiNW was extracted from the MBE system. During the subsequent annealing and electrode deposition steps, the oxide layer was gradually accumulated on the surface. To verify the quality of the oxide layer, the leakage current was measured on the prepared 5-nm SiNW transistor. The result is shown in Figure 7, showing the characteristics of the leakage current versus the gate voltage. Scanning was performed from −0.5 V to 0.5 V. It can be seen that the leakage current was less than 0.1 nA, demonstrating the quality of the oxide layer.

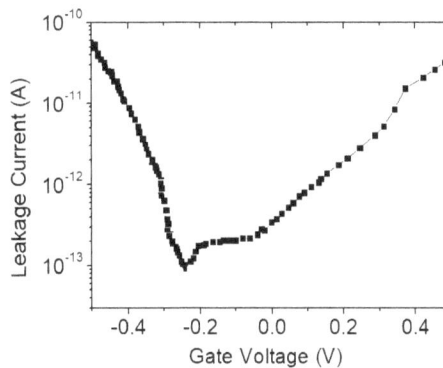

Figure 7. Leakage current versus gate voltage for the prepared SiNW transistor. Scanning was performed from −0.5 V to 0.5 V.

To investigate the states of the trapped charges in the fabricated 5-nm SiNW transistor, the hysteresis behavior was characterized on the device. Hysteresis is the shift in the threshold voltage during the forward (+V to −V) and the reverse gate voltage sweep (−V to +V), at a constant drain voltage bias. The trapping is typically manifested as hysteresis behavior during transistor IV scans. Figure 8 shows the measured hysteresis loop of the transistor at the drain voltage V_{SD} = 40 mV. The threshold voltages in the down-sweep stage were higher than the voltages in the up-sweep stage, suggesting the negative polarity of the trapped charges in the device. The trapping state scatters the charge carriers, and hence it decreases the carrier mobility. Further research work on the trapped charges is necessary to investigate the source and the locations in the device.

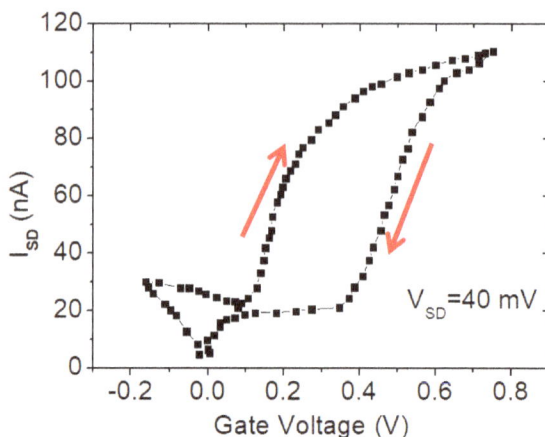

Figure 8. Transfer characteristic of the 5-nm SiNW transistor, showing a hysteresis value of ~0.2 V at a 40 mV drain voltage.

The current of the assembled 5-nm SiNW transistor are at a nA scale, as shown in Figures 5 and 6. There are obvious noises on the IV curves in Figure 6. The origin of the noises is currently unclear. There were two possible sources of the noises, which included the experimental set-up and the fabricated device. Further investigation should be conducted to clarify the origin and to improve the quality of the transistor.

To investigate the doping densities of the grown nanowires, the capacitance-voltage technique [12] was used to measure the doping profiles of the samples, with and without plasma assistance. The 20-nm SiNWs were used for the characterization. The results are shown in Figure 9, using the six different doping densities. Four results were summarized as corresponding to one single doping level, which included the chemical doping densities with plasma assistance (Line 1), the chemical doping densities without plasma assistance (Line 2), the electrical doping densities with plasma assistance (Line 3), and the electrical doping densities without plasma assistance (Line 4). It could be seen that there were obvious discrepancies among the four lines for one single doping level. As shown in Figure 9a, the doping level was $N_D = 1 \times 10^{16}$ cm^{-3}. For the nanowire prepared using plasma assistance (Line 1), its chemical doping concentration on the surface was 1.2×10^{16} cm^{-3}. It drops at the center, to around 5×10^{15} cm^{-3}. For the nanowire prepared without plasma assistance (Line 2), its chemical doping concentration on the surface was 1×10^{16} cm^{-3}. Its center density was 5×10^{14} cm^{-3}. Comparing the data in Line 1 and Line 2, it could be seen that the plasma assistance was helpful for the doping distribution, assisting the distribution more homogeneously. For the five other doping concentrations, the same tendency could be found in which the chemical doping profiles assisted by plasma showed more homogeneity than those without plasma assistance.

The actual electrical doping densities for the sample with $N_D = 1 \times 10^{16}$ cm^{-3} are shown in Figure 9a. Line 3 corresponds to the electrical doping profile with plasma. It showed the same tendency as that of the chemical doping profile (Line 1). The surface concentration was higher than that at the center. At the same position, the electrical density was smaller than the chemical density, demonstrating that partial dopants were not involved in the electrical transportation. However, it showed that the electrical doping concentrations with the plasma (Line 3) were obviously higher than concentrations without plasma (Line 4). This means that the plasma was helpful in improving the electrical doping concentrations in the grown SiNWs. The same conclusion could be made for the other samples with different chemical concentrations, as shown in Figure 9b–f.

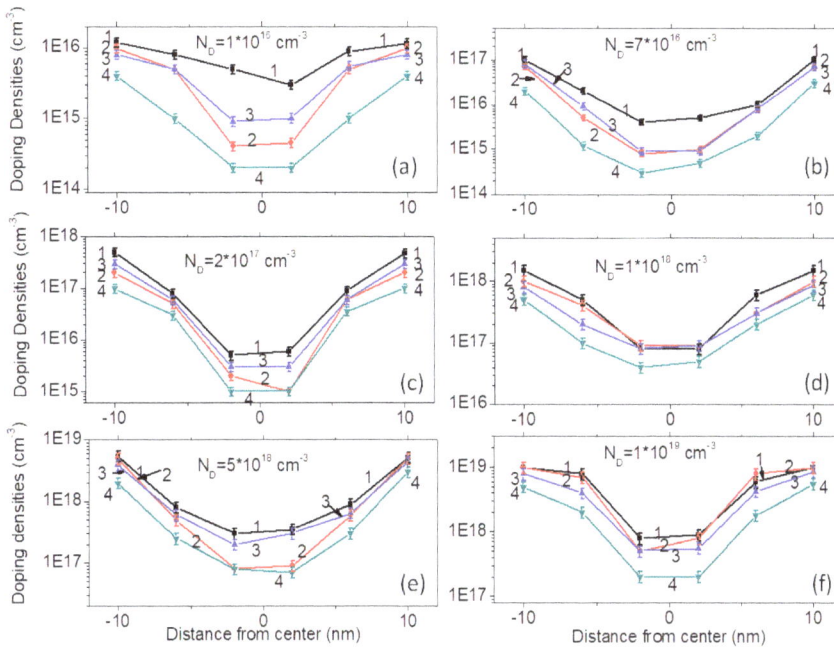

Figure 9. Doping densities of the grown SiNWs with diameter 20 nm. Line "1" is the chemical doping densities with plasma assistance. Line "2" is the chemical doping densities without plasma assistance. Line "3" is the electrical doping densities with plasma assistance. Line "4" is the electrical doping densities without plasma assistance. (a) $N_D = 1 \times 10^{16}$ cm^{-3}; (b) $N_D = 7 \times 10^{16}$ cm^{-3}; (c) $N_D = 2 \times 10^{17}$ cm^{-3}; (d) $N_D = 1 \times 10^{18}$ cm^{-3}; (e) $N_D = 5 \times 10^{18}$ cm^{-3}; (f) $N_D = 1 \times 10^{19}$ cm^{-3}.

For the silicon nanowire, p- or n-type nanowires can be manufactured when suitable impurities are doped in it. However, the question of how the electrical conductivity depends on the doping level remains largely open. A prominent part of the doping atoms has been blocked because of the existence of surface states, trapped charges, or defects, et al. [9,11]. The segregation of the doping atoms appears more obvious when the diameter gets smaller. This means that the modulation of the doping concentration on the nanowire's conductivity is not as sensitive as anticipated. For the potential application of SiNWs with a smaller diameter, the efficiency of the doping concentration on the modulation of conductivity should be increased.

Most of the related works have used high impurity concentrations, even near or above the Mott density corresponding to the metal–nonmetal transition in the bulk semiconductor [18]. However, a high concentration could induce surface segregation. This may be one of the main reasons for the low conductivity. In this paper, six different concentrations ranging from 1×10^{16} cm^{-3} to 1×10^{19} cm^{-3} were considered, in which the wide range of concentrations can help us to not only observe the incorporation effect more clearly, but the surface segregation as well. The merits of this approach have been verified because the conductivities of the nanowires can be measured at a medium level with a smaller diameter.

The mechanism of the plasma assisted incorporation in this paper is still under investigation. The measured doping profiles in Figure 9 show that the surface concentrations of the samples with plasma assistance were higher than those without plasma assistance, indicating that the side wall conductance was enhanced by the plasma. I. Amit et al. [19,20] and E. Koren et al. [21] found that the side wall conductance significantly contributed to the conductivity of the nanowires. The results in this paper, as shown in Figure 9, reveal one important contribution of the plasma on the doping profiles. That is, the surface profiles with plasma are improved in terms of the doping concentrations. The side wall conductance of the SiNWs with plasma assistance, either the chemical concentrations or the electrical concentrations, are both improved.

From the above results, two interesting conclusions can be reached. Firstly, a large part of the doping impurities are incorporated into the wires using the plasma assisting method. A relative number of the segregation atoms still exist because the calculated electron mobility was smaller than that in the bulk material. However, the comparison between groups A and B showed obvious improvements. Secondly, all the nanowires in group A had measurable conductivities, even down to a 5 nm diameter. Therefore, the plasma assisting approach is effective in solving the impurity segregation issue in small nanowires down to 5 nm [14]. In this paper, the transistor assembled with 5 nm SiNW using a bottom-up approach was demonstrated as showing good transistor characteristics. Another 5 nm transistor based on the SiNW was reported by N. Singh et al. [7].

5. Conclusions

The method of growing SiNWs using plasma was reported in this paper. Different concentrations ranging from 1×10^{16} cm^{-3} to 1×10^{19} cm^{-3} of the SiNWs with diameters from 5 nm to 20 nm were fabricated and characterized. Compared with the SiNWs without plasma assistance, obvious improvements in the conductivities could be observed in the SiNWs with plasma assistance. By preparing the SiNW transistor based on the 5 nm nanowire, it showed good transistor characterizations. The device's I_{ON}/I_{OFF} ratio can be as high as 10^4. This work is helpful for future work on SiNW transistors with diameters in several nanometers.

Author Contributions: Y.J. and J.-P.W. conceived and designed the experiments; W.W. performed the experiments; W.W. and Z.W. analyzed the data; Y.J. wrote the paper.

Funding: This work was funded by the National Science Foundation of China (NSFC) Contract No.61774078.

Conflicts of Interest: The authors declare no conflict of interest.

References

1. Cui, Y.; Lieber, C.M. Functional nanoscale electronic devices assembled using silicon nanowire building blocks. *Science* **2001**, *291*, 851–853. [CrossRef] [PubMed]
2. Tian, B.; Zheng, X.; Kempa, T.J.; Fang, Y.; Yu, N.; Yu, G.; Huang, J.; Lieber, C.M. Coaxial silicon nanowires as solar cells and nanoelectronic power sources. *Nature* **2007**, *449*, 885–889. [CrossRef] [PubMed]
3. Cui, Y.; Zhong, Z.; Wang, D.; Wang, W.U.; Lieber, C.M. High performance silicon nanowire field effect transistors. *Nano Lett.* **2003**, *3*, 149–152. [CrossRef]
4. Patolsky, F.; Zheng, G.; Lieber, C.M. Fabrication of silicon nanowire devices for ultrasensitive, label-free, real-time detection of biological and chemical species. *Nat. Protoc.* **2006**, *1*, 1711–1724. [CrossRef]
5. Hannon, J.; Kodambaka, S.; Ross, F.; Tromp, R. The influence of the surface migration of gold on the growth of silicon nanowires. *Nature* **2006**, *440*, 69–71. [CrossRef]
6. Chung, S.-W.; Yu, J.-Y.; Heath, J.R. Silicon nanowire devices. *Appl. Phys. Lett.* **2000**, *76*, 2068–2070. [CrossRef]
7. Singh, N.; Agarwal, A.; Bera, L.; Liow, T.; Yang, R.; Rustagi, S.; Tung, C.H.; Kumar, R.; Lo, G.Q.; Balasubramanian, N.; et al. High-performance fully depleted silicon nanowire (diameter/spl les/5 nm) gate-all-around CMOS devices. *IEEE Electron. Device Lett.* **2006**, *27*, 383–386. [CrossRef]
8. Ma, D.; Lee, C.; Au, F.; Tong, S.; Lee, S. Small-diameter silicon nanowire surfaces. *Science* **2003**, *299*, 1874–1877. [CrossRef]
9. Allen, J.E.; Perea, D.E.; Hemesath, E.R.; Lauhon, L.J. Nonuniform nanowire doping profiles revealed by quantitative scanning photocurrent microscopy. *Adv. Mater.* **2009**, *21*, 3067–3072. [CrossRef]
10. Wang, D.; Chang, Y.-L.; Wang, Q.; Cao, J.; Farmer, D.B.; Gordon, R.G.; Dai, H. Surface chemistry and electrical properties of germanium nanowires. *J. Am. Chem. Soc.* **2004**, *126*, 11602–11611. [CrossRef]
11. Zhang, S.; Hemesath, E.R.; Perea, D.E.; Wijaya, E.; Lensch-Falk, J.L.; Lauhon, L.J. Relative influence of surface states and bulk impurities on the electrical properties of Ge nanowires. *Nano Lett.* **2009**, *9*, 3268–3274. [CrossRef] [PubMed]

12. Perea, D.E.; Hemesath, E.R.; Schwalbach, E.J.; Lensch-Falk, J.L.; Voorhees, P.W.; Lauhon, L.J. Direct measurement of dopant distribution in an individual vapour–liquid–solid nanowire. *Nat. Nanotechnol.* **2009**, *4*, 315–319. [CrossRef] [PubMed]
13. Björk, M.T.; Schmid, H.; Knoch, J.; Riel, H.; Riess, W. Donor deactivation in silicon nanostructures. *Nat. Nanotechnol.* **2009**, *4*, 103–107. [CrossRef] [PubMed]
14. Rosenberg, R.; Shenoy, G.; Heigl, F.; Lee, S.-T.; Kim, P.-S.; Zhou, X.-T.; Sham, T.K. Effects of in situ vacuum annealing on the surface and luminescent properties of ZnS nanowires. *Appl. Phys. Lett.* **2005**, *86*, 263115. [CrossRef]
15. Das Kanungo, P.; Zakharov, N.; Bauer, J.; Breitenstein, O.; Werner, P.; Goesele, U. Controlled in situ boron doping of short silicon nanowires grown by molecular beam epitaxy. *Appl. Phys. Lett.* **2008**, *92*, 263107. [CrossRef]
16. Khanal, D.; Yim, J.W.; Walukiewicz, W.; Wu, J. Effects of quantum confinement on the doping limit of semiconductor nanowires. *Nano Lett.* **2007**, *7*, 1186–1190. [CrossRef] [PubMed]
17. Fernández-Serra, M.; Adessi, C.; Blase, X. Surface segregation and backscattering in doped silicon nanowires. *Phys. Rev. Lett.* **2006**, *96*, 166805. [CrossRef]
18. Diarra, M.; Niquet, Y.-M.; Delerue, C.; Allan, G. Ionization energy of donor and acceptor impurities in semiconductor nanowires: Importance of dielectric confinement. *Phys. Rev. B* **2007**, *75*, 045301. [CrossRef]
19. Amit, I.; Givan, U.; Connel, J.G.; Paul, D.F.; Hammond, J.S.; Lauhon, L.J.; Rosenwaks, Y. Spatially resolved correlation of active and total doping concentrations in VLS grown nanowires. *Nano Lett.* **2013**, *13*, 2598–2604. [CrossRef]
20. Amit, I.; Jeon, N.; Lauhon, L.J.; Rosenwaks, Y. Impact of dopant compensation of graded p-n junctions in Si Nanowires. *ACS Appl. Mater. Interfaces* **2016**, *8*, 128–134. [CrossRef]
21. Koren, E.; Berkovitch, N.; Rosenwaks, Y. Measurement of active dopant distribution and diffusion in individual silicon nanowires. *Nano Lett.* **2010**, *10*, 1163–1167. [CrossRef] [PubMed]

micromachines

MDPI

Article

High Performance Drain Engineered InGaN Heterostructure Tunnel Field Effect Transistor

Xiaoling Duan *, Jincheng Zhang *, Jiabo Chen, Tao Zhang, Jiaduo Zhu, Zhiyu Lin and Yue Hao

Wide Bandgap Semiconductor Technology Disciplines State Key Laboratory, School of Microelectronics, Xidia University, Xi'an 710071, China; jbchen@stu.xidian.edu.cn (J.C.); zhangtao9204@sina.com (T.Z.); jdzhu@xidian.edu.cn (J.Z.); zylin@xidian.edu.cn (Z.L.); yhao@xidian.edu.cn (Y.H.)
*   Correspondence: duanxiaoling@xidian.edu.cn (X.D.); jchzhang@xidian.edu.cn (J.Z.); Tel.: +86-29-8820-1759 (X.D.); +86-29-8820-1446 (J.Z.)

Received: 27 December 2018; Accepted: 18 January 2019; Published: 21 January 2019

Abstract: A drain engineered InGaN heterostructure tunnel field effect transistor (TFET) is proposed and investigated by Silvaco Atlas simulation. This structure uses an additional metal on the drain region to modulate the energy band near the drain/channel interface in the drain regions, and increase the tunneling barrier for the flow of holes from the conduction band of the drain to the valence band of the channel region under negative gate bias for n-TFET, which induces the ambipolar current being reduced from 1.93×10^{-8} to 1.46×10^{-11} A/µm. In addition, polar InGaN heterostructure TFET having a polarization effect can adjust the energy band structure and achieve steep interband tunneling. The average subthreshold swing of the polar drain engineered heterostructure TFET (DE-HTFET) is reduced by 53.3% compared to that of the nonpolar DE-HTFET. Furthermore, I_{ON} increases 100% from 137 mA/mm of nonpolar DE-HTFET to 274 mA/mm of polar DE-HTFET.

Keywords: drain engineered; tunnel field effect transistor (TFET); polarization; ambipolar; subthreshold; ON-state

1. Introduction

Tunnel field effect transistors (TFETs) have been considered as attractive alternative replacements to metal-oxide-semiconductor field effect transistors (MOSFETs) for low power applications [1–3], due to the conductive mechanism of band to band tunneling, realizing a steep subthreshold swing (SS) (less than 60 mV/dec at room temperature), good immunity against Short Channel Effects (SCEs) and high ON-state current (I_{ON}) to OFF-state current (I_{OFF}) ratio (I_{ON}/I_{OFF}) [1,4,5]. Although TFETs have various benefits, there are still some problems to be solved, such as the ambipolar behavior [6,7], low ON-state current (I_{ON}) [8], and less-than-idea SS. To solve these problems, different techniques such as the use of high-k dielectric materials [9], heterojunction engineering [10–13], source pocket based devices [14,15], junction-less concept based devices [16–18], and narrow bandgap materials have been investigated to boost I_{ON}. Drain doping profile investigation [7], gate-drain electrode gap control [19], the hetero-dielectric box concept [20], and heterojunction engineering have been developed to restrain ambipolar behavior.

III Nitride is the direct bandgap semiconductor, and its bandgap can be modulated from 0.7 eV (InN) to 6.2 eV (AlN), while the natural polarization effect will facilitate the formation of a steep energy band at the tunneling junction of the heterostructure, inducing a steep tunneling junction, small subthreshold swing and large ON-state current [21]. Recently, it has been reported that III Nitride TFETs exhibit superior device characteristics, showing the great potential of the application in the low power field [21–24]. However, the research on III Nitride TFET devices is just beginning, and it is worth further study.

In this paper, the drain engineering method uses workfunction engineering on an additional metal to modulate the energy band on the drain/channel interface in the drain regions and increase the tunneling barrier for the flow of holes from the conduction band of the drain to the valence band of the channel region under negative gate bias for n-TFET, which will reduce the ambipolar current of InGaN heterojunction TFETs (HTFETs). In addition, the improvement mechanism of the polarization effect on the subthreshold and ON-state characteristics of polar InGaN HTFETs are investigated at length.

The remaining paper is organized as follows: Section 2 presents the device structure, simulation models and material parameters. Section 3 is dedicated to results and discussions, including the drain engineered TFET used to suppress the ambipolar current of InGaN HTFET, and the performance comparison and analysis of polar and nonpolar HTFETs. Finally, Section 4 concludes the paper with some important findings.

2. Device Structure and Simulation Parameters

Figure 1 shows the cross-sectional views of InGaN nonpolar heterostructure TFET (HTFET), InGaN nonpolar drain engineered heterostructure TFET (DE-HTFET), and InGaN polar DE-HTFET, respectively. The InGaN HTFET contains a source-side channel to improve the ON-state current and reduce the subthreshold swing. This work has been reported in reference [24]. The channel length (L_{ch}) and the channel thickness (t) are set as 50 nm, and 10 nm, respectively. The drain metal and the gate (L_{GD}) are 100 nm. The gate dielectric material is Al_2O_3 with a thickness (t_{ox}) of 3 nm. The doping concentrations in the source, the channel, and the drain region are p+ 9.9×10^{19} cm^{-3}, n 1×10^{15} cm^{-3}, and n+ 1.2×10^{19} cm^{-3}, respectively. The workfunction of gate metal is 5.1 eV. The proposed nonpolar DE-HTFET contains an additional drain engineering metal with the length of 10 nm and a dielectric thickness of 1.2 nm to modulate the energy band in the drain/channel region and finally reduce the ambipolar current. Along with this, the polar DE-HTFET is proposed to modulate the energy band at the source/channel tunneling junction, and will further improve the device characteristics.

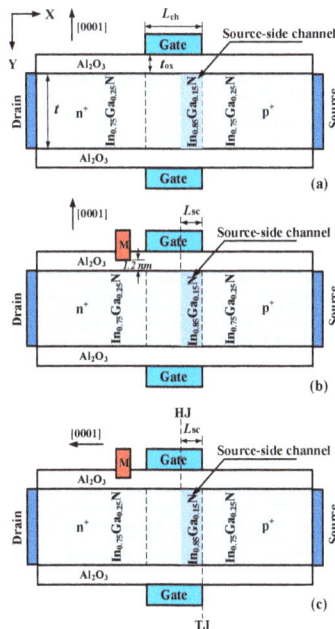

Figure 1. Cross-sectional views of the nonpolar (**a**) conventional, (**b**) drain engineered InGaN tunnel field effect transistor (TFET), and (**c**) polar drain engineered InGaN TFET.

Simulations are carried out using a 2D Silvaco Atlas simulator (5.19.20.R, Silvaco Inc., Santa Clara, CA, USA) [25]. The nonlocal band to-band tunneling (BTBT) is incorporated in the simulation for the calculation of the tunneling rate of charge carriers. The Shockley–Read–Hall (SRH) carrier recombination model, bandgap narrowing (BGN) model, constant low field mobility model, and field-dependent mobility model at high electric fields are activated using the SRH, BGN and FLDMOB parameters, respectively. The polarization effect in the polar TFET is simulated by a fixed polarization charge at the heterojunction interface [26], and the density of polarization surface charge (σ_{pol}) will be discussed in Section 3.2. Other material parameters used in the simulations are presented in Table 1.

In the electrical characteristics analysis below, I_{ON} is defined to be the drain current I_D at $V_G = V_D$ = 1 V. The average sub-threshold swing (SS_{avg}) is obtained from the I_D-V_G curve, it is given by

$$SS_{avg} = (V_{TH} - V_{OFF}) / (logI_{V_{TH}} - logI_{OFF}) \qquad (1)$$

where the threshold voltage V_{TH} is defined to be the gate voltage (V_G) at current of 1×10^{-7} A/mm, and V_{OFF} is the gate voltage at I_D of 10^{-18} A/mm.

Table 1. Material parameters used in simulations [21].

Parameters	$In_{0.75}Ga_{0.25}N$	$In_{0.85}Ga_{0.15}N$
Band gap E_g (eV)	1.1125	0.9265
Hole effective mass (m_0)	0.295	0.273
Electron effective mass (m_0)	0.1025	0.0895
Static dielectric constant ε_r	12.75	13.05
Electron mobility μ_e (cm^2/Vs)	1050	1050
Hole mobility μ_h (cm^2/Vs)	20	20

3. Results and Discussions

3.1. Drain Engineered HTFET to Suppress Ambipolar Current

3.1.1. Impact of Drain Engineering Metal Position on Electrical Characteristic of DE-HTFET

Drain engineered HTFET using an additional drain engineering metal (M) to modulate the energy band can increase the tunneling barrier for the flow of holes on the drain/channel interface, and reduce the ambipolar current at the negative V_G bias. The M position will obviously affect the electrical characteristic of DE-HTFET. The following simulations are discussed with the different M position at the M workfunction (φ_M) of 5.15 eV.

Figure 2 displays the I-V curves of HTFET without M and DE-HTFET with the different gate to M space (L_{GM}). HTFET without drain engineering metal shows an ambipolar current of 1.93×10^{-8} A/μm. Overall, $I_{ambipolar}$ of DE-HTFET by drain metal engineering has been obviously reduced compared with that of HTFET without M. As L_{GM} increases from 5 nm to 20 nm, $I_{ambipolar}$ reduces firstly and then increases. At L_{GM} of 15 nm, DE-HTFET obtains the smallest $I_{ambipolar}$ (1.46×10^{-11} A/μm), while the ON-state current (I_{ON}) is almost not degraded.

To gain a better insight into the variation of $I_{ambipolar}$ and I_{ON}, energy band profiles for V_G of -1 V and 1 V are shown in Figure 3a,b, respectively. Figure 3a shows the energy band of DE-HTFET near the drain/channel junction at V_G of -1 V. With drain metal engineering, the tunneling distance of DE-HTFET at the negative gate bias is larger than that of HTFET. The tunneling distance of DE-HTFET firstly increases and then decreases with the increase of L_{GM}. DE-HTFET with $L_{GM} = 15$ nm has the largest tunneling distance in the L_{GM} range from 5 nm to 20 nm. The larger tunneling distance will contribute to a lower electric field (E) and thereby a smaller tunneling rate, as illustrated with Kane's formula [27]:

$$P_{tun} \sim \frac{E^2 m_r^{1/2}}{E_g^{1/2}} \exp\left(-\frac{C_2 m_r^{1/2} E_g^{3/2}}{E}\right), \qquad (2)$$

where E is the electric field, C_2 is a constant and m_r is effective mass. This can well explain the variation tendency of $I_{ambipolar}$. Thus, DE-HTFET with L_{GM} of 15 nm has the smallest $I_{ambipolar}$. Figure 3b shows the energy band of DE-HTFET at the ON-state ($V_G = V_D = 1$ V). It is obvious that L_{GM} variation has almost no influence over the tunneling distance at the source/channel junction, therefore I_{ON} is almost not degraded. Considering the aforementioned factors, L_{GM} of 15 nm is selected in the following discussions.

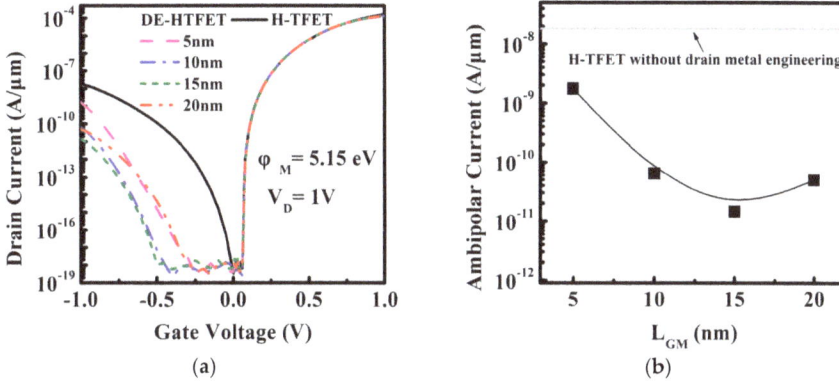

Figure 2. (a) Transfer characteristic of HTFET and DE-HTFET with varied L_{GM} at V_D of 1 V and φ_M of 5.15 eV, and (b) ambipolar current with different L_{GM} extracted from the I-V curve.

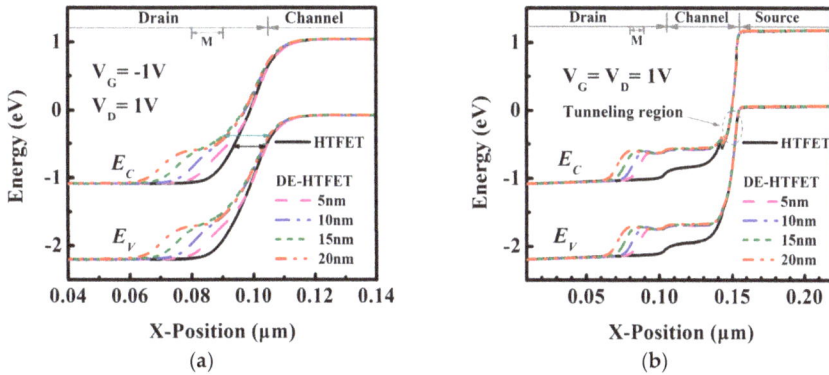

Figure 3. (a) Energy band diagram of DE-HTFET near the drain/channel tunneling junction with $L_{GM} = 15$ nm at V_G of -1 V, V_D of 1 V, and (b) Energy band diagram of DE-HTFET at ON-state.

3.1.2. Impact of Drain Engineering Metal Workfunction (φ_M) on Electrical Characteristic of DE-HTFET

In this section, in order to find the optimal drain engineering metal, the impacts of the various of workfunction are studied in detail. The function of the drain engineering metal is to increase the tunneling barrier of the drain/channel junction at the negative V_G bias, so a relatively high work function of the drain engineering metal may be a good choice. Here, electrical characteristics of DE-HTFET with drain engineering metal $\varphi_M = 5.0, 5.15$, and 5.3 eV are simulated, respectively, and the results are shown in Figure 4. Based on the results, it is indicated that $I_{ambipolar}$ presents an exponential decrease trend with the increment of φ_M, and I_{ON} stays in the same level. However, there is also a balance between $I_{ambipolar}$ and I_{ON}, that is to say, a higher workfunction is not always better. I_{ON} also reduces with the increment of φ_M, although it is in the same level. As φ_M increases from 5.0 eV to 5.15 eV, I_{ON} decreases slightly. However, at φ_M of 5.3 eV, I_{ON} decreases significantly, which is not desirable.

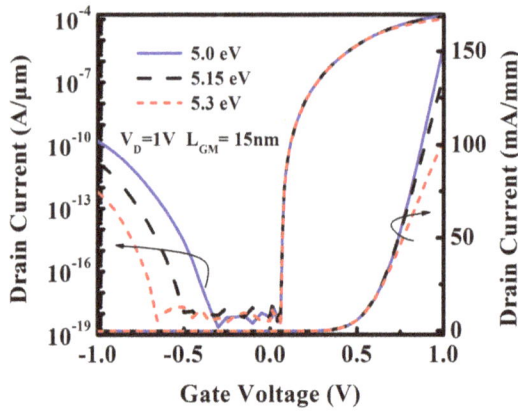

Figure 4. Transfer characteristic of DE-HTFET with various φ_M at V_D of 1 V.

To gain a better insight into the variation of $I_{ambipolar}$ and I_{ON}, Figure 5 exhibits the energy band diagram of DE-HTFET with φ_M = 5.0 eV and 5.3 eV for VG of −1 V and 1 V, respectively. Figure 5a shows that the tunneling distance in the tunneling window for DE-HTFET with φ_M = 5.3 eV is larger than that for DE-HTFET with φ_M = 5.0 eV, which results in the lower BTBT tunneling rate at the drain/channel interface for DE-HTFET with higher φ_M, therefore, lower $I_{ambipolar}$ is obtained for higher φ_M. Figure 5b shows the energy band profile at the ON-state. An energy peak is created near the drain/channel interface, which will act as a barrier for electrons tunneling from the source to the drain region at the positive bias [28]. Therefore, I_{ON} of DE-HTFET with φ_M = 5.3 eV significantly degrades as shown in Figure 4. To make a trade-off between $I_{ambipolar}$ and I_{ON}, φ_M of 5.15 eV is selected as the last choice.

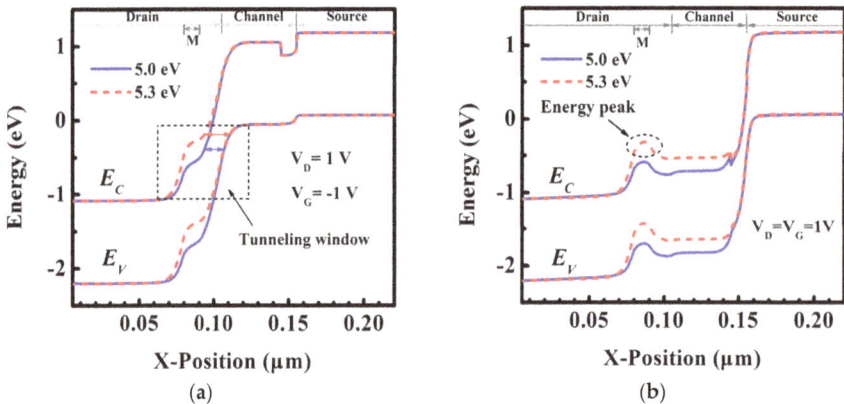

Figure 5. Energy band diagram at the line 1 nm away from the interface of InGaN/Al$_2$O$_3$ for DE-HTFET with φ_M = 5.0 eV and 5.3 eV as (a) V_D = 1 V, V_G = −1 V, and (b) V_D = V_G = 1 V.

3.2. Polar Heterostructure DE-HTFET to Improve Device Performances

The heterostructures in TFETs studied above are all along the nonpolar plane, so no polarization charge exists at the interface of heterojunction. However, studies show that the polarization engineering in the III-nitride heterostructure can further adjust the energy band structure and achieve the interband tunneling [21]. It needs the heterostructure growing along the c-axis. Because of spontaneous polarization and piezoelectric polarization along the c-axis, large amounts of fixed polarization charge can be induced at the heterojunction interface. The polarization charge generated by polarization engineering can lead to a large internal electric field near the heterojunction interface, and increase the tunneling rate. On the other hand, polarization also changes the energy structure near the tunneling junction and helps improve TFET performance.

Herein, the polar and non-polar DE-HTFETs are compared. Figure 1c shows the schematic diagram of polar DE-HTFETs. The growth direction is along [0001], that is, the negative direction of X-axis in Figure 1. In this case, the polarization charge at the interface of III-Nitride heterojunction will be taken into account in the simulations. In particular, the polarization charge near the tunneling junction may have important effects on the device characteristics of InGaN DE-HTFET. By calculation, the density of polarization surface charge (σ_{pol}) is $-1.27134 \times 10^{13}/cm^2$ at the position of TJ, and $1.27134 \times 10^{13}/cm^2$ at the position of HJ. The negative value of σ_{pol} represents a negative polarization charge, and vice versa.

The polarization changes the energy structure near the tunneling junction. Figure 6 shows that the polar DE-HTFET has a steeper energy band near the tunneling junction, which facilitates the better subthreshold and ON-state characteristics. Figure 7a displays the transfer characteristic of nonpolar and polar DE-HTFETs at L_{GM} of 15 nm and V_D of 1 V. I_D-V_G curve indicates a smaller threshold voltage for the polar DE-HTFET, which will be more beneficial to the low-power applications. Besides this, the average subthreshold swing (SS_{avg}) of the polar DE-HTFET is reduced by 53.3% compared to that of the nonpolar DE-HTFET. Also, I_{ON} increases 100% from 137 mA/mm of nonpolar DE-HTFET to 274 mA/mm of polar DE-HTFET. Figure 7b shows the point subthreshold swing (SS_{point}) of TFETs with different I_D extracted from the I_D-V_G curve. Obviously, SS_{point} of the polar DE-HTFET is smaller than that of the nonpolar TFET at any I_D value.

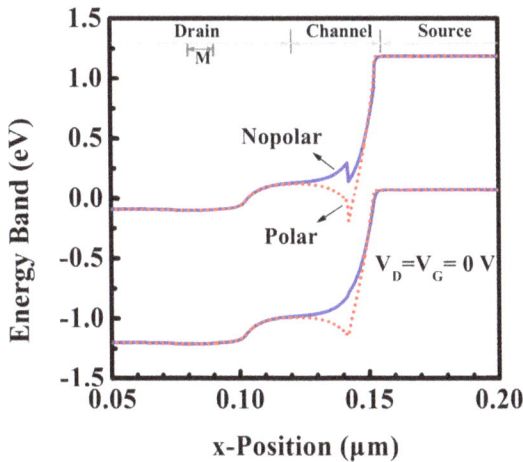

Figure 6. Energy band diagram in the thermal state.

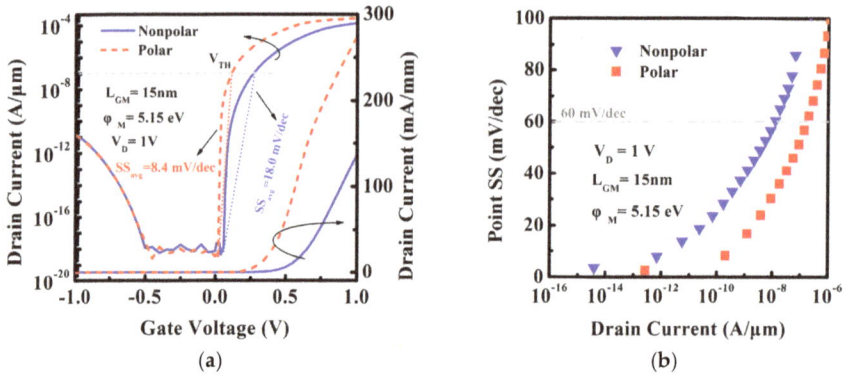

Figure 7. (**a**) Transfer characteristic and (**b**) point subthreshold swing for the nonpolar and polar InGaN TFET at V_D of 1 V.

To further investigate the mechanism of performance improvement for polar DE-HTFET, the electric field and e-BTBT rate are discussed as follows. Figure 8a exhibits that the transverse electric field intensity of the channel near the tunneling junction of polar DE-HTFET is higher than that of the non-polar device under the effect of polarization at the ON-state. Therefore, the e-BTBT rate at ON-state between the source and channel of the polar DE-HTFET is much larger than that of the nonpolar DE-HTFET as shown in Figure 7b, which is consistent with Kane's Formula (1). This is the reason why the polar InGaN DE-HTFETs have the larger I_{ON}. On the other hand, the e-BTBT rate with various of V_G is shown in Figure 8b. When $V_G = 0$ V, the tunneling rate of both TFETs is almost 0; when V_G increases to 0.1 V, the tunneling rate of polar DE-HTFET increases to 1.14×10^{29} cm^{-3}s^{-1}, however the tunneling rate of non-polar DE-HTFET is only 4.49×10^{24} cm^{-3}s^{-1}. The significant increase of the tunneling rate with the increase of V_G for polar DE-HTFETs in the subthreshold region makes polar DE-HTFETs have the lower subthreshold swing.

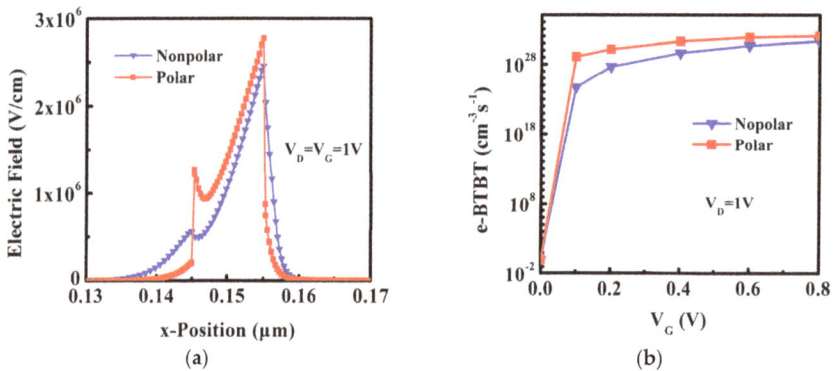

Figure 8. (**a**) Electric field in the ON-state, and (**b**) electron tunneling rate with various V_G.

4. Conclusions

In summary, a drain engineered InGaN HTFET is investigated by Atlas simulation in this paper. Firstly, with the trade-off between $I_{ambipolar}$ and I_{ON}, $I_{ambipolar}$ is reduced from 1.93×10^{-8} A/μm of DE-HTFET to 1.46×10^{-11} A/μm of HTFET by studying the impact of the drain engineering metal position and workfunction on device performances. The decreased $I_{ambipolar}$ results from the energy band modulation by the drain engineering metal and workfunction near the drain/channel junction in the drain region. Secondly, the polarization effect induces a large internal electric field near the heterojunction interface, changes the energy structure near the tunneling junction, and helps improve TFET performance. SS_{avg} of the polar InGaN DE-HTFET is 8.4 mV/dec, which is reduced by 53.3% compared to that of the nonpolar InGaN DE-HTFET. Also, I_{ON} increases 100% from 137 mA/mm of nonpolar DE-HTFET to 274 mA/mm of polar DE-HTFET. Therefore, the structure of polar InGaN DE-HTFET embodies much more promising in low power applications.

Author Contributions: X.D. provided the concept, designed the structures, performed the simulations, and wrote the manuscript; J.Z. (Jincheng Zhang) and Y.H. gave valuable suggestions on the manuscript; All authors discussed the results and approved the final manuscript.

Funding: This work was supported by the National Key Research and Development Program (Grant No. 2016YFB0400100) and the Fundamental Research Funds for the Central Universities (Grand No. JB181104).

Conflicts of Interest: The authors declare no conflict of interest.

References

1. Ionescu, A.M.; Riel, H. Tunnel field effect transistors as energy efficient electronic switches. *Nature* **2011**, *479*, 329–337. [CrossRef] [PubMed]
2. Avci, U.E.; Morris, D.H.; Young, I.A. Tunnel field-effect transistors: Prospects and challenges. *J. Electron. Dev. Soc.* **2015**, *3*, 88–95. [CrossRef]
3. Koswatta, S.O.; Lundstrom, M.S.; Nikonov, D.E. Performance comparison between p-i-n tunneling transistors and conventional MOSFETs. *IEEE Trans. Electron Devices* **2007**, *56*, 456–465. [CrossRef]
4. Luisier, M.; Klimeck, G. Performance comparisons of tunneling field-effect transistors made of InSb, Carbon, and GaSb–InAs broken gap heterostructures, In Proceedings of the IEEE International Electron Devices Meeting (IEDM), Baltimore, MD, USA, 7–9 December 2009; pp. 1–4.
5. Nirschla, T.; Henzler, S.; Fischer, J.; Fulde, M.; Bargagli-Stoffi, A.; Sterkel, M.; Sedlmeir, J.; Weber, C.; Heinrich, R.; Schaper, U.; et al. Scaling properties of the tunneling field effect transistor (TFET): Device and circuit. *Solid-State Electron.* **2006**, *50*, 44–51. [CrossRef]
6. Vijayvargiya, V.; Vishvakarma, S.K. Effect of drain doping profile on double-gate tunnel field-effect transistor and its influence on device RF performance. *IEEE Trans. Nanotechnol.* **2014**, *13*, 974–981. [CrossRef]
7. Loan, S.A.; Alharbi, A.G.; Rafat, M. Ambipolar leakage suppression in electron–hole bilayer TFET: Investigation and analysis. *J. Comput. Electron.* **2018**, *17*, 977–985.
8. Beneventi, G.B.; Gnani, E.; Gnudi, A.; Reggiani, S.; Baccarani, G. Dual-metal-gate InAs tunnel FET with enhanced turn-on steepness and high ON-current. *IEEE Trans. Electron. Dev.* **2014**, *61*, 776–784. [CrossRef]
9. Choi, W.Y.; Lee, W. Hetero-gate-dielectric tunneling field-effect transistors. *IEEE Trans. Electron. Dev.* **2010**, *57*, 2317–2319. [CrossRef]
10. Chauhan, S.S.; Sharma, N. Enhancing analog performance and suppression of subthreshold swing using hetero-junctionless double gate TFETs. *Superlattices Microstruct.* **2017**, *112*, 257–261. [CrossRef]
11. Peng, Y.; Han, G.; Wang, H.; Zhang, C.; Liu, Y.; Wang, Y.; Zhao, S.; Zhang, J.; Hao, Y. InN/InGaN complementary heterojunction-enhanced tunneling field-effect transistor with enhanced subthreshold swing and tunneling current. *Superlattices Microstruct.* **2016**, *93*, 144–152. [CrossRef]
12. Wang, H.; Liu, Y.; Liu, M.; Zhang, Q.; Zhang, C.; Ma, X.; Zhang, J.; Hao, Y.; Han, G. Performance improvement in novel germanium–tin/germanium heterojunction-enhanced p-channel tunneling field-effect transistor. *Superlattices Microstruct.* **2015**, *83*, 401–410. [CrossRef]
13. Wang, Y.; Liu, Y.; Han, G.; Wang, H.; Zhang, C.; Zhang, J.; Hao, Y. Theoretical investigation of GaAsBi/GaAsN tunneling field-effect transistors with type-II staggered tunneling junction. *Superlattices Microstruct.* **2017**, *106*, 139–146. [CrossRef]
14. Raad, B.R.; Sharma, D.; Kondekar, P.; Nigam, K.; Baronia, S. DC and analog/RF performance optimisation of source pocket dual work function TFET. *Int. J. Electron.* **2017**, *104*, 1992–2006. [CrossRef]
15. Chang, H.-Y.; Adams, B.; Chien, P.-Y.; Li, J.; Woo, J.C.S. Improved subthreshold and output characteristics of source-pocket Si tunnel FET by the application of laser annealing. *IEEE Trans. Electron Devices* **2013**, *60*, 92–96. [CrossRef]

16. Kumar, M.J.; Janardhanan, S. Doping-less tunnel field effect transistor: Design and investigation. *IEEE Trans. Electron Devices* **2013**, *60*, 3285–3290. [CrossRef]

17. Bashir, F.; Loan, S.A.; Rafat, M.; Alamoud, A.R.M.; Abbasi, S.A. A high performance gate engineered charge plasma based tunnel field effect transistor. *J. Comput. Electron.* **2015**, *14*, 477–485. [CrossRef]

18. Duan, X.; Zhang, J.; Wang, S.; Li, Y.; Xu, S.; Hao, Y. A High-Performance Gate Engineered InGaN Dopingless Tunnel FET. *IEEE Trans. Electron Devices* **2018**, *65*, 1223–1229. [CrossRef]

19. Abdi, D.B.; Kumar, M.J. PNPN tunnel FET with controllable drain side tunnel barrier width: Proposal and analysis. *Superlattices Microstruct.* **2015**, *86*, 121–125. [CrossRef]

20. Sahay, S.; Kumar, M.J. Controlling the drain side tunneling width to reduce ambipolar current in tunnel FETs using heterodielectric BOX. *IEEE Trans. Electrons Devices* **2015**, *62*, 3882–3885. [CrossRef]

21. Li, W.; Sharmin, S.; Ilatikhameneh, H.; Rahman, R.; Lu, Y.; Wang, J.; Yan, X.; Seabaugh, A.; Klimeck, G.; Jena, D.; et al. Polarization-Engineered III-Nitride Heterojunction Tunnel Field-Effect Transistors. *IEEE J. Explor. Solid-State Comput. Devices Circuits* **2015**, *1*, 28–34. [CrossRef]

22. Li, W.; Cao, L.; Lund, C.; Keller, S.; Fay, P. Performance projection of III-nitride heterojunction nanowire tunneling field-effect transistors. *Phys. Status Solidi A* **2016**, *213*, 905–908. [CrossRef]

23. Cho, M.S.; Kwon, R.H.; Seo, J.H.; Yoon, Y.J.; Jang, Y.I.; Won, C.H.; Kim, J.-G.; Lee, J.; Cho, S.; Lee, J.-H.; et al. Electrical performances of InN/GaN tunneling field-effect transistor. *J. Nanosci. Nanotechnol.* **2017**, *17*, 8355–8359. [CrossRef]

24. Duan, X.; Zhang, J.; Wang, S.; Quan, R.; Hao, Y. Effect of graded InGaN drain region and 'In' fraction in InGaN channel on performances of InGaN tunnel field-effect transistor. *Superlattices Microstruct.* **2017**, *112*, 671–679. [CrossRef]

25. *ATLAS Device Simulation Software*, Silvaco Int.: Santa Clara, CA, USA, 2013.

26. Duan, X.; Zhang, J.; Xiao, M.; Zhao, Y.; Ning, J.; Hao, Y. Groove-type channel enhancement-mode AlGaN/GaN MIS HEMT with combined polar and nonpolar AlGaN/GaN heterostructures. *Chin. Phys. B* **2016**, *25*, 087304. [CrossRef]

27. Verhulst, A.S.; Vandenberghe, W.G.; Maex, K.; Groeseneken, G. Boosting the on-current of a n-channel nanowire tunnel field-effect transistor by source material optimization. *J. Appl. Phys.* **2008**, *104*, 064514. [CrossRef]

28. Raad, B.R.; Sharma, D.; Kondekar, P.; Nigam, K.; Yadav, D.S. Drain work function engineered doping-less charge plasma TFET for ambipolar suppression and RF performance improvement: A proposal, design, and investigation. *IEEE Trans. Electron Devices* **2016**, *63*, 3950–3957. [CrossRef]

micromachines

MDPI

Article

Design and Characterization of Semi-Floating-Gate Synaptic Transistor

Yongbeom Cho [1], Jae Yoon Lee [1], Eunseon Yu [1], Jae-Hee Han [2], Myung-Hyun Baek [3], Seongjae Cho [1,*] and Byung-Gook Park [3,*]

[1] Department of Electronics Engineering, Gachon University, Gyeonggi-do 13120, Korea;
    jj2928@naver.com (Y.C.); ldhh1015@nate.com (J.Y.L.); yesemic@naver.com (E.Y.)
[2] Department of Energy IT, Gachon University, Gyeonggi-do 13120, Korea; jhhan388@gachon.ac.kr
[3] Department of Electrical and Computer Engineering, Seoul National University, Seoul 08826, Korea;
    applewhisky90@gmail.com
*   Correspondence: felixcho@gachon.ac.kr (S.C.); bgpark@snu.ac.kr (B.-G.P.); Tel.: +82-31-750-8722 (S.C.);
    +82-2-880-7270 (B.-G.P.)

Received: 14 November 2018; Accepted: 2 January 2019; Published: 7 January 2019

Abstract: In this work, a study on a semi-floating-gate synaptic transistor (SFGST) is performed to verify its feasibility in the more energy-efficient hardware-driven neuromorphic system. To realize short- and long-term potentiation (STP/LTP) in the SFGST, a poly-Si semi-floating gate (SFG) and a SiN charge-trap layer are utilized, respectively. When an adequate number of holes are accumulated in the SFG, they are injected into the nitride charge-trap layer by the Fowler–Nordheim tunneling mechanism. Moreover, since the SFG is charged by an embedded tunneling field-effect transistor existing between the channel and the drain junction when the post-synaptic spike occurs after the pre-synaptic spike, and vice versa, the SFG is discharged by the diode when the post-synaptic spike takes place before the pre-synaptic spike. This indicates that the SFGST can attain STP/LTP and spike-timing-dependent plasticity behaviors. These characteristics of the SFGST in the highly miniaturized transistor structure can contribute to the neuromorphic chip such that the total system may operate as fast as the human brain with low power consumption and high integration density.

Keywords: semi-floating gate; synaptic transistor; neuromorphic system; spike-timing-dependent plasticity (STDP); highly miniaturized transistor structure; low power consumption

1. Introduction

In 2016, AlphaGo, one of the results of artificial intelligence (AI) won the Go game against top-ranked Go players [1]. Because Go had been considered suitable only for humans, as it requires not only intelligence but also experience, this achievement led to a media sensation. Why has AI now attracted the public's attention, and why has AI research become so active once again? One of the reasons is the efficiency of the AI system. Currently, as AI technology develops, a method for increasing operation speed by using a graphic card in parallel is being adopted [2]. Hence, although the amount of necessary computation is large, replicating and mimicking activities that humans would carry out for human mental activities have become possible. So far, these operations have been realized in the software technologies in the von Neumann architecture. However, in order to imitate the human brain with higher resemblance, which performs great deal of mental activities with very small amount of power consumption, power efficiency should be considered more importantly now and in the future calling for the hardware-driven neuromorphic system. In other words, efficiency should be supported not only by algorithms but also by the hardware. In this respect, studies on neuromorphic chips that integrate software and hardware are attracting particular interest [3–6].

In this study, we have focused on synaptic cells shown in Figure 1, which is thought to be closely related to experiences in human mental activity and accumulation of them, as the element to imitate the nervous system [7–10]. This biological motivation is projected to an electronic component, synaptic transistor. In order to enable low-power and high-speed operations, a poly-Si semi-floating gate (SFG) structure with a tunneling field-effect transistor is adopted for realizing short-term potentiation (STP), and a SiN charge-trap layer is stacked on the SFG for realizing long-term potentiation (LTP) operation. Further, we have obtained the spike-timing-dependent plasticity (STDP) characteristic [11–13]. Finally, we propose a novel synaptic device that has STP/LTP capabilities with STDP operation which are the essential functions of the human biological synapse.

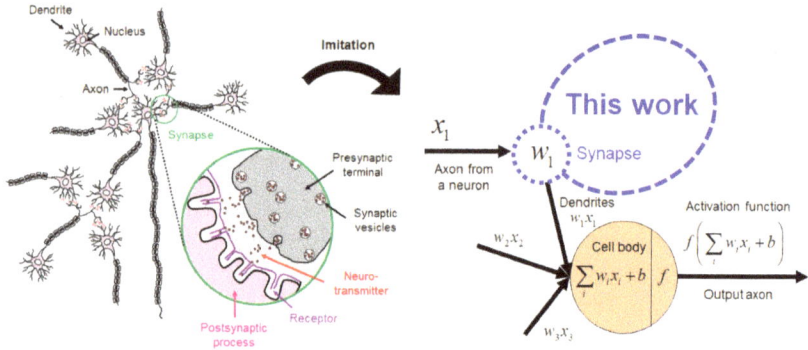

Figure 1. Biological nerve cell element targeted to imitate by the electron device in this work and its mathematical representation.

2. Device Structure and Operation Schemes

Figure 2a shows the schematic of the proposed SFG synaptic transistor (SFGST) and its circuit symbol representation. Although the proposed device is based on the integration of volatile and nonvolatile memory components in a miniaturized transistor, write/erase operations are analogously termed as potentiation and depression from the stance of new synaptic functions expected from the proposed device. As shown in Figure 2b, the $n^+/n/p^+$ (drain/channel/SFG) junction is embedded for low-power and high-speed potentiation by hole tunneling into the SFG. When the first and second gates are negatively biased and the drain is positively biased, potentiation occurs as demonstrated in Figure 2c. The device fabrication has full Si processing compatibility and higher mass producibility. Hole tunneling takes place between the drain junction of the SFGST and the first gate by the operation of a p-type tunnel field-effect transistor (TFET), by which holes are accumulated in the SFG. This accumulation of holes in the SFG lowers the threshold voltage of the SFGST and increases the channel conductivity eventually. These series of changes in carrier population and potential distribution make up the potentiation process. When the biases on the two terminals are reversed, the holes are discharged from the SFG by drift and diffusion due to the turn-on of the diode part residing between the drain junction of the SFGST and the SFG. Equivalently, it can be understood as electrons are charged in the SFG and the threshold voltage of the SFGST is elevated. These series of carrier and potential redistributions make the depression process happen. Figure 3 and Table 1 show the mesh structure and the device parameters of the simulated device by technology computer-aided design (TCAD) [14]. The meshes are weaved more densely in the SFG, nitride charge-trap layer, and near the tunneling sites for achieving higher accuracy in this simulation. In order to obtain the TCAD simulation results with higher accuracy and credibility, multiple physical models including Fowler–Nordheim (FN) tunneling model, band-to-band tunneling model, nitride charge-trap model, concentration-dependent generation-recombination model, and concentration/temperature-dependent mobility models have been activated simultaneously for respective simulation tasks. The band-to-band tunneling model has been adjusted with the correction factors empirically suggested by Hurks [15].

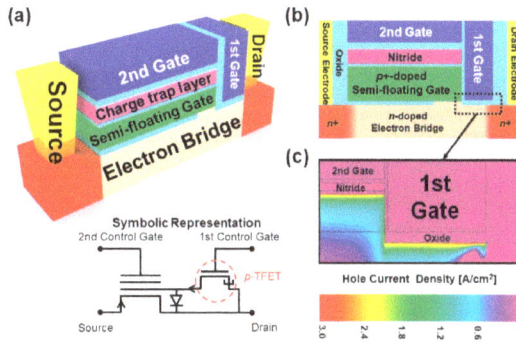

Figure 2. Device structure and potentiation process. (**a**) Aerial view of the proposed synaptic device and its circuit symbol representation; (**b**) Cross-sectional view of the device; (**c**) Contour of hole current density during the potentiation through band-to-band tunneling.

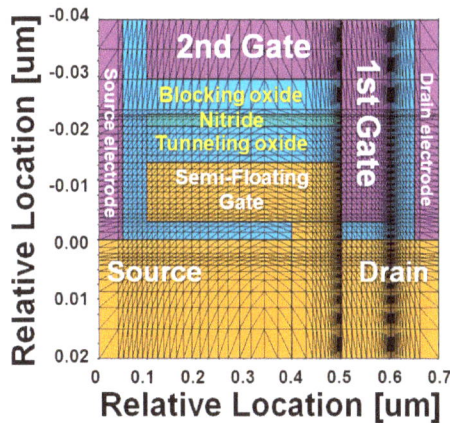

Figure 3. Mesh structure of the simulated device with notations of the terminals.

Table 1. Critical dimensions and process parameters in the designed semi-floating-gate synaptic transistor (SFGST).

Region	Length (nm)	Thickness (nm)	Doping Concen Tration (cm^{-3})
1st Gate	100	37	p-type 1×10^{20}
2nd Gate	400	10	p-type 1×10^{20}
SFG	400	10	p-type 1×10^{18}
Source junction	100	20	n-type 1×10^{20}
Channel	500	20	n-type 1×10^{17}
Drain junction	100	20	n-type 1×10^{20}
Gate oxide	-	3	-
Tunneling oxide	-	6	-
Nitride	-	2	-
Blocking oxide	-	6	-

3. Synaptic Operation Characteristics

3.1. Short-Term and Long-Term Potentiation Operations

SFG is partially connected to the channel at the end unlike the commonly used floating gates which is isolated from the channel. By using the SFG, holes can be easily stored by the tunneling current and erased by the drift and diffusion mechanisms. Here, the holes accumulated in the SFG region by proper operation voltages but vanish if there is no hold bias. This characteristic can be adopted for realizing the STP operation. However, when input pulses are successively provided before the holes vanish, the total charges in the SFG increase with time. The number of holes in the SFG increases as the pulses with short time interval are successively applied and the number of newly generated holes is larger than that of holes disappearing by either diffusion or recombination. The higher energy states allowed in the SFG region were mostly vacant due to smaller occupation probabilities since they are located in the tail region of the Fermi–Dirac distribution but now they are occupied by the holes accumulated in a large number in the SFG. The holes in the higher energy states have higher probability of Fowler–Nordheim (FN) tunneling into the nitride charge-trap layer through the tunneling oxide energy barrier deformed to a triangular shape under a high electric field. Thus, the FN tunneling has the predominance in the region of large amount of holes accumulated in the SFG as shown in Figure 4a. These characteristics make distinction between STP and LTP. Figure 4b shows the actual results of the simulated total charges in the SFG (left) and nitride (right) regions after successive potentiating pulses. Here, the charges in the specified regions identify the total net charges which have been extracted by integration of current over a period of time for an operation. It should be reasonable to have an individual look into the electron and hole densities in order to investigate the time-varying amount of stored charges in case of conventional floating-gate (FG) memory devices. The proposed device in this work equips an SFG but there should be conduction of electrons and hole into and out of the floating gate according to the relation among potential distributions over the diode and TFET regions linked to the floating gate. Thus, total net charge might make a more practical sense in this case and Figure 4b conveys the total net charges vs. number of potentiation pulses. Here, a negative value implies that the electrons have the predominance in population, and inversely, a positive one reveals the predominance of holes. It is confirmed that more than three pulses are required for the transition from STP to LTP at the bias condition of $V_{GS1} = V_{GS2} = -1.5$ V with a pulse width and interval of 1 μs. It is expected that an increased number of pulses will be required for STP to transit into LTP as the tunneling oxide (TO) becomes thicker. The hole current density after a specific number of pulses is shown in Figure 5. Here, it is Figure 5 that qualitatively demonstrates the directions of carrier movements over the short- and long-term potentiation processes. Holes are injected by the operation of p-type TFET functional region near the drain for potentiation. The holes are injected into the SFG and a part of them occupying higher energy states in the Fermi–Dirac distribution after accumulation of significant amount of holes tunnel into the nitride layer.

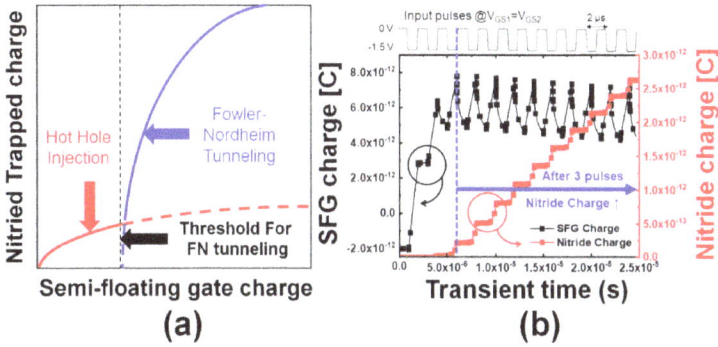

Figure 4. Total charges in the semi-floating gate (SFG) and the nitride regions. (**a**) Qualitative explanation of required holes accumulated in the SFG region to meet the condition of increased probability of injection into the nitride charge-trap layer by Fowler–Nordheim (FN) tunneling; (**b**) Total charges in the SFG (left) and nitride (right) regions after series of potentiating pulses as a function of time obtained by technology computer-aided design (TCAD) simulation.

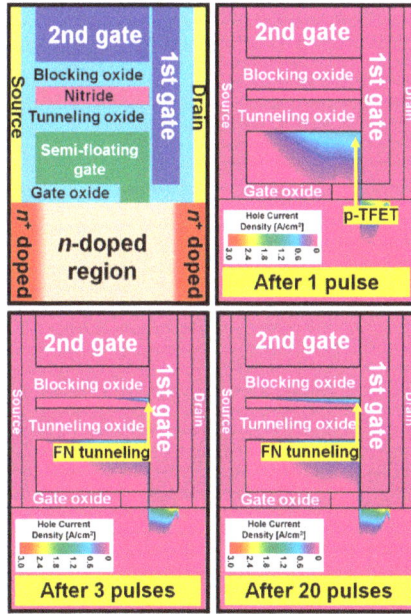

Figure 5. Hole current density after successive potentiation pulses, $V_{GS1} = V_{GS2} = -1.5$ V. As the number of pulses increases, the holes at the Fermi distribution tail accumulated in the SFG region see the triangular energy barrier and become more probable for injection into the nitride by FN tunneling.

Figure 6a shows the hole distributions in the simulated synaptic transistor after 1 and 20 potentiation pulses are applied. The electric field across the blocking oxide, nitride charge-trap layer, tunneling oxide, and SFG along the cutline A-A′ is investigated in Figure 6b. It is confirmed by Figure 6a,b that the charge-trap layer between tunneling oxide and blocking oxide layers has relatively larger population of holes injected from the SFG by tunneling. Consequently, the electric field across the cutline is increased, with the reference at point A′, owing to the holes trapped in the nitride layer.

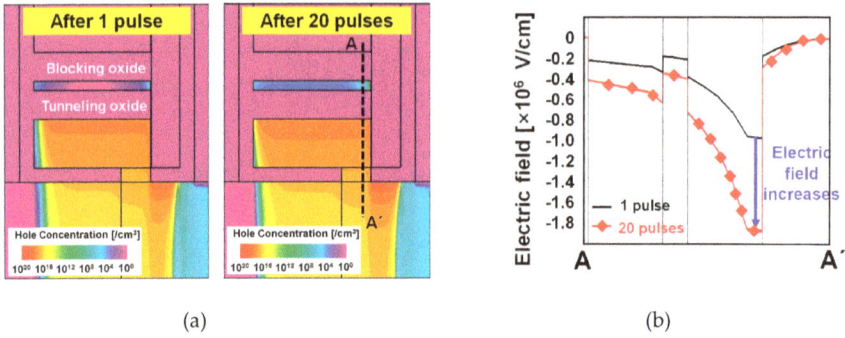

Figure 6. Pulse-number-dependent carrier distribution and electric field. (**a**) Distribution of hole concentration in the SFG and nitride charge-trap layer and (**b**) electric field along the cutline A-A' in (**a**) after 1 and 20 potentiation pulses.

Figure 7 shows the retention characteristics under a constant read bias condition, $V_{GS1} = V_{GS2} = V_{DS} = 0.5$ V, after different numbers of pulses are provided. When the number of pulses is 0, 1, and 3, the drain current decreases as time passes, and then, converges into the initial state due to the semi-floating structural characteristic. In contrast, when the number of pulses is more than 3, such as 10, 20, and 50, the drain current converges into a higher value than that of the initial state. In particular, when the number of pulses exceeds 50, the current levels of STP and LTP are almost the same. As previously mentioned, this is due to the trapped charges in the nitride layer and the electric field repulsing the holes downward. Further, in both the short-term and long-term potentiation cases, multi-level states can be realized. Hence, it is confirmed that SFGST can distinguish between STP and LTP with multi-level current states, which is the essence for mimicking synaptic operation which modulates the connectivity strength by frequencies. While performing the STP operations, the accumulated holes vanish by recombination and diffusion conduction. When the holes are still existing in the SFG, the threshold voltage of the SFGST is elevated and the sensing current increases in accordance. On the other hand, the number of holes in the SFG decreases as time passes without successive pulses with a short enough interval time and the sensing current goes back to the original low level as the result. Thus, the pulse time shorter than times for recombination and travelling by diffusion can only fluctuate the SFG potential and the sensing current. However, in the existence of a large amount of holes accumulated in the SFG, without being provided with a long enough pulsing time to be released from the SFG, the holes become very probable to occupy higher energy states, see a lower effective energy barrier toward the nitride charge-trap layer, and can subsequently tunnel into nitride even at a smaller tunneling electric field necessitating a small voltage. Once the holes are trapped in the nitride, the sensing current is semi-permanently decided and invariant with time as shown in Figure 7. The additional pulses under the LTP condition contribute to increasing the amount of holes trapped in the nitride and determine the level of constant current, which eventually modulates the electrical conductivity of the SFGST and presents the multi-level states.

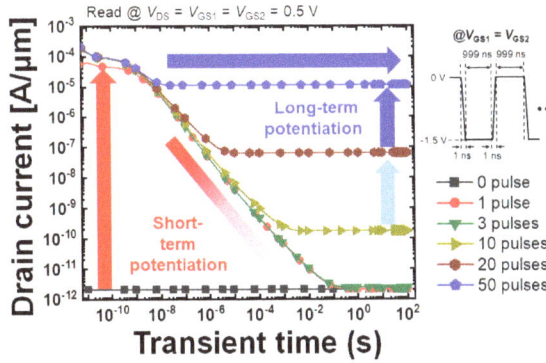

Figure 7. Retention characteristics under a constant read bias condition, $V_{GS1} = V_{GS2} = V_{DS} = 0.5$ V after potentiating operations with different numbers of potentiation pulses.

3.2. Spike-Timing-Dependent Plasticity (STDP)

In order to utilize the SFGST as a synaptic device capable of STDP operation, the array architecture for realizing the artificial spike neural network (SNN) hardware based on the proposed SFGST device with the full accommodation of the operation bias schemes needs to be proposed, as demonstrated in Figure 8. Since the second gate and the drain are tied together, the device is operated as a three-terminal device. Here, the operating condition that the biases of the first and the second gates are opposite to each other makes it possible to operate the SFGST realizing the STDP behavior. Figure 9 shows the simulated transient STDP characteristics of the SFGST for a single triangular spike. If the pre-neuron signal comes in earlier than the post-neuron signal, the time difference has a positive value, and vice versa. It is confirmed that the SFGST follows the Hebbian learning rule successfully. The change in weight increases as $|\Delta t|$ decreases, and decreases as $|\Delta t|$ increases.

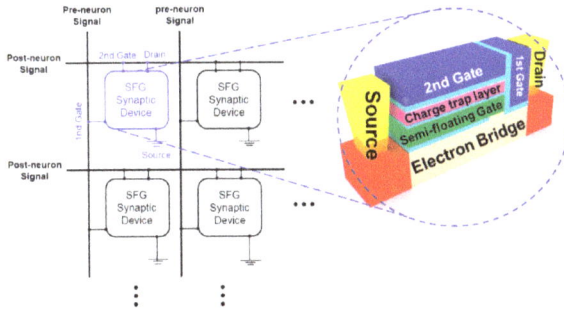

Figure 8. Array architecture for the artificial spike neural network (SNN) based on the proposed SFGST device with the full accommodation of the developed bias schemes.

This can be also verified by Figure 10, which shows the variation in the threshold voltage by the potentiated or depressed SFG. As briefly mentioned earlier, the variation in the threshold voltage becomes larger as $|\Delta t|$ gets larger. Finally, Figure 11 shows the simulated learning operations as a function of number of potentiation pulses. From the results demonstrating that the given pre- and post-neuron signals potentiate the SFGST making the distinction between STP and LTP, it is confirmed that the designed SFGST is fully functional as a synaptic device.

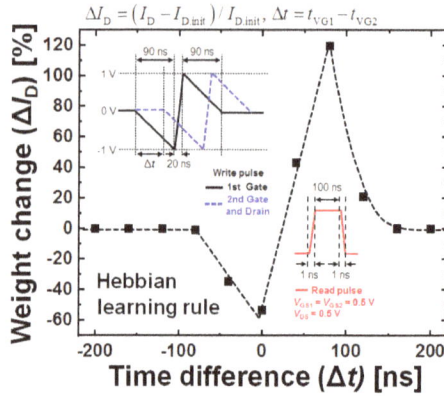

Figure 9. Simulated spike-timing-dependent plasticity (STDP) characteristics of the SFGST after a single triangular spike. Following the Hebbian learning rule [16], the synaptic change is determined by Δt.

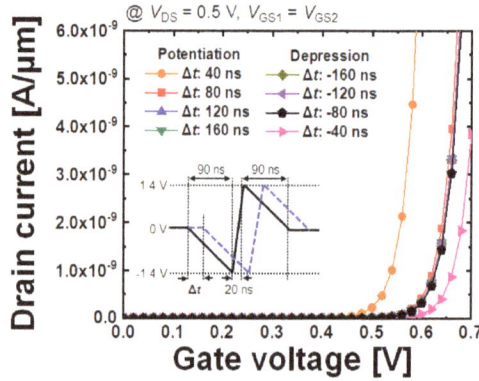

Figure 10. Variation in threshold voltage by the potentiated or depressed SFG. After two pulses are fed with time difference (pre- and post-input signals), potentiation and depression take place under the conditions of $\Delta t > 0$ and $\Delta t < 0$, respectively. The shorter time interval between the pre-and post-input signals, the larger becomes the variation in the threshold voltage.

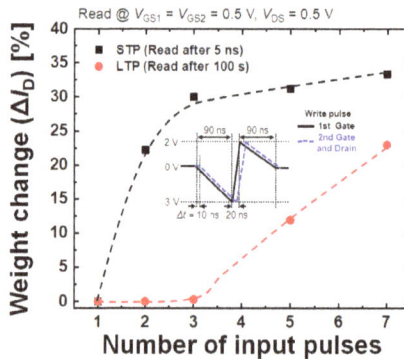

Figure 11. Simulated learning operations of the SFGST according to the number of potentiation pulses.

4. Conclusions

In this work, a novel synaptic transistor featuring the semi-floating gate and charge-trap layer has been proposed and designed, and its essential synaptic operations have been verified through TCAD simulation. The SFGST performs both STP and LTP operations discriminable by the number of potentiation pulses. Also, it is confirmed that multiple states, i.e., multiple conductance values can be obtained in the LTP, which corresponds to modulation in the biological synaptic connectivity representing the synaptic weight. Based on the STP and LTP operation capabilities, STDP operation has been verified and the presumable array architecture into which the proposed synaptic transistor and the operation schemes are converged has been proposed. The proposed miniaturized transistor embedding both volatile and nonvolatile memory components can be a promising intelligent component realizing the hardware-driven neuromorphic system which is mainly based on the semiconductor technology with full Si processing compatibility.

5. Patents

(1)   Seongjae Cho and Yongbeom Cho, "Synaptic Semiconductor Device and Neural Networks Using the Same,"

-         Korean patent filed, 10-2017-0152803, 16 November 2017
-         United States patent filed, 15/892,658, February 2018.

(2)   Byung–Gook Park and Seongjae Cho, "Neuron circuit and synapse array integrated circuit architecture and fabrication method of the same,"

-         Korean patent filed, 10-2017-0062097, 19 May 2017.
-         United States patent filed, 15/895,255, 13 February 2018.

Author Contributions: Y.C. and S.C. conceived the device structure and wrote the manuscript. Y.C., J.Y.L., and E.Y. performed the device simulations. J.-H.H. exchanged constructive discussions with S.C. and confirmed the biological analogies. M.-H.B. helped the simulation task and checked the practicability of idea by evaluating the process viability of the proposed device structure. S.C. made the direction of manuscript and prepared the steady-state and the transient simulation strategies. B.-G.P. conceived the hardware-driven neuromorphic system based on the conventional well-matured Si complementary metal-oxide semiconductor (CMOS) processing, initiated the overall research project, and confirmed the validities of the simulated synaptic operations towards the artificial spike neural network.

Funding: This work was supported by Nano Material Technology Development Program through the National Research Foundation of Korea (NRF) funded by the Ministry of Science and ICT (MSIT) (Grant No. NRF-2016M3A7B4910348) and by Mid-Career Researcher Program through NRF funded by the MSIT (Grant No. NRF-2017R1A2B2011570).

Conflicts of Interest: The authors declare no conflict of interest.

References

1.   Lee, C.S.; Wang, M.H.; Yen, S.J.; Wei, T.H.; Wu, I.C.; Chou, P.C.; Chou, C.H.; Wang, M.W.; Yan, T.H. Human vs. Computer Go: Review and Prospect [Discussion Forum]. *IEEE Comput. Intell. Mag.* **2016**, *11*, 67–72. [CrossRef]

2.   Silver, D.; Huang, A.; Maddison, C.J.; Guez, A.; Sifre, L.; Schrittwieser, J.; Antonoglou, I.; Panneershelvam, V.; Lanctot, M.; Dieleman, S.; et al. Mastering the game of Go with deep neural networks and tree search. *Nature* **2016**, *529*, 484–489. [CrossRef] [PubMed]

3.   Ishiwara, H. Proposal of adaptive-learning neuron circuits with ferroelectric analog-memory weights. *Jpn. J. Appl. Phys.* **1993**, *32*, 442–446. [CrossRef]

4.   Kuzum, D.; Yu, S.; Wong, H.-S.P. Synaptic electronics: Materials, devices and applications. *Nanotechnology* **2013**, *24*, 382001. [CrossRef] [PubMed]

5. Nishitani, Y.; Kaneko, Y.; Ueda, M.; Fujii, E.; Tsujimura, A. Dynamic observation of brain-like learning in a ferroelectric synapse device. *Jpn. J. Appl. Phys.* **2013**, *52*, 04CE06. [CrossRef]

6. Kim, H.; Cho, S.; Sun, M.-C.; Park, J.; Hwang, S.; Park, B.-G. Simulation study on silicon-based floating body synaptic transistor with short- and long-term memory functions and its spike timing-dependent plasticity. *J. Semicond. Technol. Sci.* **2016**, *16*, 657–663. [CrossRef]

7. Selkoe, D.J. Alzheimer's disease is a synaptic failure. *Science* **2002**, *298*, 789–791. [CrossRef] [PubMed]

8. Lüscher, C.; Isaac, J.T. The synapse: Center stage for many brain diseases. *J. Physiol.* **2009**, *15*, 727–729. [CrossRef] [PubMed]

9. Barker, R.A.; Cicchetti, F.; Neal, M.J. *Neuroanatomy and Neuroscience at a Glance*, 4th ed.; Wiley-Blackwell: Hoboken, NJ, USA, 2012; ISBN 978-1-118-36852-7.

10. Duman, R.S.; Aghajanian, G.K.; Sanacora, G.; Krystal, J.H. Synaptic plasticity and depression: New insights from stress and rapid-acting antidepressants. *Nat. Med.* **2016**, *22*, 238–249. [CrossRef] [PubMed]

11. Song, S.; Miller, K.D.; Abbot, L.F. Competitive Hebbian learning through spike-timing-dependent synaptic plasticity. *Nat. Neurosci.* **2000**, *3*, 919–926. [CrossRef] [PubMed]

12. Kwon, M.-W.; Kim, H.; Park, J.; Park, B.-G. Integrate-and-fire neuron circuit and synaptic device using floating body MOSFET with spike timing-dependent plasticity. *J. Semicond. Technol. Sci.* **2015**, *15*, 658–663. [CrossRef]

13. Choi, H.-S.; Wee, D.-H.; Kim, H.; Kim, S.; Ryoo, K.-C.; Park, B.-G.; Kim, Y. 3-D floating-gate synapse array with spike-time-dependent plasticity. *IEEE Trans. Electron Devices* **2018**, *65*, 101–107. [CrossRef]

14. *ATLAS User's Manual*; Silvaco International Inc.: Santa Clara, CA, USA, 2016.

15. Hurkx, G.A.M.; Klaassen, D.B.M.; Knuvers, M.P.G. A new recombination model for device simulation including tunneling. *IEEE Trans. Electron Devices* **1992**, *39*, 331–338. [CrossRef]

16. Abbott, L.F.; Nelson, S.B. Synaptic plasticity: Taming the beast. *Nat. Neurosci.* **2000**, *3*, 1178–1183. [CrossRef] [PubMed]

micromachines

MDPI

Article

Process Variability—Technological Challenge and Design Issue for Nanoscale Devices

Jürgen Lorenz [1,*], Eberhard Bär [1], Sylvain Barraud [2], Andrew R. Brown [3], Peter Evanschitzky [1], Fabian Klüpfel [1] and Liping Wang [3]

[1] Fraunhofer Institut für Integrierte Systeme und Bauelementetechnologie, Schottkystrasse 10, 91058 Erlangen, Germany; eberhard.baer@iisb.fraunhofer.de (E.B.); peter.evanschitzky@iisb.fraunhofer.de (P.E.); Fabian.Kluepfel@iisb.fraunhofer.de (F.K.)

[2] CEA, LETI, MINATEC campus and Univ. Grenoble Alpes, 38054 Grenoble, France; sylvain.barraud@cea.fr

[3] Synopsys Northern Europe Ltd., Glasgow G3 8HB, UK; Andrew.Brown@synopsys.com (A.R.B.); Liping.Wang@synopsys.com (L.W.)

* Correspondence: juergen.lorenz@iisb.fraunhofer.de; Tel.: +49-9131-761-210

Received: 8 December 2018; Accepted: 15 December 2018; Published: 23 December 2018

Abstract: Current advanced transistor architectures, such as FinFETs and (stacked) nanowires and nanosheets, employ truly three-dimensional architectures. Already for aggressively scaled bulk transistors, both statistical and systematic process variations have critically influenced device and circuit performance. Three-dimensional device architectures make the control and optimization of the device geometries even more important, both in view of the nominal electrical performance to be achieved and its variations. In turn, it is essential to accurately simulate the device geometry and its impact on the device properties, including the effect caused by non-idealized processes which are subject to various kinds of systematic variations induced by process equipment. In this paper, the hierarchical simulation system developed in the SUPERAID7 project to study the impact of variations from equipment to circuit level is presented. The software system consists of a combination of existing commercial and newly developed tools. As the paper focuses on technological challenges, especially issues resulting from the structuring processes needed to generate the three-dimensional device architectures are discussed. The feasibility of a full simulation of the impact of relevant systematic and stochastic variations on advanced devices and circuits is demonstrated.

Keywords: process simulation; device simulation; compact models; process variations; systematic variations; statistical variations; FinFETs; nanowires; nanosheets

1. Introduction

Aggressively scaled transistors are affected by three kinds of process variations. Most frequently and since long discussed in the literature are statistical process variations which are caused by the granularity of matter, such as Random Dopant Fluctuations (RDF). However, as summarized earlier [1] a diversity of systematic process variations is caused by non-idealities of process equipment, like inhomogeneity of gas flow or temperature distributions, or imperfect control of parameters like the distance between the last lens and the wafer in lithography, the so-called defocus. Moreover, layout effects caused, e.g., by pattern density also affect the results of various process steps [1]. Especially for aggressively scaled three-dimensional devices such as FinFETs or nanowires as shown in Figure 1, not only systematic variations of simple geometrical parameters such as the gate length must be considered, but also three-dimensional shapes may vary, critically affecting device performance. These requirements are being addressed and met in the cooperative project SUPERAID7 [2] funded by the European Commission within the Horizon 2020 programme.

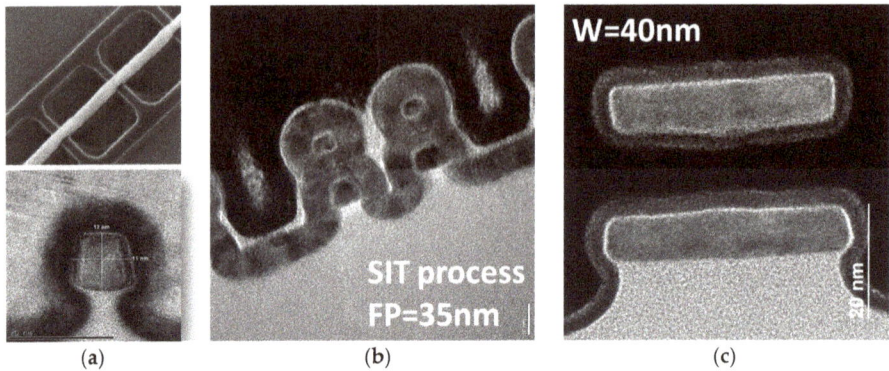

Figure 1. Example of transistor as addressed in SUPERAID7: (**a**) Trigate; (**b**) stacked nanowire; (**c**) stacked nanosheet.

Whereas the impact of statistical variations such as Random Dopant Fluctuations (RDF), Metal Gate Granularity (MGG), and Line Edge Roughness (LER) have been frequently and since long discussed in the literature (e.g., in [3–5]), the effects of systematic process variations have so far got much less attention. Some publications with involvement of authors of this paper referred to bulk transistors [1,6–9]. Two examples for the impact of patterning processes on FinFET transistors were presented earlier [10–12], in the latter case also discussing the impact on a static random-access memory (SRAM) cell. However, these papers did not discuss the key differences between the different patterning processes in terms of variability and impact on design. Furthermore, the general case where not only two or three geometrical parameters of the transistor, e.g., fin width and gate length, are changed but also the shape of a fin or of nanowires was not considered

In the following, the main processes available for the patterning of three-dimensional nanoscale transistors are discussed concerning their simulation and their variability, and a method to extract compact models which include both systematic and stochastic variations which may also affect the shape of the transistor is presented.

2. Materials and Methods

This paper reports about results obtained in the cooperative project SUPERAID7 [2], funded within the Horizon 2020 programme of the European Union, and partly about related background work from Fraunhofer IISB, and from partners as cited. Within SUPERAID7 a software system for the simulation of the impact of systematic and statistical process variations on advanced More Moore devices and circuits down to the 7 nm node and below has been developed, including especially interconnects. Besides enhanced and new software tools this needs improved physical models and extended compact models. In terms of software integration and application, SUPERAID7 and this paper are partly based on the SENTAURUS simulation system from Synopsys [13], which is for academic use also available via EUROPRACTICE [14]. Topography simulation for this paper has been performed with the tools Dr.LiTHO [15], ANETCH [16], and DEP3D [17], which are available under license from Fraunhofer IISB, and also with ViennaTS which is available from TU Wien as Open Source [18].

3. Results

3.1. Nanoscale Patterning Processes and Their Variations

Besides the physical properties of the semiconductor material used, which among others limit carrier mobility, the patterning of the transistors has increasingly become challenging. This is both due to the ever-smaller feature sizes needed and the complex three-dimensional geometries employed for nanoscale transistors in order to achieve good channel control. In the following, this development is outlined especially in view of process variations and design implications.

3.1.1. Lithography

For traditional bulk CMOS, mainly gates and their spacers had to be patterned. In corresponding simulations of lithography, etching, and deposition mainly, the resulting footprint of gate and spacer was relevant, because this defined the physical gate length L and width W. In turn, variations in these patterning processes led to variations of L and W, which could readily be considered during the extraction of compact models [8]. In most cases, standard optical lithography was used, partly including water immersion [19]. Such lithography steps are subject to variations of the distance between the last lens of the optical system and the photoresist to be developed (which can be transferred into the so-called focus position or focus) and of the illumination dose used. These lead to variations of the feature size printed, the so-called critical dimensions (CD). Generally, best resolution is obtained for so-called dense lines, that is, a regular pattern of lines and spaces of the same width (duty factor 1:1). However, the situation encountered during circuit fabrication usually deviates from this situation. In Figure 2, the CD of the intensity distribution of the light just above the resist (the so-called aerial image) is shown for 65 nm lines and 130 nm spaces (duty factor 1:2). The figure illustrates how the variations of focus and dose lead to variations of the CD of the aerial image, which would then during resist development be transferred into variations of the CD of the resist structures. Moreover, the distribution of the CDs is shifted from the nominal 65 nm to larger values. Figure 3 shows the dependence of the CD on focus and dose: Here, the central curve shows the combinations of focus and dose which lead to the nominal CD of 65 nm. The two other curves show the combinations of focus and dose which result in a CD increased (upper line) or decreased (lower line) by 10%, respectively. A standard task in the lithography community is then to optimize the lithography process in terms of its stability against variations of dose and focus. Here, the so-called process window is maximized, which is the largest rectangle in the space of focus and dose which results in CDs which deviate not more than 10% from the targets values. With the duty factor of 1:2 shown in Figure 3a, the process window is rather small. For the ideal case of dense lines with a duty factor of 1:1, shown in Figure 3b, the process window is much larger. This illustrates that both the resolution and also the variability of a lithography process critically depend not only on the size but also on the layout of the features or circuits to be printed.

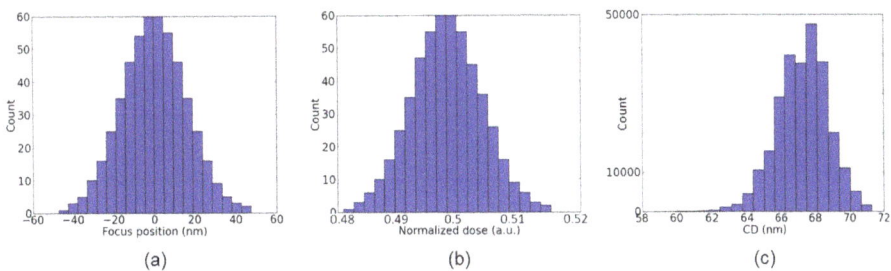

(a) (b) (c)

Figure 2. Impact of variations of focus and dose on a 193 nm dry lithography process with adapted illumination for 65 nm lines/130 nm spaces (duty factor 1:2): Variation of (**a**) focus and (**b**) threshold considered; (**c**) resulting variation of CD.

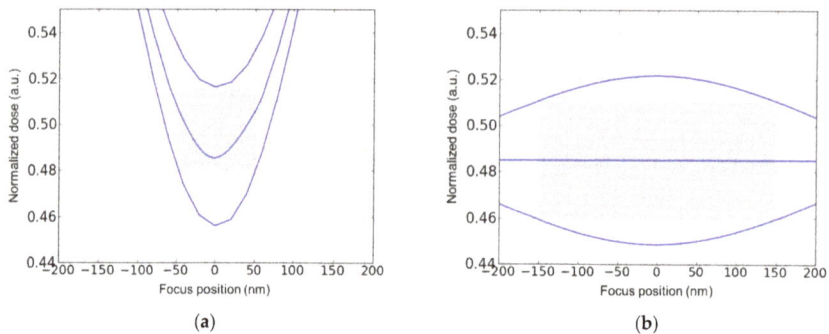

Figure 3. Aerial image-based process window on a 193 nm dry lithography system with adapted illumination as in Figure 2, for (**a**) 65 nm lines/130 nm spaces (duty factor 1:2). (**b**) 65 nm dense lines: 65nm lines and spaces with a duty factor of 1:1.

Another important non-ideality which challenges aggressive patterning steps is the printing of defects. These include especially defects in the multilayer mirrors used for masks and instead of lenses in an Extreme Ultraviolet (EUV) lithography step, or defects of the mask structures [20]. Equipment and mask makers generally try to compensate mirror defects [20]. Contamination with particles during the handling of masks can be compensated by the use of pellicles which move the particles out of the mask area and in turn make sure that they do not print [21]. The effect of defects of the mask itself should be considered as a process variation insofar as the patterns generated on the wafer may deviate from the shapes which would be generated without a defect. Figure 4 shows an example for mask defect printing as frequently simulated in the lithography community. However, the impact on device level has so far not been considered systematically. In order to study this effect, the subsequent etching of the underlying layer to be patterned by the lithography step must be considered. This needs the intimate coupling of lithography and etching simulation, which is one of the core topics of the SUPERAID7 project. Figure 5 shows how the image of the mask defect shown in Figure 4 is transferred into the underlying layer for the two extreme cases of anisotropic and isotropic etching. The etching simulator ANETCH [16] used for this example can be calibrated to the specific etching process applied and therefore enables the assessment of defect transfer to patterned layers.

Shrinking device dimensions have led to more complex device architectures in order to maintain sufficient electrostatic control of the channel, low leakage and high drive currents. In order to extend CMOS scaling to its limits, first buried oxide layers were introduced for the so-called SOI (silicon-on-insulator) transistors. This has been followed by transistor architectures such as FinFETs, (stacked) nanowires and (stacked) nanosheets. These small and truly three-dimensional structures lead to considerable challenges for the patterning processes and for their simulation. As outlined below, they also generate different problems in terms of variability and impact on design. The problem starts from the basic limitation of optical lithography, where the minimum half pitch printed is given by $d = k_1 \cdot \lambda / NA$, where k_1 is the technology factor with minimum value of 0.25 for dense lines, λ is the wavelength of the light used, and NA is the numerical aperture, equal to the sine of the opening angle of the last lens times the refractive index of the medium between lens and wafer (1 for air, and 1.44 for water). With 193 nm being the minimum wavelength at which lenses are sufficiently transparent, a minimum half pitch for dense lines of about 34 nm results in case of water immersion lithography.

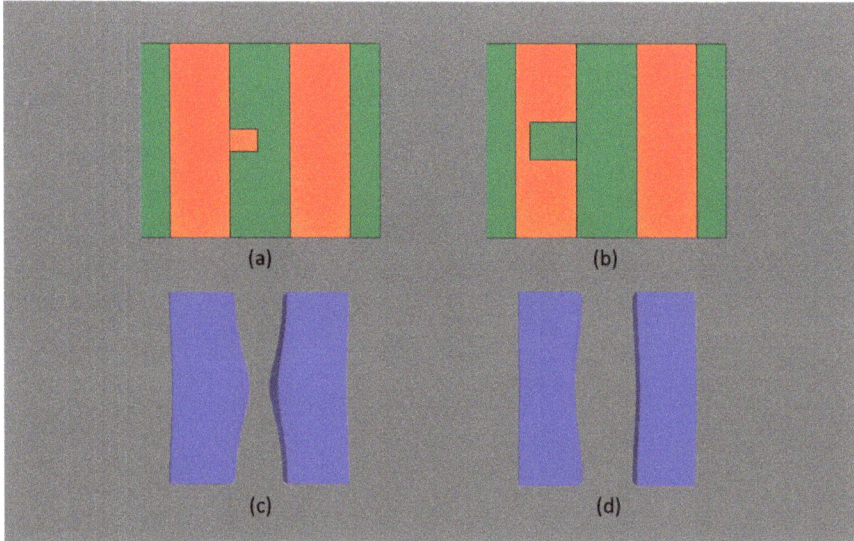

Figure 4. Example for defect printing in 193 nm dry optical lithography. Mask defects considered (65 nm dense lines/spaces with (**a**) 30 nm extrusion and (**b**) 50 nm intrusion; (**c**) and (**d**) resulting resist geometries for (**a**) and (**b**), respectively, for lithography with ideal focus and dose.

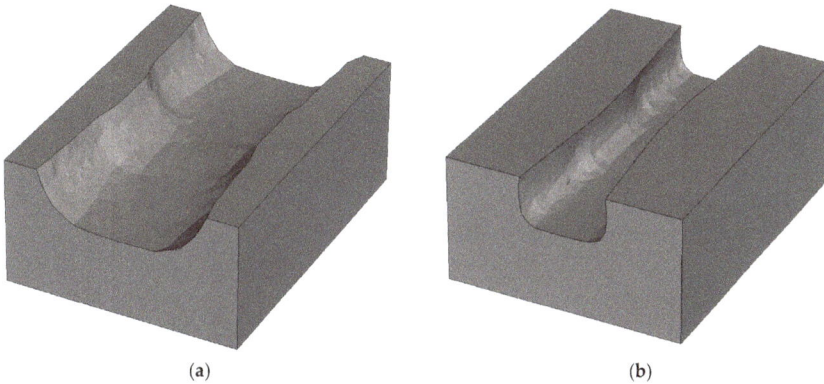

Figure 5. Simulation of oxide etching using the resist shape of Figure 4c as a mask. (**a**) Isotropic etching. (**b**) Anisotropic etching. To allow better visualization of the etched material, the resist mask is not shown. The lateral width of the simulation domain is 190 nm.

The favorite industrial approaches for the patterning of nanoscale devices smaller than that limit are various kinds of double or even quadruple patterning. In Litho-Etch-Litho-Etch (LELE) [22] and Litho-Freeze-Litho-Etch (LFLE) [23] the number of features is doubled and the pitch size halved, by exposing the wafer twice with two different masks (e.g., the mask shifted by half the pitch for the second illumination), employing some memory process of the photoresist and finally developing the resist after the second illumination. Consequently, focus and especially dose variations are not correlated between these two incremental lithography steps. This has severe consequences on circuit level: e.g., for an SRAM cell the densest structures are the polysilicon gate lines, see Figure 6. As shown therein, the three transistors T1, T2, and T6 are patterned with one illumination step, whereas the transistors T3, T4, and T 5 are patterned with the other illumination step. In turn, focus and dose variations correlate within each of the two triples of transistors, but especially in terms of dose not between the two triples. This has severe consequences for variability-aware circuit simulation. Considering dose variations in the double patterning step for example, one value of the varying dose must be used for the first triple and a (likely different) value for the second triple. In turn, the dose variations may, e.g., cause an increase of the CD for the first triple and a decrease for the second one, leading to negative consequences for circuit performance. This was discussed in more detail elsewhere [10].

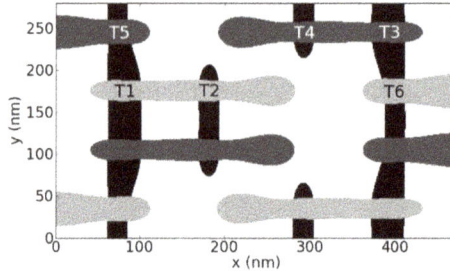

Figure 6. Top-view of lithography simulation of an SRAM cell. Black vertical features are the active silicon areas, the horizontal light grey and dark grey features are the polysilicon gate areas generated with the first and the second incremental lithography step, respectively. Transistors T1 to T6 are explained in the text.

A different situation holds for Self-Aligned Double Patterning (SADP): Here, first a carbon layer is patterned by optical lithography. Then, spacers are created by deposition followed by back etching, see Figure 7. In result, the pattern density is doubled compared with the initial pattern of the carbon lines. The CDs of the final spacers are nearly independent of the lithography step, and are largely defined by the deposition and etching processes used. Variations of the CDs depend on the parameters of the etching and deposition processes [24]. Because etching and deposition partly depend on the open view angle of the surface element towards the reactor volume, a layout dependence of the CD is introduced: The CD of outer lines is different from the CD of inner lines. An important target of process optimization is the tuning of the deposition and etching processes in order to minimize their variability across the wafer and the layout impact explained above.

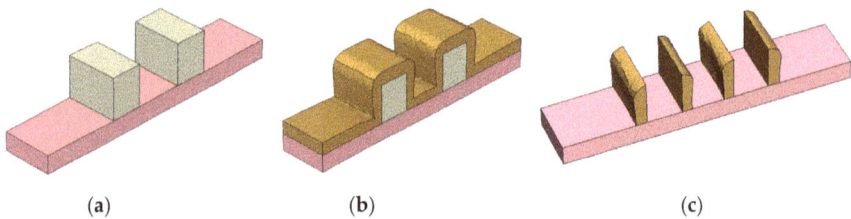

(a)   (b)   (c)

Figure 7. Example for the simulation of Self-Aligned Double Patterning: (**a**) carbon lines patterned by optical lithography; (**b**) layer deposition; (**c**) spacer pattern after back etching.

For current FinFET transistors, usually SADP is applied for the patterning of the fins as dense lines, whereas LELE is used for gate patterning, because it is more flexible in terms of the layouts.

Another design issue appears in case of EUV. Here, soft X-rays with 13.5 nm wavelength are used instead of laser light. Because lenses are not sufficiently transparent for light below 193 nm, in EUV instead of refractive optics, reflective optics are used, consisting of multilayer mirrors and new absorber materials [20]. Whereas dose and defocus continue to be major sources of systematic process variability, EUV leads to an additional design issue, which has so far not been addressed when discussing process flows for the fabrication of advanced transistors. Because the reflected light must be separated from the incoming light, illumination of all mirrors and especially of the mask is not vertical but tilted by a so-called "chief ray angle" of about 5.3°, see Figure 8. In turn, features generate an asymmetric shadowing relative to the plane of incidence, resulting in a position shift, a CD increase and a telecentricity error. The first two effects can be compensated by the so-called "Optical Proximity Correction" (OPC) techniques. The third effect introduces an additional source of variations since it causes a focus dependent position of the features. Furthermore, the absorber features are thick compared to the illumination wavelength resulting in phase deformations causing mask-induced aberration like effects. Finally, a new class of defects, the multilayer defects, introduce new issues in the field of defect detection and repair.

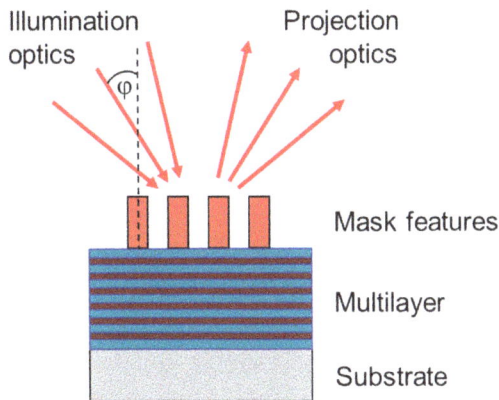

Figure 8. Principle of illumination of an EUV mirror and mask. φ denotes the chief ray angle.

In summary, various kinds of systematic, layout-dependent and defect-induced variations influence the patterns generated in a lithography or multiple patterning process. There are major even qualitative differences in terms of variability between optical lithography, SADP and other double patterning processes. Etching and deposition, which are key components of SADP, are discussed in the next chapter.

3.1.2. Deposition and Etching

In semiconductor technology, a large variety of deposition and etching processes is employed. Deposition processes differ among others in terms of the properties of the deposited layers and the layer conformality. The latter for instance critically influences the filling of contact holes and trenches. Etching processes differ, e.g., in terms of selectivity concerning the materials to be etched, etch-induced damage, and the degree of isotropy and anisotropy. Equipment simulation tools such as CFD-ACE+ [25] and Q-VT [26], which model the electrical, thermal, fluid-dynamical, and plasma properties of the reactor can be used to predict the fluxes and energies of different species (ions, neutrals) above the flat or patterned wafer. Using these as boundary conditions, feature-scale simulation allows one to simulate the evolving geometry during deposition or etching. To this end, the interaction of the different species with the substrate is considered and local values of the different fluxes are extracted which allow one to determine local etching or deposition rates. Examples for this approach are given in [27] for deposition and in [28] for etching. The discussion of the large variety of deposition and etching processes is beyond the scope of this paper.

Deposition and etching processes are subject to several sources of systematic variability. Depending on the process in question, these include the inhomogeneity of gas flow and temperature in the etching or deposition reactor, the non-homogeneous emission of metal atoms, e.g., from a sputter target, or the finite size of a sputter target. These sources lead to variations of deposition and etching rates across the wafer. Furthermore, non-vertical incidence of contributing species can lead to layout-dependent deposition or etching rates, as discussed above for the SADP process. An overview of sources of variability in deposition and etching processes is given elsewhere [1].

DEP3D and ANETCH allow the simulation of these deposition and etching processes, provided that physical/chemical models suitable for the process in question have been implemented, parameters have been calibrated, and the required boundary conditions above the wafer are known, e.g., from equipment simulation. As an example, Figure 9 shows cross-sections of three-dimensional simulations of a contact hole etching process for two different angular distributions of the rate determining species: Figure 9a shows the result for purely physical sputter etching with a highly directional ion flux, Figure 9b depicts the result for chemical dry etching, that is, for an isotropic angular distribution of the etching species. Figure 10 shows simulation results for long-throw sputter deposition into a contact hole structure for different positions on the wafer. The asymmetry of the deposited layer for the off-axis positions of the contact hole is clearly visible.

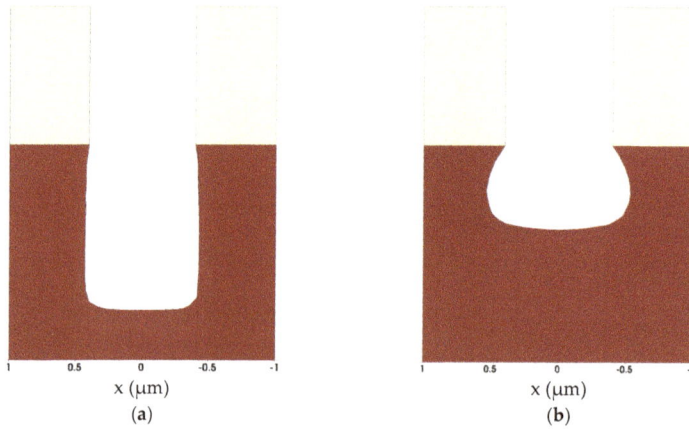

Figure 9. Example for the simulation of contact hole etching: (**a**) sputter etching with highly directional ion flux; (**b**) chemical dry etching with an isotropic angular distribution of the etching species. The material to be etched (for instance oxide) is shown in dark red. The mask which is assumed to be not affected by the etching process is depicted in yellow.

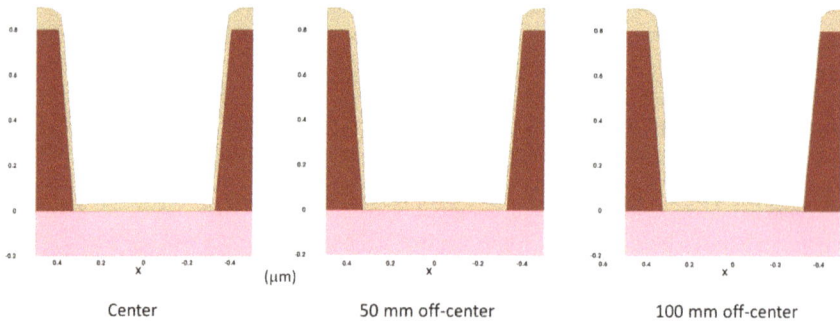

Figure 10. Example for the simulation of long-throw sputter deposition into a contact hole structure for different positions on the wafer. Ballistic transport of the metal atoms in the reactor is assumed, the target diameter and the distance between target and substrate were set to 300 mm and 150 mm, respectively. The step coverage (= ratio of the thickness at the sidewall bottom to the thickness on top) for the feature position at the center, 50 mm off-center, and 100 mm off-center is here given for sidewall left | sidewall right: 0.16 | 0.16, 0.16 | 0.12, and 0.20| 0.05, respectively.

3.1.3. Integrated Topography Simulation

As outlined above, advanced nanoscale devices generally employ truly three-dimensional geometries. Furthermore, the approximation of these structures using rectangular shapes is no more possible, because the electrical performance depends on details, such as taper angles of FinFETs or corner rounding. Moreover, as discussed above already basic features such as line width depend on the three-dimensional interim structures generated during the patterning flow, e.g., in the form of non-vertical and/or non-straight resist edges. In turn, one of the key tasks in the SUPERAID7 project has been the development of an integrated three-dimensional topography simulator, which combines Dr.LiTHO [15], ANETCH [16], DEP3D [17], which all output a triangle-based surface representation, and the level-set based etching and deposition simulator ViennaTS [18]. To this end, DEP3D and ANETCH have been seamlessly integrated into the Python [29] simulation framework of Dr.LITHO. The ViennaTS level-set-method-based tool has been exposed to the Python programming language to allow the usage as a Python module. A common geometry conversion engine has been developed to handle the different data representations used in the tools, including surface and volume meshes, structured and unstructured grids. The physical models from DEP3D, ANETCH and ViennaTS are available in the integrated topography simulator and enable the simulation of a large variety of deposition and etching processes. Figure 11 shows the architecture of the integrated topography simulator.

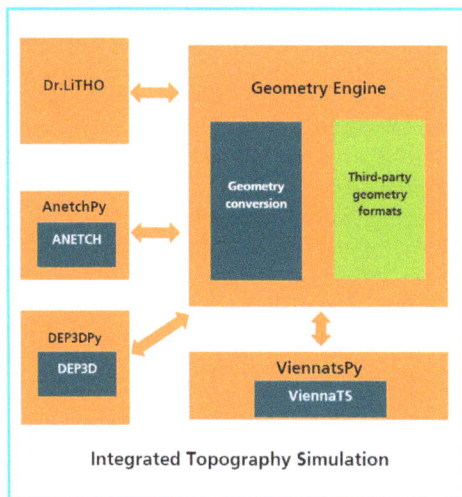

Figure 11. Integrated 3D topography simulation, combining Dr.LiTHO, DEP3D, ANETCH, and ViennaTS.

Figure 12 depicts an example for the combined simulation of lithography and etching: As layout, an SRAM cell pattern on polysilicon level has been used. The lithography simulation was performed with Dr.LiTHO assuming 193 nm water immersion with a strong off-axis illumination and unpolarized light. Dr.LiTHO provides the resist profile as a triangulated surface. To reduce the number of surface elements for the etching simulation, the footprint of the resist was extracted, smoothed and the 3D resist region was generated with steep sidewalls. Figure 12a shows one polysilicon line forming part of the SRAM layout. To study the effect of varying parameters on the etched profile, it is efficient to simulate etching of a slice located at the position where, according to the SRAM layout, the gate electrode located is above the active region. The etching simulations have been carried out with ANETCH using coupling to equipment simulation thus allowing one to study the effect of the feature position on the wafer. Details are provided elsewhere [28]. Briefly, an inductively-coupled plasma reactor is simulated, which is operated at a pressure of 1.5 Pa and powered with 1500 W, and the substrate is biased with 200 V. Cross sections of the etched gate electrode for two different positions on the wafer are shown in Figure 12b,c respectively. It can be seen that the shape particularly at the foot of the gate electrode significantly depends on the position on the wafer.

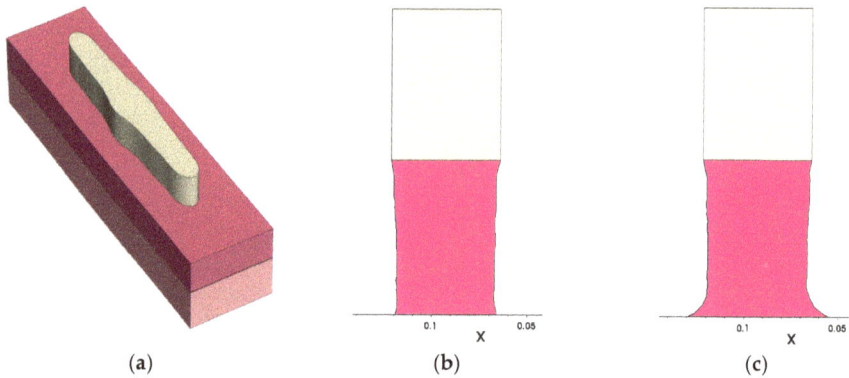

(a)   (b)   (c)

Figure 12. Coupled simulation of lithography and etching: (**a**) 3D simulation of the resist line geometry (yellow) above the polysilicon (purple); (**b**) simulated etched gate electrode for a position close to the center of the wafer (distance to center axis = 0.14 cm); (**c**) Simulated etched gate electrode for a position at the wafer rim (distance to center axis = 9.7 cm). The unit for the x axis in (**b**) and (**c**) is micron.

3.2. Integration with other Tools

In terms of process simulation and systematic process variations, the SUPERAID7 project has focused on topography simulation and the variations affecting the topography processes lithography, etching, and deposition: These processes and variations raise new challenges especially for three-dimensional nanoscale transistors as addressed in SUPERAID7 and in this paper.

For the simulation of the other process steps, especially ion implantation and dopant activation/diffusion, the well-established process simulator Sentaurus Process [13] has been used. To this end, results from SUPERAID7 topography simulation are transferred to Sentaurus Process via the DF-ISE file format of Synopsys. More specifically, the exported results are used by Sentaurus Process to update an initial structure by means of Boolean operations. Device simulation has been carried out with Sentaurus Device [13], and statistical device simulation with Garand of GSS (now belonging to Synopsys). To this end the SUPERAID7 topography simulation tools have been integrated with Sentaurus Process, the Sentaurus Workbench, and the statistical device simulator Garand. This overall hierarchical approach implemented in SUPERAID7 to simultaneously simulate the impact of systematic and stochastic variations on devices and circuits has been discussed elsewhere [30].

3.3. Compact Model Extraction

Because conventional compact models are not suitable for the three-dimensional devices as studied in SUPERAID7, a novel compact model (LETI-NSP) based on surface potential has been developed by CEA/Leti and used for the compact model extraction work in SUPERAID7. This model has among others been presented at IEDM 2016 [31].

Within SUPERAID7 the hierarchical approach developed at GSS (now Synopsys) to extract statistical compact models [8] has been used and extended. In the traditional approach illustrated in Figure 13 (without the red extension "and/or process parameters"), first a so-called "uniform" compact model is extracted from device simulations without taking process variations into account. Then, process corners are defined for the process results (e.g., channel length L and width W) generated by the systematic variations (e.g., focus F and dose D in a lithography step) to be considered. Following this, an extended compact model is extracted from device simulations at these process corners, using a first group of compact model parameters defined during this extraction. Finally, this compact model is further extended into a statistical compact model, extracted from statistical device simulations carried out at each of the process corners. This compact model then allows one to calculate for relevant electrical device data A (e.g., the threshold voltage or the leakage current) the probability distribution P(A,s) that certain values of A occur. This distribution depends on values s (e.g., L and W) of the process results generated by the systematic variations of process input (or equipment) data r (e.g., F and D) considered in the study. In order to get the final distribution of A, process simulations are carried out to calculate the dependence s(r) of the process results on the (equipment) parameter r. The final distribution P_A of the electrical device data is then given as:

$$P_A = \int P(A,s(r)) \cdot f(r) \, dr \tag{1}$$

where f(r) is the probability distribution of the (equipment) parameter r, e.g., the distributions of focus F and dose D.

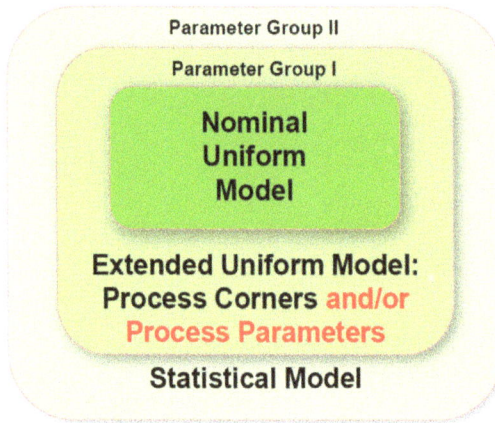

Figure 13. Extraction and generation of hierarchical compact model aware of systematic and statistical process variations.

Here, in general A is a vector, because various electrical properties are considered. As soon as more than one statistical variation occurs, also r is a vector. It should be noted that in case of more than one statistical variation, the components of r should be statistically independent. Otherwise it would be necessary to find the root cause of their dependence, and express this by relating them to some (hidden) parameters which cause the (partial) correlation.

Since one statistical variation may influence several process results (e.g., the channel length of one transistor and the channel width of another), s(r) is generally a matrix. This also allows the proper description of correlations, which arise in the case that an input parameter r influences more than one (intermediate or final) result parameter. The same holds for P(A,s).

In examples studied earlier (e.g., in [8]), process corners referred to the minimum and maximum values of two or three discrete parameters affected by systematic process variations, such as channel length and width, influenced among others by focus and dose variations in lithography. However, for three-dimensional nanoscale devices systematic process variations might not only lead to variations of such discrete parameters, but also change the shape of the device, e.g., the channel cross section. Such variations can hardly be described with the variation of just one or two geometrical parameters. In turn, the hierarchical compact model extraction strategy has been extended in SUPERAID7 as illustrated in Figure 13 (red text): The extended uniform compact model is not only based on process corners (e.g., channel length and width), but may also be based on the minimum and maximum values of statistically varying process parameters (e.g., focus F and dose D in lithography). In turn, compact model extraction is then not based only on the results of device simulation, but also on the result of coupled process and device simulation, needed in order to trace the impact of the parameter r on the device properties. If the extended uniform compact model is based on statistically varying process parameters r only, Equation (1) will simplify to:

$$P_A = \int P(A,r) \cdot f(r) \, dr \qquad (2)$$

The mixed case where both process corners s and statistically varying process parameters r are used in the extraction of the extended compact model must be treated with some care, because only variables can be used in the compact model extraction which do not depend on each other and can therefore be treated independently. In turn, it must be differentiated between varying process parameters r_1 which influence process corners s, and varying process parameters r_2 which do not influence process corners s. The probability distribution of an electrical device performance parameter A is then given by

$$P_A = \int P(A,s(r_1), r_2) \cdot f_1(r_1) \cdot f_2(r_2) \, dr_1 \, dr_2 \qquad (3)$$

where $f_1(r_1)$ and $f_2(r_2)$ are the probability distributions of the varying process parameters r_1 and r_2, respectively.

The selection of the process corners and/or the varying process parameters to be considered in the extraction of the group I parameters (see Figure 13) depends on an a-priori assessment of the relevant variations. This assessment can be based on technological knowledge on which variations are most important, and/or on numerical simulation where each potentially relevant variation is considered alone. Then, the most relevant systematical variations are identified and used in the compact extraction process outlined above.

Generally, several different systematic variations may influence a device. Tracing all their combinations in parallel through the full simulation of the process sequence would require far too much effort. In turn, suitable design-of-experiment approaches must be used to limit the number of simulation splits. A simple and efficient approach is to first disregard variations which are due to expert knowledge already known to have negligible impact. Next, the impact of the remaining variations should be considered one by one, to identify a small number of most important variations. These variations could then either be considered in parallel, or more refined design-of experiment techniques could be employed to skip some or several elements of the full matrix of simulation splits.

3.4. Example for Compact Model Extraction

Figure 14 shows a TEM micrograph of a nanowire transistor investigated in the benchmarks carried out in the SUPERAID7 project. The process flow included among others the deposition of a sacrificial SiGe layer on top of an SOI substrate, plus an upper Si layer. Self-Aligned Double patterning was used to create Si/SiGe/Si fins, and optical lithography for gate patterning. For this device architecture and the process flow used, ten potentially relevant systematical variations were identified. Their relative impact on the transistor performance was studied with coupled process and device simulation. Among these parameters, the three most important ones were identified [30]: (1) The SiGe mole fraction x_{Ge} which influences the etching of the inner spacers; (2) the fin SADP deposition factor d_{sadp} which gives the relative variation of the deposition thickness from the nominal thickness in the SADP spacer creation process; (3) the defocus F_{Gate} of the lithography step used to pattern the gate. Both for n-type NMOS and p-type PMOS MOSFETs, F_{gate} had the largest impact on both saturation current and leakage current.

Coupled process and device simulation was then carried out for the process corners of these three statistical variations. Figure 15 shows as an example the impact of these variations on NMOS and PMOS, respectively. The horizontal axis shows the minimum, the nominal and the maximum value of the varying parameters. Except for the parameter studied for each of the curves, the parameters have their nominal values. For instance, for the blue curve in Figure 15, x_{Ge} is varied whereas d_{sadp} and F_{gate} have their nominal values. The V-shape for the impact of F_{gate} is an intrinsic feature of the dependence of CD on defocus in lithography, discussed above in connection with Figure 2. For d_{sadp} a technologically appropriate variation was assumed. Its impact was, however, smaller than the impact of numerical discretization errors. In turn, in further work, a larger variation will be assumed here for the extraction of the compact model, whereas the realistic value can and will then be introduced via the probability distribution f(d_{sadp}) in Equation (2). Based on these variations, the extended compact model was extracted, using the Mystic [13] tool from Synopsys. Figure 16 shows as an example the comparison between the saturation current simulated by TCAD and the results from the extended compact model extracted. Further work includes among others the updated extraction of the dependence on the variations of X_{Ge} and the final extraction of the statistical compact model, as indicated in Figure 13. The result will be reported elsewhere.

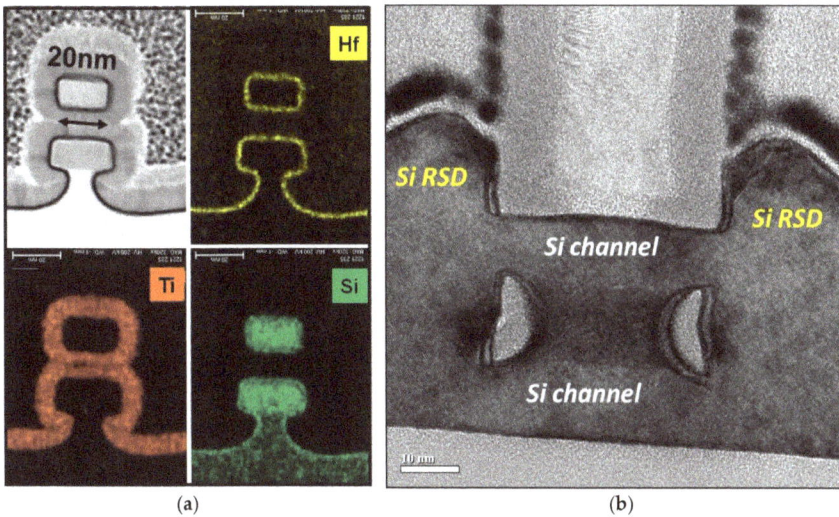

Figure 14. Example of nanowire transistors considered in this work: (**a**) cross section, including identification of Hf, Ti, and Si layers; (**b**) cut along source-drain direction, with SiN inner spacers.

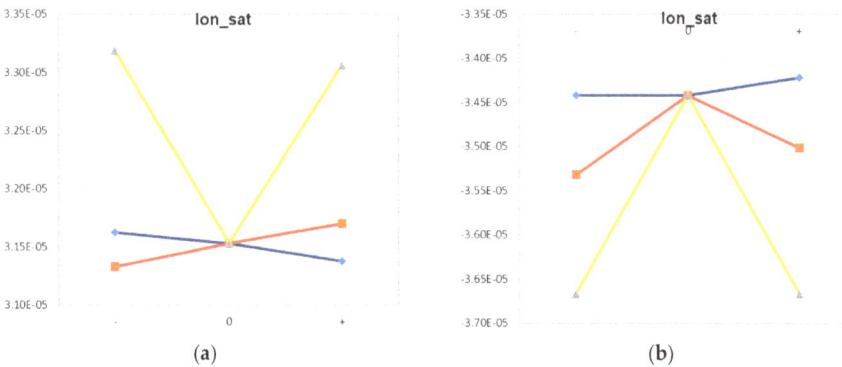

Figure 15. Example for extraction of extended compact model: Results for saturation current of (**a**) NMOS; (**b**) PMOS. Key: blue—x_{Ge}; orange—d_{sadp}; yellow—F_{gate}. The horizontal axis shows the minimum, the nominal and the maximum value of the varying parameters.

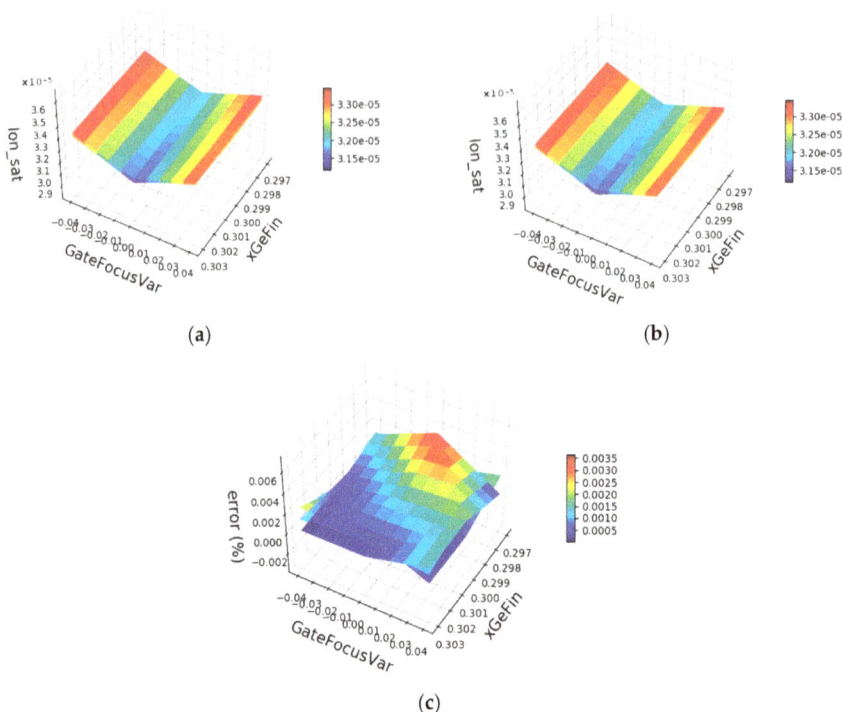

Figure 16. Example for extraction of extended compact model: Comparison between results (**a**) from TCAD; (**b**) from the extended compact model for the saturation current (color scale in Ampere of a nanowire NMOS; (**c**) relative error (color scale in %) of the compact model compared with TCAD.

4. Discussion

Especially for three-dimensional nanoscale transistors, the details of the geometries generated during device fabrication are important both for the nominal device performance and for its stability against systematic variations caused by process equipment and/or by layout effects. Other than for statistical variations which can be simulated with stand-alone statistical device simulation, there is a huge variety of potential systematic variations, which must be addressed with various equipment simulation tools and process simulation modules. In turn, it is indispensable for optimized device and circuit manufacturing to well identify and characterize the sources of systematic variability, to sort out those with the largest impact, depending on the device and circuit in question, and to quantify their joint impact on the final nanoelectronic product. The impacts of systematic and statistical variability must be simulated and minimized in parallel, by adapting process flows, but potentially also device architectures and even fabrication equipment. In this paper, a holistic approach to use coupled equipment, process and device simulation combined with the extraction of variation-aware compact models has been presented together with some examples. The approach is partly based on well-established tools used in industry, partly on additional modules and extensions which are compatible to such standard frameworks. The suggestions made are intended to support industry in their efforts to optimize the stability of their products against all kinds of process variations.

Author Contributions: J.L. coordinated the acquisition of the funding and managed the overall SUPERAID7 research project. P.E. and E.B. carried out the lithography and topography simulation. S.B. defined benchmarks and provided experimental data. F.K., A.B., and L.W. carried out benchmark simulations. J.L. wrote the paper, with figures provided by all co-authors. All authors critically reviewed the manuscript.

Funding: The research leading to these results has received funding from the European Union's Horizon 2020 research and innovation programme under grant agreement No. 688101 SUPERAID7.

Acknowledgments: The authors want to acknowledge the discussions with and the contributions of all members of the SUPERAID7 team at Fraunhofer IISB, CEA/Leti, University of Erlangen-Nuremberg, University of Glasgow, Synopsys and TU Wien, especially of the work package leaders A. Asenov and C. Millar who are not among the co-authors.

Conflicts of Interest: The authors declare no conflict of interest.

References

1. Lorenz, J.K.; Bär, E.; Clees, T.; Jancke, R.; Salzig, C.P.J.; Selberherr, S. Hierarchical Simulation of Process Variations and Their Impact on Circuits and Systems: Methodology. *IEEE Trans. Electron Devices* **2011**, *58*, 2218–2226. [CrossRef]
2. Horizon 2020 Project SUPERAID7. Available online: www.superaid7.eu (accessed on 18 October 2018).
3. Asenov, A.; Brown, A.R.; Davies, J.H.; Kaya, S.; Slavcheva, G. Simulation of intrinsic parameter fluctuations in decananometer and nanometer-scale MOSFETs. *IEEE Trans. Electron Devices* **2003**, *50*, 1837–1852. [CrossRef]
4. Takeuchi, K.; Nishida, A.; Hiramoto, T. Random Fluctuations in Scaled MOS Devices. In Proceedings of the 2009 International Conference on Simulation of Semiconductor Processes and Devices (SISPAD), San Diego, CA, USA, 9–11 September 2009; pp. 79–85.
5. Brown, A.R.; Idris, N.M.; Watling, J.R.; Asenov, A. Impact of Metal Grain Granularity on Threshold Voltage Variability: A Full-Scale Three-Dimensional Statistical Simulation Study. *IEEE Electron Device Lett.* **2010**, *31*, 1199–1201. [CrossRef]
6. Lorenz, J.K.; Bär, E.; Clees, T.; Evanschitzky, P.; Jancke, R.; Kampen, C.; Paschen, U.; Salzig, C.P.J.; Selberherr, S. Hierarchical Simulation of Process Variations and Their Impact on Circuits and Systems: Results. *IEEE Trans. Electron Devices* **2011**, *58*, 2227–2234. [CrossRef]
7. Lorenz, J.; Baer, E.; Burenkov, A.; Evanschitzky, P.; Asenov, A.; Wang, L.; Wang, X.; Brown, A.R.; Millar, C.; Reid, D. Simultaneous Simulation of Systematic and Stochastic Process Variations. In Proceedings of the 2014 International Conference on Simulation of Semiconductor Processes and Devices (SISPAD), Yokohama, Japan, 9–11 September 2014; pp. 289–292.
8. Wang, X.; Reid, D.; Wang, L.; Burenkov, A.; Millar, C.; Cheng, B.; Lange, A.; Lorenz, J.; Baer, E.; Asenov, A. Variability-Aware Compact Model Strategy for 20-nm Bulk MOSFETs. In Proceedings of the 2014 International Conference on Simulation of Semiconductor Processes and Devices (SISPAD), Yokohama, Japan, 9–11 September 2014; pp. 293–296.
9. Wang, X.; Reid, D.; Wang, L.; Burenkov, A.; Millar, C.; Lorenz, J.; Asenov, A. Hierarchical Variability-Aware Compact Models of 20 nm Bulk CMOS. In Proceedings of the International Conference on Simulation of Semiconductor Processes and Devices (SISPAD), Washington, DC, USA, 9–11 September 2015; pp. 325–328.
10. Evanschitzky, P.; Burenkov, A.; Lorenz, J. Double Patterning: Simulating a Variability Challenge for Advanced Transistors. In Proceedings of the International Conference on Simulation of Semiconductor Processes and Devices (SISPAD), Glasgow, Scotland, UK, 3–5 September 2013; pp. 105–108.
11. Baer, E.; Burenkov, A.; Evanschitzky, P.; Lorenz, J. Simulation of process variations in FinFET transistor patterning. In Proceedings of the 21st International Conference on Simulation of Semiconductor Processes and Devices (SISPAD), Nuremberg, Germany, 6–8 September 2016; pp. 299–302.
12. Wang, X.; Reid, D.; Wang, L.; Millar, C.; Burenkov, A.; Evanschitzky, P.; Baer, E.; Lorenz, J.; Asenov, A. Process Informed Accurate Compact Modelling of 14-nm FinFET Variability and Application to Statistical 6T-SRAM Simulations. In Proceedings of the 21st International Conference on Simulation of Semiconductor Processes and Devices (SISPAD), Nuremberg, Germany, 6–8 September 2016; pp. 303–306.
13. Synopsys. TCAD. Available online: https://www.synopsys.com/silicon/tcad.html (accessed on 18 October 2018).
14. Europractice Software Service. Available online: http://www.europractice.stfc.ac.uk/welcome.html (accessed on 18 October 2018).
15. IISB R&D lithography simulator Dr.LiTHO. Available online: www.drlitho.com (accessed on 18 October 2018).
16. IISB etching simulator ANETCH. Available online: https://www.iisb.fraunhofer.de/en/research_areas/simulation/topography/ANETCH1.html (accessed on 18 October 2018).

17. IISB deposition simulator DEP3D. Available online: https://www.iisb.fraunhofer.de/en/research_areas/simulation/topography/dep3d.html (accessed on 18 October 2018).

18. TU Wien topography simulator. Available online: http://www.iue.tuwien.ac.at/software/viennats/ (accessed on 18 October 2018).

19. Kwok-Kit Wong, A. *Optical Imaging in Projection Microlithography*; SPIE Press: Bellingham, WA, USA, 2005; pp. 126–127, ISBN 978-0-8194-5829-5.

20. Bakshi, V. *EUV Lithography*; SPIE Press: Bellingham, WA, USA, 2009; pp. 325–381, ISBN 978-0-8194-6964-9.

21. Lapedus, M. Next EUV Challenge: Pellicles, Semiconductor Engineering. Available online: www.semiengineering.com (accessed on 15 November 2018).

22. Dusa, M.; Arnold, B.; Finders, J.; Meiling, H.; van Ingen Schenau, K.; Chen, A.C. The lithography technology for the 32 nm HP and beyond. *Proc. SPIE* **2008**, *7028*, 702810. [CrossRef]

23. Bae, Y.C.; Liu, Y.; Cardolaccia, T.; McDermott, J.C.; Trefonas, P. Materials for single-etch double patterning process: surface curing agent and thermal cure resist. *Proc. SPIE* **2009**, *7273*, 727306. [CrossRef]

24. Baer, E.; Lorenz, J. The Effect of Etching and Deposition Processes on the Width of Spacers Created during Self-Aligned Double Patterning. In Proceedings of the 2018 International Conference on Simulation of Semiconductor Processes and Devices (SISPAD), Austin, TX, USA, 24–26 September 2018; pp. 236–239.

25. *Simulator CFD-ACE+*; ESI Group: Paris, France, 2018.

26. *Q-VT Plasma Processing Simulator*; Quantemol Ltd.: London, UK, 2018.

27. Baer, E.; Evanschitzky, P.; Lorenz, J.; Roger, F.; Minixhofer, R.; Filipovic, L.; de Orio, R.L.; Selberherr, S. Coupled simulation to determine the impact of across wafer variations in oxide PECVD on electrical and reliability parameters of Through-Silicon Vias. *Microelectron. Eng.* **2015**, *137*, 141–145. [CrossRef]

28. Baer, E.; Kunder, D.; Evanschitzky, P.; Lorenz, J. Coupling of Equipment Simulation and Feature-Scale Profile Simulation for Dry-Etching of Polysilicon Gate Lines. In Proceedings of the 15th International Conference on Simulation of Semiconductor Processes and Devices (SISPAD 2010), Bologna, Italy, 6–8 September 2010; pp. 57–60.

29. Python. Available online: www.python.org (accessed on 15 November 2019).

30. Lorenz, J.K.; Asenov, A.; Baer, E.; Barraud, S.; Kluepfel, F.; Millar, C. Process Variability for Devices at and beyond the 7 nm Node. *J. Solid State Sci. Technol.* **2018**, *7*, 595–601. [CrossRef]

31. Rozeau, O.; Martini, S.; Poiroux, T.; Triozon, F.; Barraud, S.; Lacord, J.; Niquet, Y.M.; Tabone, C.; Coquand, E.; Augende, E.; Vinet, M.; Faynot, O.; Barbe, J.-Ch. Physical Compact Model for Stacked-planar and Vertical Gate-All-Around MOSFETs. In Proceedings of the International Electron Devices Meeting (IEDM), San Francisco, CA, USA, 3–7 December 2016.

micromachines

MDPI

Article

Empirical and Theoretical Modeling of Low-Frequency Noise Behavior of Ultrathin Silicon-on-Insulator MOSFETs Aiming at Low-Voltage and Low-Energy Regime

Yasuhisa Omura [1,2]

[1] Department of Electrical, Electronics and Information Engineering, Kansai University, Yamate-cho, Suita 564-8680, Japan; omuray@kansai-u.ac.jp; Tel.: +81-663-681-121
[2] Academic Collaboration Associate (ACA), Isehara 259-1135, Japan

Received: 4 November 2018; Accepted: 18 December 2018; Published: 22 December 2018

Abstract: This paper theoretically revisits the low-frequency noise behavior of the inversion-channel silicon-on-insulator metal-oxide-semiconductor field-effect transistor (SOI MOSFET) and the buried-channel SOI MOSFET because the quality of both Si/SiO_2 interfaces (top and bottom) should modulate the low-frequency fluctuation characteristics of both devices. It also addresses the low-frequency noise behavior of sub-100-nm channel SOI MOSFETs. We deepen the discussion of the low-frequency noise behavior in the subthreshold bias range in order to elucidate the device's potential for future low-voltage and low-power applications. As expected, analyses suggest that the weak inversion channel near the top surface of the SOI MOSFET is strongly influenced by interface traps near the top surface of the SOI layer because the traps are not well shielded by low-density surface inversion carriers in the subthreshold bias range. Unexpectedly, we find that the buried channel is primarily influenced by interface traps near the top surface of the SOI layer, not by traps near the bottom surface of the SOI layer. This is not due to the simplified capacitance coupling effect. These interesting characteristics of current fluctuation spectral intensity are explained well by the theoretical models proposed here.

Keywords: low-frequency noise; silicon-on-insulator; MOSFET; inversion channel; buried channel; subthreshold bias range; low voltage; low energy; theoretical model

1. Introduction

It has long been considered that there are two possible explanations for the low-frequency noise (LFN) exhibited by metal-oxide-semiconductor field-effect transistors (MOSFETs). They are carrier density fluctuations due to interface traps near the oxide/semiconductor interface [1–3], and carrier mobility fluctuations [4–6]. Theoretical models have been proposed to comprehensively understand such LFN characteristics (frequently, the 1/f noise) [7–12]. Hooge introduced a specific parameter (the so-called Hooge parameter) to characterize the 1/f noise [4]. Related to these theories, the quantum 1/f noise model was proposed by Peter H. Handel [7,8]. However, the physical origins of 1/f noise are not simple and remain controversial [13–15] because it is anticipated that the carrier density fluctuation and the carrier mobility fluctuation may be correlated in some cases [11]. In addition, it is considered that the difficulty of understanding the 1/f noise behavior stems from the fact that the Hooge parameter depends on device material and structure [9,15,16].

Although some people have challenged a deeper understanding of the mechanisms of 1/f noise [9,15], clear separation of the aforementioned noise sources, such as carrier density fluctuation and carrier mobility fluctuation, remains rather unclear [15].

The LFN characteristics of various buried-channel MOSFETs (BC-MOSFETs) [17] and various inversion-channel MOSFETs (IC-MOSFETs) [18] have already been discussed in detail based on Hooge's idea; it was clarified phenomenologically and theoretically that a useful interpretation of the aspects of the Hooge parameter may be possible depending on how the two fluctuation modes (the carrier density fluctuation and the carrier mobility fluctuation) are correlated [17,18]. The gate voltage dependence of the Hooge parameter was explained well by correlating the carrier mobility fluctuation to the carrier density fluctuation. The proposed model gives a valid fundamental physical basis for interpreting various aspects of the Hooge parameter [16,19,20], which suggests that the LFN characteristics of metal-oxide-semiconductor (MOS) devices are not so easily classified

The conventional scaling concept of semiconductor devices has run into the barrier of the cooling limits of very large integrated circuits; the multi-core technology design of integrated circuits and low-power device technology has been proposed because the down-scaling of semiconductor devices is still a goal to permit the greater integration of devices on chips. Such chips must offer low-voltage operation to suppress power dissipation because various sensor networks are created for the purpose of health monitoring and others. Many such sensor devices have to work without any battery. Therefore, high-performance semiconductor devices that can work in a low-voltage condition are needed. Since the low-voltage operation of MOSFETs degrades the signal-to-noise ratio, we have to identify and reduce the various noise sources. In this sense, the physical origin of LFN is now attracting attention because the random telegraph noise (RTN) exhibited by IC-MOSFET memories will be a key determiner of scaled device performance [21–25]. Some articles have theoretically addressed LFN behavior [11,26,27], but very few papers have paid attention to the LFN behavior in the subthreshold bias range [28,29]. In addition, only simplified and qualitative expressions for LFN behavior have been given [30–33]. Although the conventional model uses just the capacitance coupling effect to express the relation between LFN behavior and both Si/SiO_2 interfaces [34], we demonstrate here that this simple understanding is incomplete.

In this paper, the LFN behavior of silicon-on-insulator (SOI) IC-MOSFETs and SOI BC-MOSFETs is theoretically revisited from the viewpoint of attaining SOI MOSFETs that can support low-voltage and low-energy applications [35,36]. The SOI layer has two interfaces, and each interface influences the carrier transport and thus LFN. Fortunately, we can investigate the carrier transport of both electrons and holes in the SOI wafer easily by making all SOI layers have the same polarity when MOSFET devices are fabricated on the wafer. In addition, many large scale integration circuits (LSIs) assume this combination from the point of low fabrication cost and design feasibility of threshold voltage operation. Accordingly, this configuration was chosen in this paper. Basically, we concentrate the discussion on the current fluctuation of devices in the subthreshold bias range, and this paper assumes that the LFN behaviors of devices can be characterized by the carrier density fluctuation. This paper also examines on aspects of the drain current fluctuations of short-channel inversion-channel and buried-channel SOI MOSFETs with ultrathin p-Si bodies [37–40] because it is anticipated that differences in the conduction property will change the noise behavior. First, aspects of the drain current noise behavior are analyzed experimentally in order to categorize the dominant noise sources like interface traps. In addition, physics-based models of the current fluctuation in the subthreshold regime are proposed. Theoretical expressions for LFN behaviors are calculated straightforwardly based on Langevin's method because we assume the trap-related carrier density fluctuation. The models are validated by measured results and some new findings are discussed. This work will contribute to advances in the device physics of future nano-wire MOSFETs.

2. Experiments

A schematic view of the SOI MOS device structures used in the experiments is shown in Figure 1 [41]; the SOI layer (t_S) is 30 nm thick, the gate oxide layer (t_{OX}) is 7 nm thick, the buried-oxide (BOX) layer (t_{BOX}) is 80 nm thick, the body-doping concentration (N_A) is 5×10^{17} cm^{-3} (n-ch MOSFET) or 4×10^{17} cm^{-3} (p-ch MOSFET) of acceptor Boron atoms, and the gate length (L_G) is 0.1 m or 1.0 m.

The n-ch MOSFET and p-ch MOSFET have different doping levels (N_A) such that the absolute values of their threshold voltages are nominally the same. The channel length (L_{eff}) of the 0.1-μm-long gate device is 40 nm, while that of the 1.0-μm-long gate device is 0.95 m. Primary device parameters are summarized in Table 1. The SOI substrates used here were fabricated in the 1990s. Therefore, not only the SOI layer/buried oxide layer interface quality, but also the gate oxide/SOI layer interface quality is only mediocre. As a result, the trap density is larger than expected as is mentioned later.

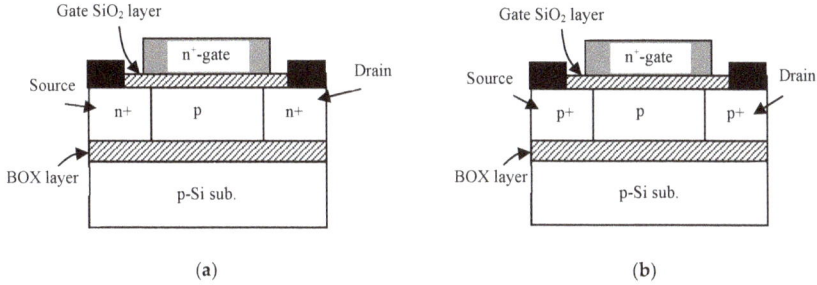

(a)   (b)

Figure 1. Schematic silicon-on-insulator (SOI) device structures used in experiments. (a) n-type inversion-channel (IC) metal-oxide-semiconductor field-effect transistors (MOSFET), (b) p-type buried-channel (BC) MOSFET.

Table 1. Parameters of fabricated devices.

Devices	t_S	t_{ox}	t_{BOX}	N_A
IC-MOSFET	30 nm	7 nm	80 nm	5×10^{17} cm^{-3}
BC-MOSFET	30 nm	7 nm	80 nm	4×10^{17} cm^{-3}

Before discussing LFN behaviors, I_D vs. V_G characteristics of devices at the bias condition of LFN measurement are shown in Figures 2 and 3. Figure 2a shows the transfer characteristics of the 1-μm-long gate n-type IC-MOSFET with a 50-μm-long gate width, Figure 2b shows those of the 1-μm-long gate p-type BC-MOSFET with a 50-μm-long gate width, Figure 3a shows that of the 100-nm-long gate n-type IC-MOSFET with a 20-μm-long gate width [41], and Figure 3b shows that of the 100-nm-long gate p-type BC-MOSFET with a 20-μm-long gate width [41]. Although the 100-nm-long gate devices exhibit some slight short-channel effect [41], it does not influence the measurement.

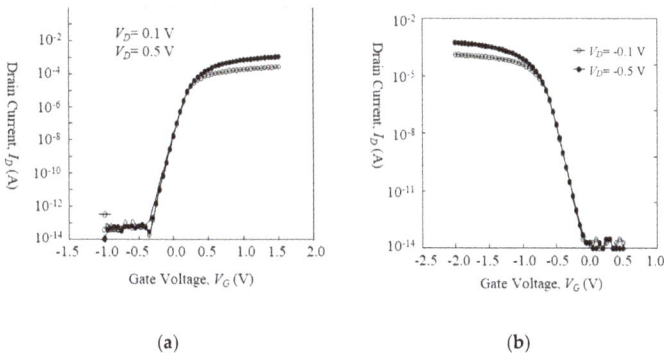

(a)   (b)

Figure 2. I_D vs. V_G characteristics of 1-μm-long gate MOSFETs. (a) n-type IC-MOSFET, (b) p-type BC-MOSFET. $V_{sub} = 0$ V.

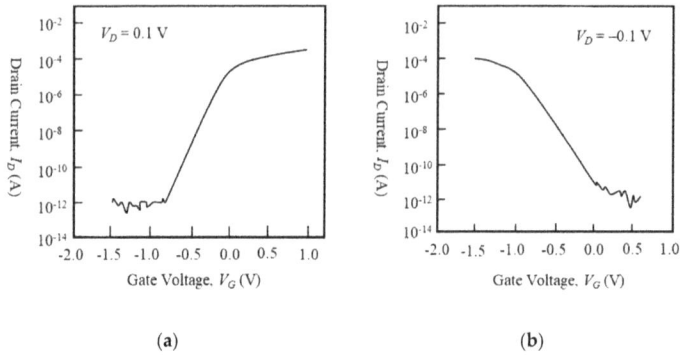

Figure 3. I_D vs. V_G characteristics of 100-nm-long gate MOSFETs. (a) n-type IC-MOSFET, (b) p-type BC-MOSFET. $V_{sub} = 0$ V.

To measure LFN characteristics, the wafer on which the semiconductor devices were fabricated was set on the vacuum chuck. The vacuum chuck was entirely covered with a metal frame to provide electromagnetic shielding; the power supply and current sensing were performed by an Agilent 4156C semiconductor parameter analyzer (Agilent Technologies, Santa Clara, CA, USA) without any preamplifier.

When measuring the drain-current fluctuation of n-channel MOSFETs, +100 mV was applied to the drain terminal. A negative bias was applied to that of the p-channel MOSFETs. The drain current fluctuation was measured from the subthreshold current range to ON-current range. A 900-second measurement of current fluctuation was carried out for every device in order to capture comprehensive sets. Although some people divide the raw data into several parts in order to average them, this paper does not apply the method because it violates the mathematical logic of Fourier transformation. The current level of each device under test was chosen to be higher than 10^{-12} A because the noise current level of the measurement system was ~10^{-13} A in the subthreshold current range.

The following discussions assume that:

$$\frac{S_{I_D}(f)}{I_D^2} = \frac{C(V_G, V_D, V_{sub})}{f^\gamma} \tag{1}$$

where $C(V_G, V_D, V_{sub})$ is a function that depends on the geometrical parameters of devices, gate voltage (V_G), drain voltage (V_D), and substrate voltage (V_{sub}). Parameter γ denotes the exponent of the frequency. In the following section, the measured drain current fluctuations is demonstrated and then theoretical models for the function $C(V_G, V_D, V_{sub})$ are proposed in the subthreshold current range of SOI MOSFETs.

3. Results and Discussion

3.1. Aspects of the Low-Frequency Noise in Long-Channel Silicon-on-Insulator Metal-Oxide-Semiconductor Field-Effect Transistor (SOI MOSFETs)

First of all, the basic aspects of the drain current fluctuation of the 1-μm-gate SOI MOSFET are investigated. Before characterizing the normalized current fluctuation power spectral intensity ($S_{I_D}(f)/I_D^2$) of the drain current, the frequency spectra of $S_{I_D}(f)$ obtained in a subthreshold current range are shown in Figure 4; Figure 4a shows data for the n MOSFET and Figure 4b for the pMOSFET. The baseline is shown to reveal that the value of $S_{I_D}(f)$ at 0.1 Hz is higher than the background noise level. Each value is different because the measurement condition of the drain current is different. Figure 4 reveals that the value in Equation (1) is larger than unity.

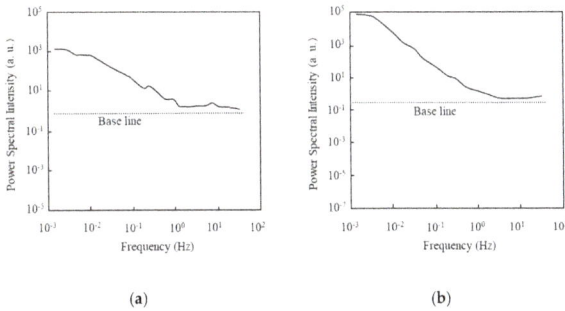

(a)   (b)

Figure 4. Power spectral intensity of 1.0-μm-long gate SOI MOSFETs at the subthreshold bias. (a) nMOSFET (V_D = 0.1 V, V_G = –0.4 V, V_{sub} = 0 V), (b) pMOSFET (V_D = –0.1 V, V_G = –0.4 V, V_{sub} = 0 V).

Normalized current fluctuation power spectral intensity ($S_{I_D}(f)/I_D^2$) of the drain current of an n-channel IC-MOSFET as a function of drain current (I_D) is shown in Figure 5 for two substrate bias conditions (V_{sub} = 0 V, –5 V). In Figure 5, values of $S_{I_D}(f)/I_D^2$ are extracted from the raw data by the parameter fitting technique and their average values are shown. Since the electric field of SOI/buried-oxide layer interface influences the drain current, numerical simulations were carried out in order to estimate the electric field of the buried-oxide layer. This electric field is estimated, at V_{sub} = 0 V, to be 2.3 × 10^5 V/cm at I_D of 10^{-12} A and 2.7 × 10^5 V/cm at I_D of 10^{-6} A. At V_{sub} = –5 V, its value is 8.0 × 10^5 V/cm at I_D of 10^{-12} A and 8.3 × 10^5 V/cm at I_D of 10^{-6} A. Since these electric field conditions suggest that the SOI/buried-oxide interface does not deplete holes even when V_{sub} = 0 V due to a small work-function difference between the SOI layer and the substrate, it is expected that electron-related traps around the SOI/buried-oxide interface don't contribute to the low-frequency noise of the front channel (electron current).

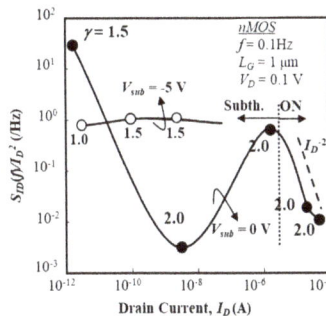

Figure 5. Normalized fluctuation power spectral intensity ($S_{I_D}(f)/I_D^2$) of drain current as a function of I_D in 1-μm gate IC nMOS. L_G =1 μm and W_G = 50 μm. Substrate voltage (V_{sub}) is 0 V and –5 V.

It is seen that $S_{I_D}(f)/I_D^2$ reveals a deep depression around I_D ~10^{-8} A for V_{sub} = 0 V. This aspect also appears in sub-micron gate devices (not shown here [29]). On the "ON" state for the gate voltage (V_G) beyond the threshold voltage, $S_{I_D}(f)/I_D^2$ is proportional to I_D^{-2} with γ = 2, which strongly suggests that the drain current (electron current) fluctuation is the primary determiner of the carrier density fluctuation due to interface traps [24,25]; it is anticipated that the carrier density fluctuation is due to the interface traps near the top surface of the SOI layer, and that most such traps are effectively shielded by the inversion layer in the "ON" state. On the other hand, $S_{I_D}(f)/I_D^2$ is basically insensitive to the drain current level (I_D) in the subthreshold bias range for V_{sub} = –5 V. The behavior of $S_{I_D}(f)/I_D^2$ for V_{sub} = –5 V suggests that the interface traps near the bottom surface of SOI layer are effectively shielded by accumulated holes. In contrast, when V_{sub} = 0 V, the interface traps near the bottom surface of SOI layer are not sufficiently shielded, and some of the electrons contributing to the subthreshold conduction are trapped near the bottom surface. Since the SOI layer thickness is less than the Debye length in this situation, it is expected according to the theoretical model proposed by V. A. Kochelap et al. [42,43] that Coulomb interactions between surface electrons and charged interface traps at the bottom surface may suppress the subthreshold current fluctuation because the surface-noise-suppression factor defined by them increases.

Normalized fluctuation power spectral intensity ($S_{I_D}(f)/I_D^2$) of the drain current of a p-channel BC-MOSFET as a function of drain current (I_D) is shown in Figure 6 for two substrate bias conditions (V_{sub} = 0 V, 5 V). In Figure 6, values of ($S_{I_D}(f)/I_D^2$) are extracted from the raw data by the parameter fitting technique and their average values are shown. Since the electric field of the SOI/buried-oxide layer interface influences the drain current, numerical simulations were carried out for the p-channel BC-MOSFET in order to estimate the electric field of the buried oxide layer. This electric field is estimated, at V_{sub} = 0 V, to be 1.7 × 10^5 V/cm at I_D of 10^{-12} A and 3.9 × 10^5 V/cm at I_D of 10^{-6} A. At V_{sub} = 5 V, it is 4.0 × 10^5 V/cm at I_D of 10^{-12} A and 4.4 × 10^5 V/cm at I_D of 10^{-6} A. When V_{sub} = 0 V, the effective buried-oxide electric field slightly lowers the threshold voltage of the p-channel BC-MOSFET. When V_{sub} = 5 V, the electric field of the buried-oxide layer depletes holes from the SOI/buried-oxide interface, which raises the threshold voltage of the p-channel BC-MOSFET; in other words, it is expected that a hole channel will be generated near the top surface of the SOI layer. As a result, it is anticipated that hole-related traps around the SOI/buried-oxide layer do not contribute to the low-frequency noise of the p-channel BC-MOSFET for V_{sub} = 5 V. Therefore, V_{sub} dependence of ($S_{I_D}(f)/I_D^2$) is reasonable.

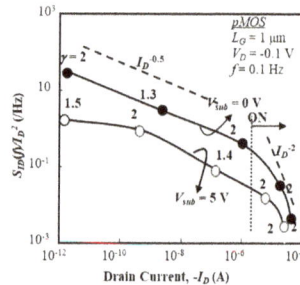

Figure 6. Normalized fluctuation power spectral intensity ($S_{I_D}(f)/I_D^2$) of drain current as a function of I_D in 1-μm gate BC pMOS. L_G = 1 μm and W_G = 50 μm. Substrate voltage (V_{sub}) is 0 V and +5 V.

It is seen that ($S_{I_D}(f)/I_D^2$) exhibits the drain current dependence of $I_D^{-0.5}$ regardless of substrate bias in the subthreshold bias range, although the magnitude of ($S_{I_D}(f)/I_D^2$) is reduced if the substrate bias is positive. This suggests that some interface traps of the buried oxide layer do not contribute to the noise because the SOI/buried-oxide layer interface is depleted for V_{sub} = 5 V. The conventional idea suggests that bulk traps of the SOI layer contribute to the current fluctuation because the buried channel width expands as the gate voltage rises, and/or that interface traps near the top surface of the SOI layer and/or near the bottom surface of the SOI layer contribute to the current fluctuation; however, this is not the case. This behavior, seen in Figure 6, is also observed in sub-micron gate devices (not shown here [29]). In the "ON" state with gate voltages (V_G) beyond the threshold voltage, $S_{I_D}(f)/I_D^2$ is proportional to I_D^{-2} with γ = 2, which strongly suggests that the drain current (hole current) fluctuation is primarily responsible for the carrier density fluctuation due to the interface traps near the top surface of the SOI layer [26,27] because the major part of the hole current consists of the hole accumulation layer near the front gate oxide layer for $V_G > V_{TH}$, where V_{TH} is the threshold voltage. It is considered that the impact of interface traps on the drain current fluctuation is almost the same as that on the inversion channel, although it is anticipated that some traps near the top surface of the SOI layer are shielded by the hole accumulation layer. This consideration is utilized in deriving the theoretical model detailed later.

Past work focused on developing theoretical models for the "ON" state [3–6,11–13,15–20,26]. A recent model [27] has been written as:

$$\frac{S_{I_D}(f)}{I_D^2} = \left(1 \pm \frac{\alpha_c \mu_{eff} C_{OX} I_D}{g_m}\right)^2 \frac{g_m^2}{I_D^2} S_{VFB}(f) \tag{2}$$

$$S_{VFB}(f) = \frac{q^2 k_B T \lambda N_t(E_F)}{W_{eff} L_{eff} C_{OX}^2 f^\gamma} \tag{3}$$

where α_c denotes the scattering factor [11], λ denotes the effective tunneling distance, and N_t denotes the effective trap density (cm^{-2}·eV^{-1}). Parameter α_c is an empirical parameter, not a physics-based parameter. For the "ON" state shown in Figures 5 and 6, Equations (2) and (3) suggest that trap density N_t is roughly constant in this gate-voltage range for both the IC-MOSFET and BC-MOSFET. This speculation is acceptable because the local Fermi level at the top surface of the SOI layer is still slightly above midgap for the n-ch IC-MOSFET and the local Fermi level at the top surface of the SOI layer is slightly below the Fermi level in the flat-band condition for the p-ch BC-MOSFET. In other words, this suggests that the distribution of N_t over the energy gap of Si definitely controls the behavior of Equation (2).

Although the noise behavior in the subthreshold bias range must be considered, no corresponding physics-based theoretical models have been proposed. This paper corrects this deficiency in a later section.

3.2. Aspects of Low-Frequency Noise in 100-nm-long Gate SOI MOSFETs

Before characterizing the normalized current fluctuation power spectral intensity ($S_{I_D}(f)/I_D^2$) of drain current, the frequency spectra of $S_{I_D}(f)$ obtained in a subthreshold current range are shown in Figure 7; Figure 7a shows the nMOSFET data and Figure 7b shows MOSFET data. Figure 7 reveals that the γ value in Equation (1) is larger than unity. Normalized fluctuation power spectral intensity ($S_{I_D}(f)/I_D^2$) of the drain current of an n-channel IC-MOSFET with 100-nm-long gate (40-nm-long channel) [44] is shown in Figure 8 as a function of drain current (I_D) for two substrate bias conditions ($V_{sub} = 0$ V, –5 V). In Figure 8, values of $S_{I_D}(f)/I_D^2$ are extracted from the raw data by the parameter fitting technique and their average values are shown. $S_{I_D}(f)/I_D^2$ is insensitive to the drain current in the subthreshold bias range regardless of the substrate bias. The behavior of $S_{I_D}(f)/I_D^2$ for $V_{sub} = 0$ V is very different from that shown in Figure 5; no depression in $S_{I_D}(f)/I_D^2$ is observed. These behaviors of the 100-nm gate device suggest the following points.

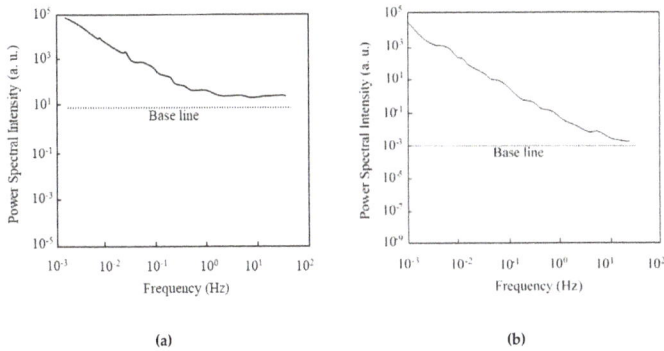

(a) (b)

Figure 7. Power spectral intensity of 100-nm-long gate SOI MOSFETs at the subthreshold bias. (a) nMOSFET ($V_D = 0.1$ V, $V_G = -0.6$ V, $V_{sub} = 0$ V), (b) pMOSFET ($V_D = -0.1$ V, $V_G = -0.4$ V, $V_{sub} = 0$ V).

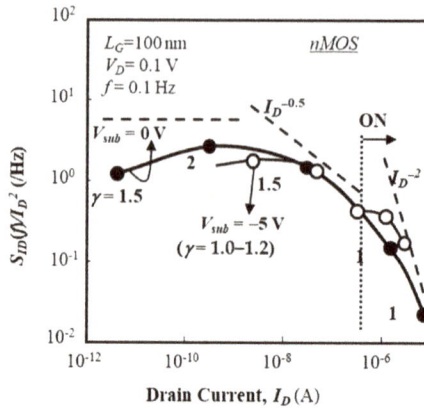

Figure 8. Normalized fluctuation power spectral intensity (S_{I_D} $(f)/I_D^2$) of drain current as a function of I_D in 100-nm gate IC nMOS. L_G = 100 nm and W_G = 20 μm. Substrate voltage (V_{sub}) is 0 V and –5 V.

(1) In the subthreshold bias range, the contribution of traps far from the top surface of the SOI layer to the current fluctuation is quite limited, which is anticipated from the insensitivity of S_{I_D} $(f)/I_D^2$ to the substrate bias. This behavior is different from that of long-channel devices, see Figure 5.

(2) Above the threshold voltage, S_{I_D} $(f)/I_D^2$ is proportional to I_D^{-2}. Equation (2), for example, suggests that the S_{VFB} factor is roughly constant above the threshold voltage, see Figure 8, when I_D is increased with a constant V_D value because the I_D value is increased when V_G is increased. This suggests that the trap density profile near the midgap is almost flat because the local Fermi level at the top surface of the SOI layer is located around the midgap.

The surface morphology of the buried oxide layer of the MOSFETs used in this experiment has a specific mesa shape aligned to the [100] direction [45,46]; the mesa scale is about 500 nm × 500 nm (in plane) as shown in Figure 9. It is considered that the local fluctuation of the surface potential rules the carrier conduction path in the subthreshold current range [47]. Therefore, it is easily anticipated that the local surface potential of the SOI layer of the MOSFET used here is modified by SOI layer thickness fluctuation [48]. In the present case, it is expected that the spatial uniformity of SOI layer thickness is limited to an area of at most 300 nm × 300 nm, which suggests that the local uniformity of interface trap density is also limited to an area of at most 300 nm × 300 nm. Therefore, the S_{I_D} $(f)/I_D^2$ behavior of the 100-nm-gate MOSFET is more insensitive to the substrate bias than that of the long-channel MOSFET.

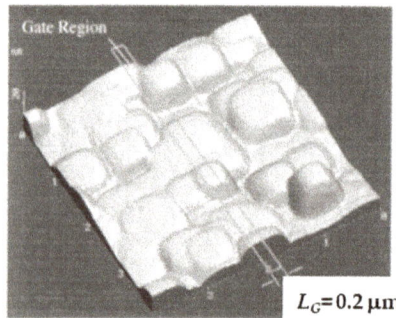

Figure 9. Atomic force microscopy (AFM) image of surface morphology of the bottom surface of the silicon layer. Reproduced with permission from [45], published by IEEE, 1996.

Next, normalized fluctuation power spectral intensity ($S_{I_D}(f)/I_D^2$) of the drain current of a p-channel BC-MOSFET as a function of drain current (I_D) is shown in Figure 10 for two substrate bias conditions (V_{sub} = 0 V, 5 V) [49]. The magnitude of $S_{I_D}(f)/I_D^2$ shows a weak dependence on I_D ($\sim I_D^{-1/2}$) in the subthreshold current range regardless of the substrate bias. The behavior of $S_{I_D}(f)/I_D^2$ is roughly the same as those shown in Figure 6, but it does show a strong dependence on I_D ($\sim I_D^{-5}$) in the "ON" state. These behaviors of the 100-nm-long gate BC-MOS device suggest the following points.

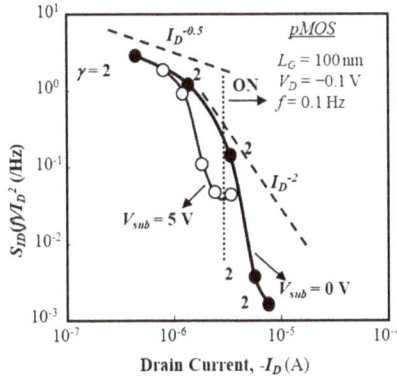

Figure 10. Normalized fluctuation power spectral intensity ($S_{I_D}(f)/I_D^2$) of drain current as a function of I_D in 100-nm gate BC pMOS. L_G = 100 nm and W_G = 20 μm. Substrate voltage (V_{sub}) is 0 V and +5 V. Reproduced with permission from [49], published by IEEE, 2017.

(1) It is anticipated that the current fluctuation in the subthreshold bias range originates from interface traps near the top surface of the SOI layer, not primarily from interface traps near the bottom interface of the SOI layer.

(2) Following Equation (2), it seems that the trap density energy profile near the top surface of the SOI layer is almost flat, but its value slightly decreases when the local Fermi level of the top surface of the SOI layer approaches the midgap from the level above the midgap because $S_{I_D}(f)/I_D^2$ decreases ($\sim I_D^{-1/2}$) as I_D increases.

(3) The contribution of interface traps existing near the bottom surface of the SOI layer to the current fluctuation is not so significant, which is supported by the fact that the magnitude of $S_{I_D}(f)/I_D^2$ is insensitive to positive substrate bias values. On the other hand, the behavior of $S_{I_D}(f)/I_D^2$ in the "ON" state reveals that the impact of interface traps near the top surface of the SOI layer on the channel current is greatly suppressed, which is suggested by the very steep decrease in $S_{I_D}(f)/I_D^2$ in the "ON" state. The channel current formed in the surface hole accumulation layer is basically not influenced by the interface traps near the top surface of the SOI layer because the hole density is very high. This suggests that the surface accumulation layer effectively shields hole traps near the valence band edge.

4. Theoretical Modeling for Subthreshold Current Fluctuations

4.1. Inversion-Channel MOSFET

Subthreshold current of the n-channel IC-MOSFET at the top surface is given by [50]:

$$I_D = I_{Sth}(\phi_{SS})\exp\left(\frac{q\phi_{SS}}{k_BT}\right)^2 \tag{4}$$

$$I_{Sth}(\phi_{SS}) = \left(\frac{q\mu_n W_{eff} V_D}{L_{eff} E_S(\phi_{SS})}\right)\left(\frac{k_B T}{q}\right)\left(\frac{n_i^2}{N_A}\right) \tag{5}$$

where ϕ_{ss} denotes the top surface potential, E_S denotes the surface electric field, and we assume V_D $< k_B T/q$. Other notations follow the conventional terminology. Here, the theoretical formulation follows Langevin's method [51]. When we assume that the noise source yields the fluctuation of the front surface potential (ϕ_{ss}), whereas the drain current fluctuation δI_D originates from $\delta\phi_{ss}$. Starting with Equation (4), δI_D is given as:

$$\delta I_D = \frac{\partial I_{Sth}(\phi_{SS})}{\partial \phi_{ss}} \exp\left(\frac{q\phi_{SS}}{k_B T}\right)\delta\phi_{SS} + I_{Sth}(\phi_{SS})\exp\left(\frac{q\phi_{SS}}{k_B T}\right)\cdot\left(\frac{q}{k_B T}\right)\delta\phi_{SS} \tag{6}$$

Then we have,

$$\frac{<\delta I_D^2>}{I_D^2} = \left(\frac{q}{k^B T} - \frac{1}{2\phi_{SS}}\right)^2 <\phi_{SS}^2> \tag{7}$$

where $<X>$ means the time averaging of the parameter X. The spectral density of drain current fluctuation is calculated as:

$$\frac{S_{I_D}(f)}{I_D^2} = \left(\frac{q}{k_B T} - \frac{1}{2\phi_{SS}}\right)^2 S_{\phi_{SS}}(f) \tag{8}$$

$$S_{\phi_{SS}}(f) = \frac{<(\delta\phi_{SS})^2>}{\Delta f} \tag{9}$$

$S_{\phi_{SS}}(f)$ corresponds to the surface potential fluctuation power spectral intensity. It is frequently thought that $S_{\phi_{SS}}(f)$ stems from the carrier density fluctuation [2,11]. One possible source of $S_{\phi_{SS}}(f)$ is the trapping-detrapping process of carriers near the top surface of the silicon layer. This expression is valid for $\phi_{ss} > 0$; that is $I_D > I_{Sth}$.

4.2. Buried-Channel MOSFET

On the other hand, the subthreshold current of the p-channel BC-MOSFET near the bottom surface or the top surface is given by [52]

$$I_D = I_{Bth}\int_{\phi_{SS}}^{\phi_{BS}} \frac{1}{\sqrt{\phi_{BS}}} \exp\left(-\frac{q\phi_B}{k_B T}\right)d\phi_B \tag{10}$$

$$I_{Bth} = \frac{\mu_n W_{eff} V_D \sqrt{2\varepsilon_S q N_A}}{L_{eff}} \tag{11}$$

where ϕ_{SS} denotes the top surface potential, ϕ_{BS} denotes the bottom surface potential, and we assume $V_D < k_B T/q$. Equation (10) can be approximated as:

$$\begin{aligned}
I_D &\approx \tfrac{1}{2}I_{Bth}(\phi_{SS} - \phi_{BS})\left[-\tfrac{1}{\sqrt{\phi_{SS}}}\exp\left(-\tfrac{q\phi_{SS}}{k_B T}\right) + \tfrac{1}{\sqrt{\phi_{BS}}}\exp\left(-\tfrac{q\phi_{BS}}{k_B T}\right)\right] \\
&= \tfrac{1}{2}I_{Bth}F(\phi_{SS})\left[-\tfrac{1}{\sqrt{\phi_{SS}}}\exp\left(-\tfrac{q\phi_{SS}}{k_B T}\right) + \tfrac{1}{\sqrt{\phi_{BS}}}\exp\left(-\tfrac{q\phi_{BS}}{k_B T}\right)\right] \\
&= I_{BS} - I_{SS}
\end{aligned} \tag{12}$$

$F(\phi_{SS})$ is given as [41],

$$\begin{aligned}
\phi_{SS} - \phi_{BS} &= \frac{C_{BOX}}{C_S + C_{BOX}}\phi_{SS} - \frac{q N_A t_S}{2(C_S + C_{BOX})} - \frac{C_{BOX} V_{SUB}^*}{C_S + C_{BOX}} \\
&= F(\phi_{SS})
\end{aligned} \tag{13}$$

and we have,

$$\frac{d\phi_{BS}}{d\phi_{SS}} = \frac{C_S}{C_S + C_{BOX}} = 1 - f_C \tag{14}$$

$$f_C = \frac{C_{BOX}}{C_S + C_{BOX}} \tag{15}$$

Parameter f_C partially represents the capacitance coupling effect. Here C_S denotes the SOI layer capacitance, C_{BOX} denotes the buried oxide layer capacitance, and V^*_{SUB} denotes the effective substrate bias.

Assuming that the current fluctuation originates from the traps near the top surface of the silicon, we can say that the surface potential fluctuation directly influences the bottom surface potential fluctuation electrostatically. The theoretical calculation is based on the same approach mentioned in Section 4.1. This argument yields the following expression for the power spectral intensity of the fluctuation of the buried-channel current.

$$
\begin{aligned}
S_{ID\_SS}(f) = {} & 4\left(\frac{f_C}{F(\phi_{SS})}\right)^2 [I_{BS} - I_{SS}]^2 S_{\phi_S}(f) \\
& + \frac{1}{4}\left[-I_{SS}\left(\frac{1}{\phi_{SS}} + \frac{2q}{k_B T}\right) + 2(1 - f_C)I_{BS}\left(\frac{1}{\phi_{BS}} + \frac{2q}{k_B T}\right)\right]^2 S_{\phi_S}(f) \\
& + f_C[I_{BS} - I_{SS}]\left[-\frac{I_{SS}}{F(\phi_{SS})}\left(\frac{1}{\phi_{SS}} + \frac{2q}{k_B T}\right) + 2(1 - f_C)\frac{I_{BS}}{F(\phi_{SS})}\left(\frac{1}{\phi_{BS}} + \frac{2q}{k_B T}\right)\right] S_{\phi_S}(f)
\end{aligned}
\tag{16}
$$

$S_{\phi_S}(f)$ corresponds to the fluctuation power spectral intensity of the top surface potential. Taking account of the fact that $I_{BS} \gg I_{SS}$, Equation (16) can be rewritten as:

$$
\frac{S_{ID\_SS}(f)}{I_D^2} \approx \left\{4\left(\frac{f_C}{F(\phi_{SS})}\right)^2 + \left[(1 - f_C)\left(\frac{1}{\phi_{BS}} + \frac{2q}{k_B T}\right)\right]^2\right\} S_{\phi_S}(f) + \left\{\frac{2f_C(1 - f_C)}{F(\phi_{SS})}\left(\frac{1}{\phi_{BS}} + \frac{2q}{k_B T}\right)^2\right\} S_{\phi_S}(f)
\tag{17}
$$

On the other hand, when it is assumed that the current fluctuation originates from the traps near not only the top surface, but also the bottom surface, of the SOI layer, we have the following expression for the power spectral intensity of the fluctuation of the buried-channel current [51].

$$
\begin{aligned}
S_{ID\_SS\_BS}(f) = {} & \left(\frac{I_D}{F(\phi_{SS})}\right)^2 (S_{\phi_S}(f) + S_{\phi_B}(f)) + I_{SS}^2\left(\frac{1}{\phi_{SS}} + \frac{2q}{k_B T}\right)^2 S_{\phi_S}(f) + I_{BS}^2\left(\frac{1}{\phi_{BS}} + \frac{2q}{k_B T}\right)^2 S_{\phi_B}(f) \\
& - 2I_D\left(\frac{I_{BS}}{F(\phi_{SS})}\right)\left(\frac{1}{\phi_{BS}} + \frac{2q}{k_B T}\right) S_{\phi_B}(f) - 2I_D\left(\frac{I_{SS}}{F(\phi_{SS})}\right)\left(\frac{1}{\phi_{SS}} + \frac{2q}{k_B T}\right) S_{\phi_S}(f) \\
\approx {} & I_D^2(S_{\phi_S}(f) + S_{\phi_B}(f)) + I_D^2\left(\frac{1}{\phi_{BS}} + \frac{2q}{k_B T}\right)^2 S_{\phi_B}(f) - 2I_D^2\left(\frac{1}{F(\phi_{SS})}\right)\left(\frac{1}{\phi_{BS}} + \frac{2q}{k_B T}\right)^2 S_{\phi_B}(f)
\end{aligned}
\tag{18}
$$

where $S_{\phi_B}(f)$ corresponds to the fluctuation power spectral density of bottom surface potential. It is assumed that $S_{\phi_B}(f) \gg S_{\phi_S}(f)$ because $I_{BS} \gg I_{SS}$. Thus we have:

$$
\frac{S_{ID\_SS\_BS}(f)}{I_D^2} \approx \left\{\frac{1}{F^2(\phi_{SS})} + \left(\frac{1}{\phi_{BS}} + \frac{2q}{k_B T}\right)^2\right\} S_{\phi_B}(f) - 2\left(\frac{1}{F(\phi_{SS})}\right)\left(\frac{1}{\phi_{BS}} + \frac{2q}{k_B T}\right) S_{\phi_B}(f)
\tag{19}
$$

4.3. Theoretical Modeling of Fluctuation Sources and Brief Examination of the Model

Following the conventional idea, it can be assumed that the fluctuation source for the current fluctuation originates from the trapping–detrapping process of Si/SiO$_2$ interface states [1,2,27]. The top surface potential fluctuation $<\phi_{SS}^2>$ can be written as [2,44]:

$$
S_{\phi_{SS}}(f) = \frac{q^2 k_B T N_t[E_F(\phi_{SS})]\eta^{\gamma - 1}}{C_{OX}^2 W_{eff} L_{eff} f^\gamma}
\tag{20}
$$

where $N_t(E_F)$ denotes the trap density at the local Fermi level, C_{OX} is the gate oxide capacitance per unit area, and η is the parameter (units of frequency) that is used in order to adjust the physical dimension of $S_{\phi_{SS}}(f)$. The local Fermi level at the top surface of the SOI layer is a function of the surface potential, ϕ_{SS}. Other than parameter λ, Equation (20) is basically the same as Equation (3).

In Figure 8, normalized fluctuation power spectral intensity ($S_{I_D}(f)/I_D^2$) of the drain current of n-channel IC-MOSFETs is almost constant and insensitive to I_D when $V_{sub} = -5$ V. Since the bottom surface of the SOI layer is electrostatically shielded by holes when $V_{sub} = -5$ V, the electron current near the top surface of the SOI layer is influenced primarily by the interface traps near the top surface of the SOI layer. As Equation (8) is not sensitive to ϕ_{SS}, Figure 8 suggests that $N_t(E_F)$ is almost flat around the midgap. This speculation is reasonable because E_F approaches the conduction band bottom via the midgap when the gate voltage approaches the threshold voltage.

Calculation results of normalized fluctuation power spectral intensity $(S_{I_D}(f)/I_D^2)$ of the drain current of the n-channel IC-MOSFET are plotted as a function of I_D by solid lines in Figure 11 for $V_{sub} = 0$ V and Figure 12 for $V_{sub} = -5$ V, where it is assumed that $N_t(E)$ is constant with the value of 4×10^{14} cm^{-2}·eV^{-1}. The effective trap density is about 1×10^{13} cm^{-2} at room temperature. Calculation results of $S_{I_D}(f)/I_D^2$ for the n-channel IC-MOSFET are insensitive to V_{sub}, and the values and behavior insensitivity to I_D well match the experimental results. Therefore, Equation (8) successfully predicts the $S_{I_D}(f)/I_D^2$ characteristics in the subthreshold bias range.

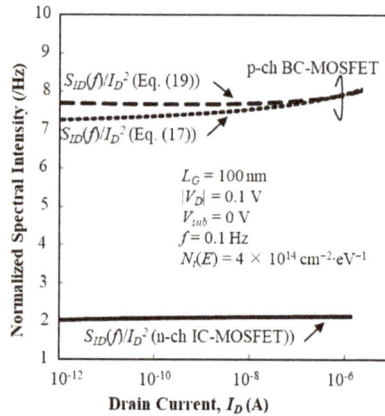

Figure 11. Calculation results of $S_{I_D}(f)/I_D^2$ as a function of I_D for n-channel IC-MOSFET. It is assumed $N_t(E) = 4 \times 10^{14}$ cm^{-2}·eV^{-1} and $V_{sub} = 0$ V.

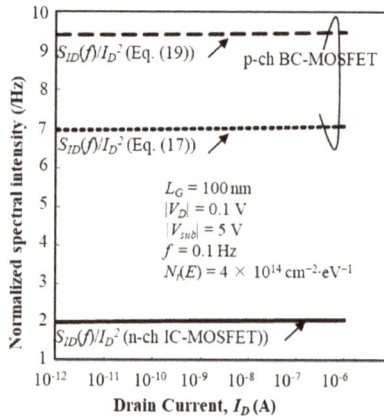

Figure 12. Calculation results of $S_{I_D}(f)/I_D^2$ as a function of I_D for p-channel BC-MOSFET. It is assumed $N_t(E) = 4 \times 10^{14}$ cm^{-2}·eV^{-1}. $V_{sub} = -5$ V for nMOSFET and $V_{sub} = +5$ V for pMOSFET.

On the other hand, we show two calculation results of the normalized fluctuation power spectral intensity $(S_{I_D}(f)/I_D^2)$ of the drain current of the p-channel BC-MOSFET as a function of I_D in Figure 11 for $V_{sub} = 0$ V and Figure 12 for $V_{sub} = 5$ V, where it is assumed that $N_t(E)$ is constant with value of 4×10^{14} cm^{-2}·eV^{-1}. The effective trap density is about 1×10^{13} cm^{-2} at room temperature.

In Figures 11 and 12, results of Equation (17) are shown by dotted lines and those of Equation (19) are shown by broken lines. $S_{I_D}(f)/I_D^2$ calculated by Equation (17) decreases with positive substrate bias as seen in Figure 12. However, $S_{I_D}(f)/I_D^2$ calculated by Equation (19) increases with positive substrate bias, see Figure 12. This assessment reveals that model Equation (17) is acceptable, which means that the current fluctuation of BC-MOSFETs is primarily ruled by the interface traps near the top surface of the SOI layer. This new finding is given by the theoretical analysis proposed in this paper. Equation (17) successfully predicts the $S_{I_D}(f)/I_D^2$ characteristics in the subthreshold bias range. Finally, we examined whether the value of $N_t(E)$ alters the substrate bias dependence of $S_{I_D}(f)/I_D^2$ assuming the adoption of Equations (17) and (19). When $N_t(E) = 4 \times 10^{13}$ cm^{-2}·eV^{-1}, $S_{I_D}(f)/I_D^2$ values are reduced to one tenth that for $N_t(E) = 4 \times 10^{14}$ cm^{-2}·eV^{-1} because the surface potential fluctuation is proportional to $N_t(E)$ (see Equation (20)). However, the substrate bias dependence of $S_{I_D}(f)/I_D^2$ does not change.

5. Conclusions

This paper elucidated the normalized drain current fluctuation spectral intensity of various long-channel and short-channel SOI MOSFETs (inversion-channel SOI MOSFETs and buried-channel SOI MOSFETs) from the viewpoint of scaling. This paper reconsidered low-frequency noise behavior in the inversion-channel SOI MOSFET and the buried-channel SOI MOSFET because it is anticipated that the quality of both Si/SiO$_2$ interfaces should modulate the low-frequency noise characteristics of both devices. Our assessments also addressed the low-frequency noise behavior of sub-100-nm-long channel SOI MOSFETs. The low-frequency noise behavior in the subthreshold bias range was discussed in some detail in order to consider device suitability for future low-voltage and low-power applications.

This paper also proposed theoretical models to explain and predict the drain current fluctuations of SOI MOSFETs. For the buried-channel device, two models were proposed; one assumes that the drain current fluctuation originates from the traps near the top surface of the SOI layer, and the other assumed that the drain current fluctuation originates from the traps near the bottom surface of the SOI layer. As expected, the analyses showed that the current fluctuation of the inversion channel SOI MOSFET is strongly influenced by interface traps near the top surface of the SOI layer because those traps are not well shielded by surface weak inversion carriers in the subthreshold bias range. However, unexpectedly, the buried channel is primarily influenced by interface traps near the top surface of the SOI layer, not by traps near the bottom surface of the SOI layer. This interesting characteristic of current fluctuation spectral intensity was well explained by the theoretical models proposed here. One theoretical expression reveals that the impact of substrate bias is not due to just the capacitance coupling effect, which contradicts the conventional model. As a result, the interface trap density of the top surface of the SOI layer should be reduced in order to improve the analog performance of SOI MOSFETs.

Funding: A part of this study is financially supported by Kansai University Grants, 2015.

Acknowledgments: The author wishes to express his thanks to Shingo Sato and the undergraduate students in the laboratory for their technical support and discussion.

Conflicts of Interest: The author declares no conflict of interest. The funders had no role in the design of the study; in the collection, analyses, or interpretation of data; in the writing of the manuscript; or in the decision to publish the results.

References

1. McWhorter, A.L. *Semiconductor Surface Physics*; Kingston, R.H., Ed.; University of Pennsylvania Press: Philadelphia, PA, USA, 1957; p. 207.
2. Sah, C.T.; Hielsher, F.H. Evidence of the Surface Origin of the 1/f Noise. *Phys. Rev. Lett.* **1966**, *17*, 956–958. [CrossRef]
3. Klaassen, F.M. Characterization of low 1/f noise in MOS transistors. *IEEE Trans. Electron. Devices* **1971**, *18*, 887–891. [CrossRef]

4. Hooge, F.N. 1/f Noise. *Physica* **1976**, *83*, 14–23. [CrossRef]
5. Kleinpenning, K.G.M.; Vandamme, L.K.J. Model for 1/f Noise in Metal-Oxide-Semiconductor Transistors. *J. Appl. Phys.* **1981**, *52*, 1594–1596. [CrossRef]
6. Jindal, R.P.; Van der Ziel, A. Phonon fluctuation model for flicker noise in elemental semiconductors. *J. Appl. Phys.* **1978**, *52*, 2884–2888. [CrossRef]
7. Handel, P.H. 1/f Noise—An "Infrared" Phenomenon. *Phys. Rev. Lett.* **1975**, *34*, 1492–1495. [CrossRef]
8. Handel, P.H. Quantum Approach to 1/f Noise. *Phys. Rev. A* **1980**, *22*, 745–757. [CrossRef]
9. Van der Ziel, A.; Handel, P.H.; Zhu, X.; Duh, K.H. A Theory of the Hooge Parameters of Solid-State Devices. *IEEE Trans. Electron Devices* **1985**, *32*, 667–671. [CrossRef]
10. Handel, P.H. Fundamental quantum 1/f noise in semiconductor devices. *IEEE Trans. Electron Devices* **1994**, *41*, 2023–2033. [CrossRef]
11. Hung, K.K.; Ko, P.K.; Hu, C.; Cheng, Y.C. A unified model for the flicker noise in metal-oxide-semiconductor field-effect transistors. *IEEE Trans. Electron Devices* **1990**, *37*, 654–665. [CrossRef]
12. Stephany, J.F. A Theory of 1/f Noise. *J. Appl. Phys.* **1998**, *83*, 3139–3143. [CrossRef]
13. Kiss, L.B.; Heszler, P. An Exact Proof of the Invalidity of 'Handel's Quantum 1/f Noise Model', based on Quantum Electrodynamics. *J. Phys. C Solid State Phys.* **1986**, *19*, 631–633. [CrossRef]
14. Nieuwenhuizen, T.M.; Frenkel, D.; van Kampen, N.G. Objection to Handel's Quantum Theory of 1/f Noise. *Phys. Rev. A* **1987**, *35*, 2750–2753. [CrossRef]
15. Van Vliet, C.M. A survey of results and future prospects on quantum 1/f noise and 1/f noise in general. *Solid State Electron.* **1991**, *34*, 1–21. [CrossRef]
16. Vandamme, E.P.; Vandamme, L.K.J. Critical Discussion on Unified 1/f Noise Models for MOSFETs. *IEEE Trans. Electron Devices* **2000**, *47*, 2146–2152. [CrossRef]
17. Omura, Y. Hooge parameter in buried-channel metal-oxide-semiconductor field-effect transistors. *J. Appl. Phys.* **2002**, *91*, 1378–1384. [CrossRef]
18. Omura, Y. Possible Unified Model for the Hooge Parameter in Inversion-Layer-Channel Metal-Oxide-Semiconductor Field-Effect Transistors. *J. Appl. Phys.* **2013**, *113*, 214508–214518. [CrossRef]
19. Park, H.S.; Van der Ziel, A. Dependence of MOSFET Noise Parameters in n-Channel MOSFETs on Oxide Thickness. *Solid State Electron.* **1982**, *25*, 313–315. [CrossRef]
20. Magnone, P.; Crupi, F.; Giusi, G.; Simoen, E.; Claeys, C.; Pantisano, L.; Maji, D.; Rao, V.R.; Srinivasan, P. 1/f noise in drain and gate current of MOSFETs with high-k gate stacks. *IEEE Trans. Device Mater. Reliab.* **2009**, *9*, 180–189. [CrossRef]
21. Nowak, E.; Kim, J.-H.; Kwon, H.Y.; Kim, Y.-G.; Sim, J.S.; Lim, S.-H.; Kim, D.S.; Lee, K.-H.; Park, Y.K.; Choi, J.-H.; et al. Intrinsic Fluctuations in Vertical NAND Flash Memories. In Proceedings of the 2012 Symposium on VLSI Technology (VLSIT), Honolulu, HI, USA, 12–14 June 2012; pp. 21–22.
22. Miki, H.; Yamaoka, M.; Frank, D.J.; Cheng, K.; Park, D.-G.; Leobandung, E.; Torii, K. Voltage and Temperature Dependence of Random Telegraph Noise in Highly Scaled HKMG ETSOI nFETs and its Impact on Logic Delay Uncertainty. In Proceedings of the 2012 Symposium on VLSI Technology (VLSIT), Honolulu, HI, USA, 12–14 June 2012; pp. 137–138.
23. Chen, J.; Hirano, I.; Tatsumura, K.; Mitani, Y. Comprehensive Investigations on Neutral and Attractive Traps in Random Telegraph Signal Noise Phenomena using (100)- and (110)-Orientated CMOSFETs. In Proceedings of the 2012 Symposium on VLSI Technology (VLSIT), Honolulu, HI, USA, 12–14 June 2012; pp. 141–142.
24. Wei, C.; Jiang, Y.; Xiong, Y.-Z.; Zhou, X.; Singh, N.; Rustagi, S.C.; Lo, G.Q.; Kwong, D.-L. Impact of Gate Electrodes on 1/f Noise of Gate-All-Around Silicon Nanowire Transistors. *IEEE Electron Devices Lett.* **2009**, *30*, 1081–1083.
25. Ioannidis, E.G.; Theodorou, C.G.; Karatsori, T.A.; Haendler, S.; Dimitriadis, C.A.; Ghibaudo, G. Drain-Current Flicker Noise Modeling in nMOSFETs From a 14-nm FDSOI Technology. *IEEE Trans. Electron Devices* **2015**, *62*, 1574–1579. [CrossRef]
26. Ghibaudo, G.; Roux, O.; Nguyen-Duc, C.; Balestra, F.; Brini, J. Improved Analysis of Low Frequency Noise in Field-Effect MOS Transistors. *Phys. Stat. Sol.* **1991**, *124*, 571–581. [CrossRef]
27. Ghibaudo, G. Impact of Device Scaling on LF Noise in CMOS Technologies. In Proceedings of the 17th International Conference Mixed Design of Integrated Circuits and Systems, Prague, Czech Republic, 1–4 June 2003; pp. 301–308.

28. Rhayem, J.; Rigaud, D.; Eya'a, A.; Valenza, M. 1/f Noise in Metal-Oxide-Semiconductor Transistors Biased in Weak Inversion. *J. Appl. Phys.* **2001**, *89*, 4192–4194. [CrossRef]
29. Ito, T.; Sato, S.; Omura, Y. Characterization of Noise Behavior of Ultrathin Inversion-Channel and Buried-Channel SOI MOSFETs in the Subthreshold Bias Range. In Proceedings of the 2014 IEEE International Meeting for Future of Electron Devices, Kansai (IMFEDK), Kyoto, Japan, 19–20 June 2014; pp. 46–47.
30. Jin, W.; Chan, P.C.H.; Fung, S.K.H.; Ko, P.K. Shot-noise-induced excess low-frequency noise in floating-body partially depleted SOI MOSFET's. *IEEE Trans. Electron Devices* **1999**, *46*, 1180–1185. [CrossRef]
31. Workman, G.O.; Fossum, J.G. Physical noise modeling of SOI MOSFETs with analysis of the Lorentzian component in the low-frequency noise spectrum. *IEEE Trans. Electron Devices* **2000**, *47*, 1192–1201. [CrossRef]
32. Gross, B.J.; Sodini, C.G. 1/f noise in MOSFETs with ultrathin gate dielectrics. In Proceedings of the 1992 International Technical Digest on Electron Devices Meeting, San Francisco, CA, USA, 13–16 December 1992; pp. 881–884.
33. Morshed, T.; Devireddy, S.P.; Rahman, M.S.; Celik-Buller, Z.; Tseng, H.-H.; Zlotnicka, A.; Shanware, A.; Green, K.; Chambers, J.J.; Visokay, M.R.; et al. A new model for 1/f noise in high-k MOSFETs. In Proceedings of the 2007 IEEE International Electron Devices Meeting, Washington, DC, USA, 10–12 December 2007; pp. 561–564.
34. Simoen, E.; Mercha, A.; Claeys, C.; Lukyanchikova, N.; Garbar, N. Critical discussion of the front-back gate coupling effect on the low-frequency noise in fully depleted SOI MOSFETs. *IEEE Trans. Electron Devices* **2004**, *51*, 1008–1016. [CrossRef]
35. Chandrakasan, A.P.; Daly, D.C.; Finchelstein, D.F.; Kwong, J.; Ramadas, Y.K.; Sinangil, M.E.; Sze, V.; Verma, N. Technologies for Ultradynamic Voltage Scaling. *Proc. IEEE* **2010**, *98*, 191–214. [CrossRef]
36. Vitale, S.A.; Wyatt, P.W.; Checka, N.; Kedzierski, J.; Keast, C.L. FESOI Process Technology for Subthreshold-Operation Ultralow-Power Electronics. *Proc. IEEE* **2010**, *98*, 333–342. [CrossRef]
37. Simoen, E.; Andrade, M.G.C.; Almeida, L.M.; Aoulaiche, M.; Caillat, C.; Jurczak, M.; Claeys, C. On the Variability of the Low-Frequency Noise in UTBOX SOI nMOS-FETs. *J. Integr. Circ. Syst.* **2013**, *8*, 71–77.
38. Van Haartman, M.; Oestling, M. Effect of channel positioning on the 1/f noise in silicon-on-insulator metal-oxide semiconductor field-effect transistors. *J. Appl. Phys.* **2007**, *101*, 034506–034509. [CrossRef]
39. Gaubert, P.; Teramoto, A.; Sugawa, S. 1/f Noise Performances and Noise Sources of Accumulation Mode Si(100) n-MOSFETs. In Proceedings of the International Conference on Solid State Devices and Materials, Sapporo, Japan, 27–30 September 2015; pp. 96–97.
40. Lee, J.-H.; Kim, S.-Y.; Cho, I.; Hwang, S.; Lee, J.-H. 1/f Noise Characteristics of Sub-100 nm MOS Transistors. *J. Semicond. Technol. Sci.* **2006**, *6*, 38–42.
41. Omura, Y.; Nakashima, S.; Izumi, K.; Ishii, T. 0.1-μm-Gate, Ultrathin-Film CMOS Devices Using SIMOX Substrate with 80-nm-Thick Buried Oxide Layer. *IEEE Trans. Electron Devices* **1991**, *40*, 675–678.
42. Kochelap, V.A.; Sokolov, V.N.; Bulashenko, O.M.; Rubi, J.M. Coulomb Suppression of Surface Noise. *Appl. Phys. Lett.* **2001**, *78*, 2003–2005. [CrossRef]
43. Kochelap, V.A.; Sokolov, V.N.; Bulashenko, O.N.; Rubi, J.M. Theory of Surface Noise under Coulomb Correlations between Carriers and Surface States. *J. Appl. Phys.* **2002**, *92*, 5347–5358. [CrossRef]
44. Omura, Y. A simple model for substrate current characteristics in short-channel ultrathin-film metal-oxide-semiconductor field-effect transistors by separation by implanted oxygen. *Jpn. J. Appl. Phys.* **1995**, *34*, 4722–4727. [CrossRef]
45. Omura, Y. Two-Dimensionally Confined Injection Phenomenon at Low Temperatures in Sub-10-nm-Thick SOI Insulated-Gate p-n-Junction Devices. *IEEE Trans. Electron Devices* **1996**, *43*, 436–443. [CrossRef]
46. Ishiyama, T.; Omura, Y. Analysis of Interface Microstructure Evolution in Separation by IMplanted OXygen (SIMOX) Wafers. *Jpn. J. Appl. Phys.* **2000**, *39*, 4653–4656. [CrossRef]
47. Arnold, E. Conduction Mechanisms in Bandtails at the SiO$_2$ Interface. *Surf. Sci.* **1976**, *58*, 60–70. [CrossRef]
48. Omura, Y.; Nagase, M. Low-Temperature Drain Current Characteristics in Sub-10-nm-Thick SOI nMOSFET's on SIMOX (Separation by IMplanted OXygen) substrates. *Jpn. J. Appl. Phys.* **1995**, *34*, 812–816. [CrossRef]
49. Omura, Y.; Sato, S. Theoretical Models for Low-Frequency Noise Behaviors of Buried-Channel MOSFETs. In Proceedings of the 2017 IEEE SOI-3D-Subthreshold Microelectronics Technology Unified Conference (S3S), Burlingame, CA, USA, 16–19 Octorber 2017.
50. Sze, S.M.; Ng, K.K. *Physics of Semiconductor Devices*, 3rd ed.; Wiley: Hoboken, NJ, USA, 2007; p. 314.

51. Kogan, S. *Electronic Noise and Fluctuations in Solids*; Cambridge University Press: Cambridge, UK, 1996; Chapter 3.

52. Omura, Y.; Ohwada, K. Threshold and Subthreshold Characteristics Theory for a Very Small Buried-Channel MOSFET Using a Majority-Carrier Distribution Model. *Solid State Electron.* **1981**, *24*, 301–308. [CrossRef]

micromachines

MDPI

Article

Remote Phonon Scattering in Two-Dimensional InSe FETs with High-κ Gate Stack

Pengying Chang *, Xiaoyan Liu *, Fei Liu and Gang Du

Institute of Microelectronics, Peking University, Beijing 100871, China; fliu003@gmail.com (F.L.);
gangdu@pku.edu.cn (G.D.)
* Correspondence: pychang@pku.edu.cn (P.C.); xyliu@ime.pku.edu.cn (X.L.); Tel.: +86-152-1038-8557 (P.C.)

Received: 15 November 2018; Accepted: 17 December 2018; Published: 19 December 2018

Abstract: This work focuses on the effect of remote phonon arising from the substrate and high-κ gate dielectric on electron mobility in two-dimensional (2D) InSe field-effect transistors (FETs). The electrostatic characteristic under quantum confinement is derived by self-consistently solving the Poisson and Schrödinger equations using the effective mass approximation. Then mobility is calculated by the Kubo–Greenwood formula accounting for the remote phonon scattering (RPS) as well as the intrinsic phonon scatterings, including the acoustic phonon, homopolar phonon, optical phonon scatterings, and Fröhlich interaction. Using the above method, the mobility degradation due to remote phonon is comprehensively explored in single- and dual-gate InSe FETs utilizing SiO_2, Al_2O_3, and HfO_2 as gate dielectric respectively. We unveil the origin of temperature, inversion density, and thickness dependence of carrier mobility. Simulations indicate that remote phonon and Fröhlich interaction plays a comparatively major role in determining the electron transport in InSe. Mobility is more severely degraded by remote phonon of HfO_2 dielectric than Al_2O_3 and SiO_2 dielectric, which can be effectively insulated by introducing a SiO_2 interfacial layer between the high-κ dielectric and InSe. Due to its smaller in-plane and quantization effective masses, mobility begins to increase at higher density as carriers become degenerate, and mobility degradation with a reduced layer number is much stronger in InSe compared with MoS_2.

Keywords: two-dimensional material; field effect transistor; indium selenide; phonon scattering; mobility; high-κ dielectric

1. Introduction

The compelling demand for higher performance and lower power consumption in complementary metal-oxide-semiconductor (CMOS) field-effect transistors (FETs) has highlighted the quest for devices and architectures based on new materials [1]. Performance boosters such as strain, high-κ dielectric, metal gate, and three-dimensional (3D) devices have enabled extraordinary improvement of performance in the past 60 years [2,3]. Recently, two-dimensional (2D) van der Waals semiconductors hold great potential for optics and electronics application due to their unique properties, including the atomic thickness, tunable bandgap, and dangling-bond-free surface, which achieves improved gate control over the channel and reduced short channel effects [4,5]. So far, many classes of 2D material-based devices have been extensively studied, such as graphene, transition metal dichalcogenides (TMDs), and black phosphorus [6–8]. Very recently, few-layer InSe has attracted much attention due to its highly promising prospect as channel material for FETs, offering small effective mass of electron ~0.14 m_0 and high electron mobility up to ~10^3 cm^2/Vs at room temperature obtained by experimental measurements [9–11]. Therefore, InSe has advantages of a similar gap as silicon, 2D nature as graphene, higher mobility than TMDs, and higher environmental stability than black phosphorus. In addition, electrostatic tunability of spin-orbit coupling in InSe has been identified, showing potential in devising III-VI based spintronic devices [12,13].

However, the charge transport properties in InSe FET have not been well understood and starve for comprehensive investigation. More recently, the ballistic performance of mono- and multi-layer InSe FET is studied by the first-principles calculation and the top of the barrier model [14], and temperature-dependent phonon-limited mobility is estimated by the physical modeling of intrinsic scattering mechanisms [15]. On the other hand, charge transport behavior is very sensitive to external surroundings, such as gaseous adsorbates from air and trapped charges in substrates [16], and their electronic performance is generally lower than their intrinsic values. Previous studies of back-gated multilayer InSe FET on various substrates (bare SiO_2, bare Al_2O_3, poly(methyl methacrylate) (PMMA)/SiO_2, and PMMA/Al_2O_3) have reported the carrier mobility ranging from 2.2 cm^2/Vs to 1055 cm^2/Vs at low operating voltage [10], while dual-gated InSe FET on hexagonal boron nitride (hBN)/SiO_2 show an excellent mobility approaching 10^3 cm^2/Vs and 10^4 cm^2/Vs at room and liquid-helium temperatures respectively [9]. It is apparently suggested that the introduction of substrate and gate dielectric has a strong effect on the electron mobility, which can be generally contributed to the extrinsic scatterings from surface roughness (SRS), interfacial Coulomb impurities (CIS), and remote phonon scatterings (RPS) [17,18]. Atomic flatness of 2D materials makes them immune to SR scattering, while CIS can be lowered or eliminated as possible by improving the fabrication process. Therefore, only remote phonon can be regarded as an intrinsic factor arising from the dielectric environment, and open questions remain as to its role in determining the electron transport in atomically-thin InSe FETs.

In this paper, the effect of remote phonons arising from the substrate and high-κ dielectric together with the intrinsic phonons of the InSe channel on electron transport is studied based on the physical modeling by self-consistently solving the Poisson and Schrödinger equations and employing the Kubo–Greenwood formula. Mobility behaviors in single-gate and dual-gate InSe FET with various gate dielectric are theoretically explored and analyzed as a function of temperature, inversion density, InSe layer number, and SiO_2 interfacial layer thickness. Acoustic phonons and optical phonons—as well as homopolar phonons—have a minor effect on electron mobility, while remote phonons and Fröhlich interaction play a comparatively major role in determining the electron transport in InSe. Compared with MoS_2, much smaller effective masses of electron in InSe give rise to a great enhancement of mobility at high density as carriers become degenerate. Simulation results in this work provide physical insight into the mobility behavior of InSe FET for carrier mobility optimization from the theoretical viewpoint.

This paper is organized as follows. Section 2 describes the device structures and simulation methods, especially the physical models of remote phonon depending on the gate stack. In Section 3, we present simulation results of mobility and corresponding explanations. Finally, the conclusion is drawn in Section 4.

2. Device Structures and Simulation Methods

Simulated device structures with 2D-layered InSe channel are shown in Figure 1, where the intrinsic channel without doping is assumed. Figure 1a shows the back-gate (single-gate) InSe FET with SiO_2 substrate as gate dielectric. Figure 1b shows the top-gate (dual-gate) InSe FETs with high-κ dielectric as top dielectric and SiO_2 substrate as back dielectric. Figure 1c shows the structure with additional SiO_2 interfacial layer (ITL) embedded between the InSe channel and high-κ dielectric compared with Figure 1b. In the case of single-gate devices, only the back gate is biased with V_{bg}, while the back gate is grounded and the top gate is biased with V_{tg} for the dual-gated devices. In this work, traditionally used high-κ dielectrics of HfO_2 and Al_2O_3 are comprehensively studied, with corresponding parameters listed in Table 1. Except for Figure 2, all the simulation results are calculated at room temperature (300 K).

We start the calculation by obtaining the electrostatic characteristic of the two-dimensional electron gas (2DEG) in InSe layer by self-consistently solving the Poisson and Schrödinger equations using the effective mass approximation with nonparabolicity correction, inherently accounting for the quantum confinement effects [19]. Particularly the energy dispersion of 2D-layered InSe is described by the thickness-dependent effective masses obtained from first-principles calculation, as shown in our previous work [14]. Next, the matrix elements and the scattering rates are calculated through the Fermi golden rule [19]. Physical models for electron mobility include the remote phonon scattering (RPS) arising from the high-κ dielectric as well as the intrinsic phonon scatterings of channel material, including the acoustic (AC) phonon-, homopolar (HO) phonon-, optical (OP) phonon- scatterings, and Fröhlich interaction (POP) [20–23]. For AC phonons, elastic and isotropic approximations are adopted. The HO and OP scatterings are treated as inelastic and isotropic process. For POP and RP scatterings, inelastic and anisotropic characteristic are considered. Once the scattering rates are obtained, the mobility is calculated by the Kubo–Greenwood formula employing the momentum relaxation time approximation. The parameters for mobility calculation in few-layer InSe are taken from our previous work [15].

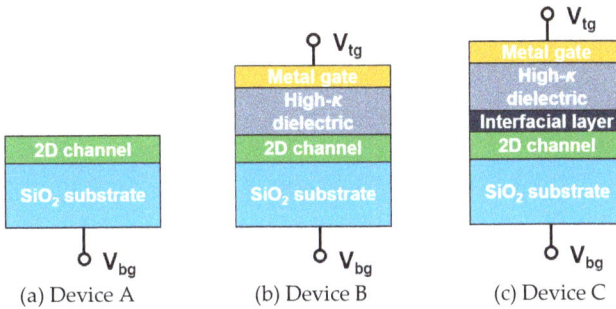

Figure 1. Simulated device structure with atomically thin InSe channel in this work. (**a**) Back-gate InSe field-effect transistors (FET) with SiO$_2$ substrate as gate dielectric. (**b**) Dual-gate InSe FET with high-κ dielectric as top-gate dielectric and SiO$_2$ substrate as back-gate dielectric. (**c**) The same structure as (**b**) with additional SiO$_2$ interfacial layer embedded between the InSe channel and high-κ dielectric. For the dual-gate structure, high-κ dielectric is covered with a metal gate.

Table 1. Parameters for the polar phonons in some high-κ materials [20]

Quantity	SiO$_2$	Al$_2$O$_3$	HfO$_2$
ε_0	3.90	12.53	22.00
ε_{int}	3.05	7.27	6.58
ε_∞	2.50	3.20	5.03
$\hbar\omega_{TO1}$	55.60	48.18	12.40
$\hbar\omega_{TO2}$	138.10	71.41	48.35

ε_0: Static (low-frequency) dielectric constant; ε_{int}: dielectric constant at an intermediate frequency; ε_∞: high-frequency dielectric constant; $\hbar\omega_{TO1}$, $\hbar\omega_{TO2}$: frequencies of the two polar phonons.

For remote phonon induced by SiO$_2$ substrate, the semi-infinite substrate is assumed, and the dispersion relationship for the remote phonon model can be written as [19]

$$\omega_{RP} = \omega_{TO1,SiO_2}\sqrt{\frac{\varepsilon_{InSe} + \varepsilon_{SiO_2,0}}{\varepsilon_{InSe} + \varepsilon_{SiO_2,\infty}}} \tag{1}$$

where $\omega_{TO1,SiO2}$ is the low-frequency phonon mode of SiO$_2$, ε_{InSe} is the dielectric constant of InSe. The potential amplitude of the remote phonon is written as

$$\frac{1}{\hat{\varepsilon}_{RP}} = \frac{1}{\varepsilon_{InSe} + \varepsilon_{SiO_2,\infty}} - \frac{1}{\varepsilon_{InSe} + \varepsilon_{SiO_2,0}} \tag{2}$$

For the remote phonon induced by top gate dielectric, high-κ dielectric covered with a metal gate is employed in the simulation. As shown in Table 1, the frequencies of two polar phonons in high-κ dielectrics such Al_2O_3 and HfO_2 show great discrepancy. Hence, for simplicity, only the low-frequency phonon mode in high-κ gate dielectric is considered [19]. For high-κ dielectric with a metal gate as shown in Figure 1b, the dispersion relationship is

$$\omega_{RP} = \omega_{TO1,HK} \frac{\left(\varepsilon_{InSe} \left(\frac{1-e^{-2qT_{HK}}}{1+e^{-2qT_{HK}}} \right) + \varepsilon_{HK,0} \right)^{1/2}}{\left(\varepsilon_{InSe} \left(\frac{1-e^{-2qT_{HK}}}{1+e^{-2qT_{HK}}} \right) + \varepsilon_{HK,int} \right)^{1/2}} \tag{3}$$

where $\omega_{TO1,HK}$ is the low-frequency phonon mode of high-κ dielectric, $\varepsilon_{HK,0}$ and $\varepsilon_{HK,int}$ are the dielectric constant at the static and intermediate frequency, T_{HK} is the thickness of top gate dielectric, and $q = |k - k'|$ is the remote phonon momentum. The effective dielectric constant depending on the frequency dependent dielectric constant of the high-κ material can be written as

$$\varepsilon_{eff} = \varepsilon_{HK}(\omega) \frac{1 + e^{-2qT_{HK}}}{1 - e^{-2qT_{HK}}} + \varepsilon_{InSe} \tag{4}$$

and then the corresponding potential amplitude is

$$\frac{1}{\hat{\varepsilon}_{RP}} = \frac{1}{\varepsilon_{eff}(\varepsilon_{HK,int})} - \frac{1}{\varepsilon_{eff}(\varepsilon_{HK,0})} \tag{5}$$

For the high-κ gate stack with a SiO_2 interfacial layer, namely ITL/high-κ/metal-gate stack as shown in Figure 1c, the dispersion relationship is [16]

$$\omega_{RP} = \omega_{TO1,HK} \left[\varepsilon_{SiO_2,0} \left(\frac{1-e^{-2qT_{HK}}}{1+e^{-2qT_{HK}}} \right) \left(\frac{1 - \frac{\varepsilon_{SiO_2,0}+\varepsilon_{InSe}}{\varepsilon_{SiO_2,0}-\varepsilon_{InSe}} e^{2qT_{ITL}}}{1 + \frac{\varepsilon_{SiO_2,0}+\varepsilon_{InSe}}{\varepsilon_{SiO_2,0}-\varepsilon_{InSe}} e^{2qT_{ITL}}} \right) - \varepsilon_{HK,0} \right]^{1/2}$$

$$\times \left[\varepsilon_{SiO_2,0} \left(\frac{1-e^{-2qT_{HK}}}{1+e^{-2qT_{HK}}} \right) \left(\frac{1 - \frac{\varepsilon_{SiO_2,0}+\varepsilon_{InSe}}{\varepsilon_{SiO_2,0}-\varepsilon_{InSe}} e^{2qT_{ITL}}}{1 + \frac{\varepsilon_{SiO_2,0}+\varepsilon_{InSe}}{\varepsilon_{SiO_2,0}-\varepsilon_{InSe}} e^{2qT_{ITL}}} \right) - \varepsilon_{HK,int} \right]^{-1/2} \tag{6}$$

where T_{ITL} is the thickness of interfacial layer. The effective dielectric constant is

$$\varepsilon_{eff}(\omega) = \varepsilon_{HK}(\omega) \left[\left(\frac{\varepsilon_{SiO_2,0}-\varepsilon_{InSe}}{2\varepsilon_{SiO_2,0}} \right)^2 e^{-2qT_{ITL}} + \left(\frac{\varepsilon_{SiO_2,0}+\varepsilon_{InSe}}{\varepsilon_{SiO_2,0}} \right)^2 e^{2qT_{ITL}} + 2\frac{\varepsilon_{SiO_2,0}^2-\varepsilon_{InSe}^2}{(2\varepsilon_{SiO_2,0})^2} \right] \cdot \frac{1+e^{-2qT_{HK}}}{1-e^{-2qT_{HK}}}$$

$$+ \frac{(\varepsilon_{SiO_2,0}+\varepsilon_{InSe})^2}{4\varepsilon_{SiO_2,0}} e^{2qT_{ITL}} - \frac{(\varepsilon_{SiO_2,0}-\varepsilon_{InSe})^2}{4\varepsilon_{SiO_2,0}} e^{-2qT_{ITL}} \tag{7}$$

Then the potential amplitude for the ITL/high-κ/metal-gate stack can be obtained through the Equations (5) and (7).

3. Results and Discussion

To begin with, we calibrate the physical models with the experimental measurement. Figure 2 shows the calculated and experimental temperature-dependent mobility at inversion density of 1.6×10^{12} cm^{-2} and 7.9×10^{12} cm^{-2} in six-layer InSe dual-gate FET. It should be noted that the experiment results are obtained from the dual-gate InSe FET with channel covered by hexagonal boron nitride (hBN) [9], which insulates InSe from the dielectric environment, leading to the absence of remote phonon scattering. From Figure 2a, considering the intrinsic scatterings by AC, HO, and OP phonon and Fröhlich interaction, the temperature-dependent electron mobility curves measured by Hall effect are reproduced successfully for T > 100 K, where phonon scatterings dominate. The excellent agreement between the simulations and experiments validate our methods and models. It should be pointed out that when temperature is down to 100 K, there is a significant discrepancy of mobility between simulations and experiments due to the fact that Coulomb scattering resulting from the channel impurities and interfacial charges is excluded, which is a dominant factor in determining the carrier mobility in the low-temperature regime.

On the other hand, if high-κ dielectric of HfO_2 is directly deposited on the InSe channel, the mobility is severely degraded from its intrinsic value, as shown in Figure 2a by solid lines. For example, at room temperature, mobility changes from 1808 to 1120 cm^2/Vs (920 to 464 cm^2/Vs) at inversion density of 7.9×10^{12} cm^{-2} (1.6×10^{12} cm^{-2}) due to the remote phonon scattering. To understand the mobility behavior in depth, Figure 2b,c plot the contributions of all the considered scattering mechanisms to the total mobility. Compared with AC, OP, and HO phonons, the remote phonon together with Fröhlich interaction plays a comparatively major role in determining the electron transport in InSe FET. This is the objective of this work to focus on the remote phonon scattering in InSe FET with high-κ gate stack in the following.

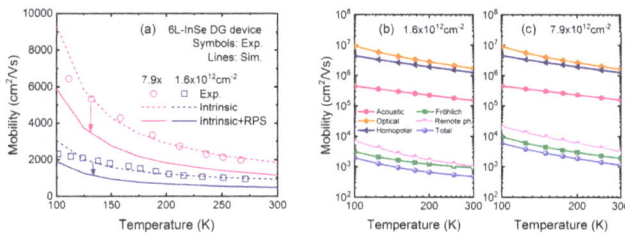

Figure 2. Temperature-dependent mobility in six-layer InSe FETs at inversion density of 1.6×10^{12} cm^{-2} and 7.9×10^{12} cm^{-2}, respectively. (**a**) Comparison between the experimental mobility (symbols) and calculated intrinsic mobility without remote phonon scattering (RPS) (dashed lines), which shows an excellent agreement. In contrast, the mobility with RPS (solid lines) is significantly degraded. (**b,c**) Contributions of each scattering mechanisms to the total mobility for different inversion density respectively. Intrinsic scatterings include acoustic (AC), optical (OP), and homopolar (HO) phonon scatterings as well Fröhlich interaction, while extrinsic scattering is remote phonon scattering arising from the HfO_2 high-κ dielectric.

The effect of remote phonon originating from the substrate and gate stack on the electron transport of few-layer InSe is shown in Figure 3. The intrinsic phonon-limited mobility in six-layer single-gate InSe FET is ~843 cm^2/Vs at low inversion density. With SiO_2 substrate employed, the mobility is degraded to ~735 cm^2/Vs due to remote phonon. In the case of dual-gate structure, when Al_2O_3 and HfO_2 are used as top-gate dielectric, the additional remote phonon further reduces the mobility to ~634 cm^2/Vs and 426 cm^2/Vs respectively. It can be seen that the HfO_2 dielectric has a much stronger influence of remote phonon than Al_2O_3 and SiO_2 dielectric since it has higher dielectric constant and softer polar vibration mode [20], as listed in Table 1. Particularly, it is worth noting that the smaller permittivity of InSe results in a stronger remote phonon coupling with electrons compared with silicon even in the SiO_2 case. Despite serious degradation due to remote phonons, the mobility of the six-layer InSe with high-κ dielectric is higher than that of silicon on insulator (SOI) device with SiO_2 dielectric at a comparative thickness [24], revealing its great potential in high-performance logic application.

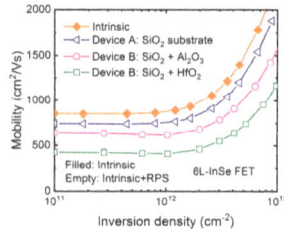

Figure 3. Calculated mobility as a function of inversion density in six-layer InSe FETs with different device structures as shown in Figure 1a,b utilizing Al_2O_3 and HfO_2 as top-gate dielectric and SiO_2 substrate as back-gate dielectric respectively. Equivalent oxide thickness (EOT) = 1 nm of high-κ dielectric is assumed. Filled symbols represent intrinsic phonon-limited mobility for benchmark, while empty symbols represent total mobility including the remote phonon scattering from substrate and high-κ dielectric.

It is also observed that in six-layer InSe, mobility is increased significantly at higher density, which is against the common sense. To confirm this behavior, mobility in 2-, 6-, 16-, and 40-layer InSe FET is calculated in Figure 4a,b, where HfO_2 and Al_2O_3 dielectric are used separately. At the same time, mobility in MoS_2 FET using same device structure is also plotted in Figure 4c,d for comparison. In the MoS_2 case, as inversion density increases, mobility monotonously decreases for thick devices as expected, and remains almost unchanged for thin devices due to strong quantum confinement. In the InSe case, at low density, mobility behavior is consistent with MoS_2. However, when inversion density is larger than ~2×10^{12} cm^{-2}, mobility quickly increases regardless of layer number or high-κ dielectric. Actually, this is also demonstrated by experimental results in [9] as shown in Figure 2a, which cannot be totally contributed to impurity scattering because at large inversion density and room temperature the screening produced by the inversion layer drastically reduce the Coulomb scattering. Therefore, this exceptional enhancement seems intrinsic for 2D-layered InSe to a great extent.

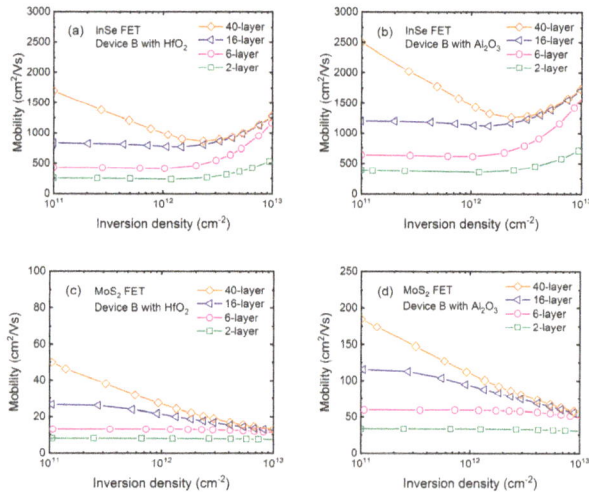

Figure 4. Mobility as a function of inversion density with layer number of 2, 6, 16, and 40 employing the dual-gate structure as shown in Figure 1b in InSe and MoS_2 FET for comparison. EOT = 1 nm of high-κ dielectric is assumed. (**a**) InSe FET with HfO_2 dielectric. (**b**) InSe FET with Al_2O_3 dielectric. (**c**) MoS_2 FET with HfO_2 dielectric. (**d**) MoS_2 FET with Al_2O_3.

To get physical insight into this mobility behavior, Figure 5a–d shows the contributions of each scattering mechanism to the total mobility in both InSe and MoS$_2$ FET with 2- and 40-layer thickness respectively. Consistent with above-mentioned results, mobility behavior is mainly governed by the remote phonon and Fröhlich interaction in all considered devices. We find that their scattering rates increase with inversion density increasing, which should reduce the mobility. In the MoS$_2$ case, carriers are always non-degenerate following the Boltzmann distribution, where the mobility is essentially determined by the relaxation times or scattering rates. On the other hand, when inversion density is larger than $\sim 2 \times 10^{12}$ cm^{-2}, carriers in InSe FET become degenerate, where subband minimum is lower than Fermi level E_F, and consequently the most influential relaxation times are those for energies close to E_F. Due to their anisotropic property, scattering rates of remote phonon and Fröhlich interaction are much smaller near E_F than those of the subband minimum, giving rise to an enhancement of mobility. This discrepancy between InSe and MoS$_2$ FET can be well understood by their effective masses. Firstly, in-plane effective mass of 0.14 m_0 in InSe is much smaller than 0.62 m_0 in MoS$_2$ [25] leading to much smaller density-of-states (DOS). In order to obtain the same density, the conduction band minimum is lower than Fermi level over a few $k_B T$, where k_B is the Boltzmann constant. Secondly, quantization effective mass of 0.08 m_0 in InSe is also much smaller than 0.49 m_0 in MoS$_2$ [26]. This is the reason that quantum confinement takes effect in 16-layer InSe, but not until the layer number is reduced to 6 layers in MoS$_2$. Actually, less subbands contributing to the carrier transport in InSe need Fermi level being higher to change the density, which makes carriers more degenerate together with the effect of small DOS.

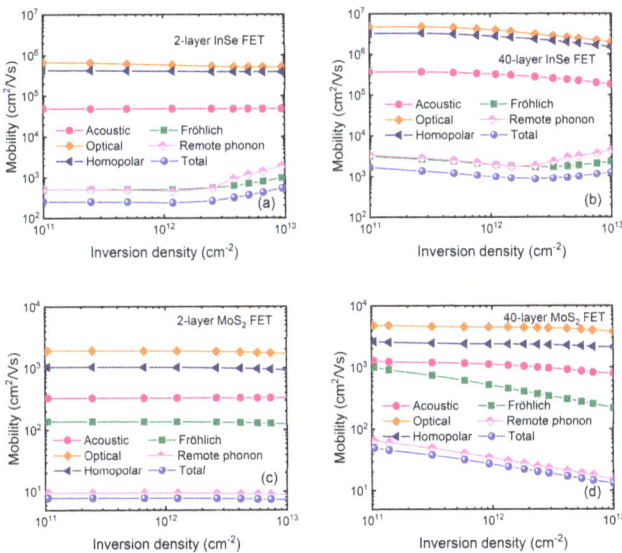

Figure 5. Contributions of each scattering process to the total mobility corresponding to Figure 4a,c for InSe and MoS$_2$ FET with HfO$_2$ dielectric respectively. (**a**) 2-layer InSe FET. (**b**) 40-layer InSe FET. (**c**) 2-layer MoS$_2$ FET. (**d**) 40-layer MoS$_2$ FET.

The dependence of mobility on number of layer (NL) in InSe FET is shown in Figure 6a, where inversion densities of 5×10^{11}, 2×10^{12}, and 8×10^{12} cm^{-2} are considered respectively. In the case of the relatively medium and high density, mobility is almost independent of channel thickness until NL ~ 15, and then drops rapidly as NL is further reduced. In the low-density case, mobility degradation occurs earlier when NL <40. It is noted that for thicker devices, mobility at 5×10^{11} cm^{-2} gradually surpasses the value at 2×10^{12} cm^{-2} and then reaches up to the value at 8×10^{12} cm^{-2}. This awkward behavior can be explained by Figure 4a,b, where mobility initially decrease and then the trend is opposite at higher density, as inversion density decreases in the 40-layer InSe FET. To get physical insight into the mobility degradation, contributions of each scattering mechanism at 5×10^{11} cm^{-2} is shown in Figure 6b. AC-, LO-, and HO-limited mobility is severely reduced when NL is less than ~15, while the degradation of remote-phonon as well as Fröhlich-limited mobility begins to decrease about NL ~40, further indicating their major role in determining the electron transport.

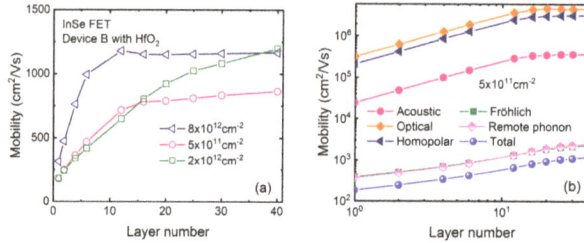

Figure 6. (**a**) Mobility as a function of number of layers ranging from 1 to 40 with different inversion density in InSe FET using HfO$_2$ as gate dielectric and SiO$_2$ as substrate, as shown in Figure 1b. (**b**) Contributions of each scattering mechanism to the total mobility at low inversion density.

Inspired by the similar situation for silicon [27,28], it is suggested that an interfacial layer can be introduced between InSe channel and high-κ dielectric. To explore the effect of the interfacial layer, Figure 7 shows the calculated mobility in InSe FET with device structure of Figure 1c capped with different high-κ dielectric. From Figure 7a, it can be seen that the interfacial layer effectively insulates the channel away from the high-κ dielectric, resulting in significant mobility enhancement in the whole range of inversion density due to weaker remote phonon coupling. As interfacial layer thickness is increased, more mobility enhancement is achieved until T_{ITL} approaches ~2 nm, when remote phonon from high-κ dielectric is totally separated, as shown in Figure 7b. Besides, a thin interfacial layer is more effective in HfO$_2$ dielectric compared with Al$_2$O$_3$ dielectric. Figure 8 shows the corresponding remote coulomb scattering (RCS)-limited mobility as a function of SiO$_2$ interfacial layer thickness, showing an exponential dependence on T_{ITL} as $\mu_{RPS} \propto \exp(2k_F T_{ITL})$, whatever the inversion density or high-κ dielectric is, with Fermi wavelength $2k_F = 1.1$ nm^{-1}. This is in agreement with theoretical predictions for a remote scattering mechanism [20,28].

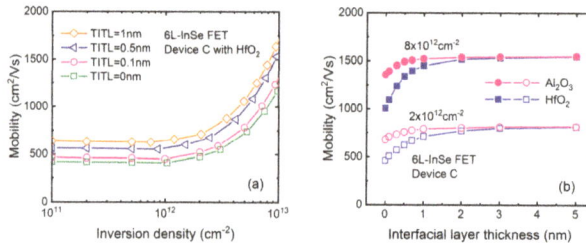

Figure 7. (**a**) Mobility as a function of inversion density featuring different SiO$_2$ interfacial layer thickness in InSe FET with dual-gate structure of Figure 1c. (**b**) Mobility as a function of SiO$_2$ interfacial layer thickness at different inversion density in both Al$_2$O$_3$- and HfO$_2$-gated InSe FET. EOT = 1 nm of high-κ dielectric is used in the simulation.

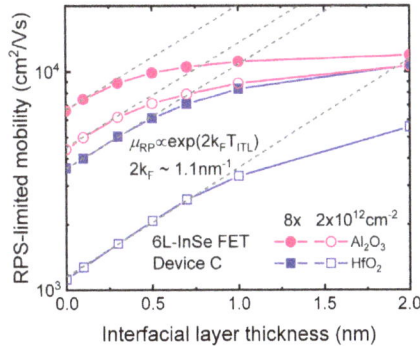

Figure 8. Remote coulomb scattering (RCS)-limited mobility as a function of SiO_2 interfacial layer thickness corresponding to total mobility in Figure 7b, showing an exponential dependence on T_{ITL} as $\exp(2k_F T_{ITL})$ with $2k_F = 1.1$ nm^{-1}, represented by the dashed lines.

4. Conclusions

Based on the self-consistent Poisson and Schrödinger equations and the Kubo–Greenwood formula, remote phonons arising from both the SiO_2 substrate and high-κ dielectrics in InSe FETs are comprehensively studied, together with the intrinsic scatterings by AC phonons, OP phonons, HO phonons, and Fröhlich interaction. It is observed that remote phonons and Fröhlich interaction plays a comparatively major role in determining the electron transport in InSe. Mobility is more severely degraded by remote phonon of HfO_2 dielectric than Al_2O_3 and SiO_2 dielectric, which can be effectively insulated by introducing a SiO_2 interfacial layer between the high-κ dielectric and InSe. Due to its smaller in-plane and quantization effective masses, mobility begins to increase at higher densities as carriers degenerate, and mobility degradation with reduced layer number is much stronger in InSe compared with MoS_2.

Author Contributions: Conceptualization, P.C.; Formal analysis, P.C.; Funding acquisition, P.C.; Investigation, P.C.; Methodology, P.C. and F.L.; Resources, X.L.; Software, P.C.; Supervision, X.L. and G.D.; Validation, P.C.; Writing—Original Draft, P.C.

Funding: This research was funded by China Postdoctoral Science Foundation (grant number 2018M630034) and National Natural Science Foundation of China (grant numbers 61804003, 61674008, 61421005).

Conflicts of Interest: The authors declare no conflict of interest.

References

1. Bernstein, K.; Cavin, R.K.; Porod, W.; Seabaugh, A.; Welser, J. Device and Architecture Outlook for Beyond CMOS Switches. *Proc. IEEE* **2010**, *98*, 2169–2184. [CrossRef]
2. Mistry, K.; Allen, C.; Auth, C.; Beattie, B.; Bergstrom, D.; Bost, M.; Brazier, M.; Buehler, M.; Cappellani, A.; Chau, R.; et al. A 45nm Logic Technology with High-κ+Metal Gate Transistors, Strained Silicon, 9 Cu Interconnect Layers, 193nm Dry Patterning, and 100% Pb-free Packaging. In Proceedings of the IEEE International Electron Devices Meeting, Washington, DC, USA, 10–12 December 2007. [CrossRef]
3. Natarajan, S.; Agostinelli, M.; Akbar, S.; Bost, M.; Bowonder, A.; Chikarmane, V.; Chouksey, S.; Dasgupta, A.; Fischer, K.; Fu, Q.; et al. A 14nm Logic Technology Featuring 2nd-Generation FinFET Transistors, Air-Gapped Interconnects, Self-Aligned Double Patterning and a 0.0588μm^2 SRAM cell size. In Proceedings of the IEEE International Electron Devices Meeting, San Francisco, CA, USA, 15–17 December 2014. [CrossRef]
4. Fiori, G.; Bonaccorso, F.; Iannaccone, G.; Palacios, T.; Neumaier, D.; Seabaugh, A.; Banerjee, S.K.; Colombo, L. Electronics based on two-dimensional materials. *Nat. Nanotechnol.* **2014**, *9*, 768–779. [CrossRef] [PubMed]

5. Novoselov, K.S.; Mishchenko, A.; Carvalho, A.; Castro Neto, A.H. 2D materials and van der Waals heterostructures. *Science* **2016**, *353*, 439. [CrossRef] [PubMed]
6. Novoselov, K.S.; Geim, A.K.; Morozov, S.V.; Jiang, D.; Zhang, Y.; Dubonos, S.V.; Grigorieva, I.V.; Firsov, A.A. Electric Field Effect in Atomically Thin Carbon Films. *Science* **2004**, *306*, 666–669. [CrossRef]
7. Schmidt, H.; Wang, S.; Chu, L.; Toh, M.; Kumar, R.; Zhao, W.; Castro Neto, A.H.; Martin, J.; Adam, S.; Ozylimaz, B.; et al. Transport Properties of Monolayer MoS$_2$ Grown by Chemical Vapor Deposition. *Nano Lett.* **2014**, *14*, 1909–1913. [CrossRef] [PubMed]
8. Li, L.; Yu, Y.; Ye, G.J.; Ge, Q.; Ou, X.; Wu, H.; Feng, D.; Cheng, X.H.; Zhang, Y. Black phosphorus field-effect transistors. *Nat. Nanotechnol.* **2014**, *9*, 372–377. [CrossRef] [PubMed]
9. Bandurin, D.A.; Tyurnina, A.V.; Yu, G.L.; Mishchenko, A.; Zólyomi, V.; Morozov, S.V.; Kumar, R.K.; Gorbachev, R.V.; Kudrynskyi, Z.R.; Pezzini, S.; et al. High Electron Mobility, Quantum Hall Effect and Anomalous Optical Response in Atomically Thin InSe. *Nat. Nanotechnol.* **2017**, *9*, 223–227. [CrossRef] [PubMed]
10. Feng, W.; Zheng, W.; Cao, W.; Hu, P. Back Gated Multilayer InSe Transistors with Enhanced Carrier Mobilities via the Suppression of Carrier Scattering from a Dielectric Interface. *Adv. Mater.* **2014**, *26*, 6587–6593. [CrossRef] [PubMed]
11. Sucharitakul, S.; Goble, N.J.; Kumar, U.R.; Sankar, R.; Bogorad, Z.A.; Chou, F.-C.; Chen, Y.-T.; Gao, X.P.A. Intrinsic Electron Mobility Exceeding 10^3 cm^2/(Vs) in Multilayer InSe FETs. *Nano Lett.* **2015**, *15*, 3815–3819. [CrossRef]
12. Premasiri, K.; Radha, S.K.; Sucharitakul, S.; Kumar, U.R.; Sankar, R.; Chou, F.-C.; Chen, Y.-T.; Gao, X.P.A. Tuning Rashba Spin-Orbit Coupling in Gated Multilayer InSe. *Nano Lett.* **2018**, *18*, 4403–4408. [CrossRef]
13. Zeng, J.; Liang, S.-J.; Gao, A.; Wang, Y.; Pan, C.; Wu, C.; Liu, E.; Zhang, L.; Cao, T.; Liu, X.; et al. Gate-tunable weak antilocalization in a few-layer InSe. *Phys. Rev. B* **2018**, *98*, 125414. [CrossRef]
14. Chang, P.; Liu, X.; Liu, F.; Du, G. First-principles based ballistic transport simulation of monolayer and few-layer InSe FET. *Jpn. J. Appl. Phys.* **2018**. under review.
15. Chang, P.; Liu, X.; Liu, F.; Du, G. Phonon-Limited Mobility in n-Type Few-Layer InSe Devices from First Principles. *IEEE Electron Devices Lett.* **2018**. [CrossRef]
16. Li, S.-L.; Tsukagoshi, K.; Orgiu, E.; Samori, P. Charge transport and mobility engineering in two-dimensional transition metal chalcogenide semiconductors. *Chem. Soc. Rev.* **2016**, *45*, 118–151. [CrossRef] [PubMed]
17. Li, S.-L.; Wakabayashi, K.; Xu, Y.; Nakaharai, S.; Komatsu, K.; Li, W.-W.; Lin, Y.-F.; Ferreira, A.A.; Tsukagoshi, K. Thickness-Dependent Interfacial Coulomb Scattering in Atomically Thin Field-Effect Transistors. *Nano Lett.* **2013**, *13*, 3546–3552. [CrossRef]
18. Zeng, L.; Xin, Z.; Chen, S.; Du, G.; Kang, J.; Liu, X. Remote phonon and impurity screening effect of substrate and gate dielectric on electron dynamics in single layer MoS$_2$. *Appl. Phys. Lett.* **2013**, *103*, 113505. [CrossRef]
19. Esseni, D.; Palestri, P.; Selmi, L. *Nanoscale MOS Transistors: Semi-Classical Transport and Applications*; Cambridge University Press: New York, NY, USA, 2011; pp. 98–103.
20. Fischetti, M.V.; Neumayer, D.A.; Cartier, E.A. Effective electron mobility in Si inversion layers in metal-oxide-semiconductor systems with a high-κ insulator: The role of remote phonon scattering. *J. Appl. Phys.* **2001**, *90*, 4587–4606. [CrossRef]
21. Poljak, M.; Jovanovié, V.; Grgec, D.; Suligoj, T. Assessment of Electron Mobility in Ultrathin-Body InGaAs-on-Insulator MOSFETs Using Physics-Based Modeling. *IEEE Trans. Electron Devices* **2012**, *59*, 1636–1643. [CrossRef]
22. Chang, P.; Liu, X.; Zeng, L.; Wei, K.; Du, G. Investigation of Hole Mobility in Strained InSb Ultrathin Body pMOSFETs. *IEEE Trans. Electron Devices* **2015**, *62*, 947–954. [CrossRef]
23. Chang, P.; Liu, X.; Du, G.; Zhang, X. Assessment of Hole Mobility in Strained InSb, GaSb and InGaSb Based Ultra-Thin Body pMOSFETs with Different Surface Orientations. In Proceedings of the IEEE International Electron Devices Meeting, San Francisco, CA, USA, 15–17 December 2014. [CrossRef]
24. Tsutsui, G.; Hiramoto, T. Mobility and Threshold-Voltage Comparison Between (110)- and (100)-Oriented Ultrathin-Body Silicon MOSFETs. *IEEE Trans. Electron Devices* **2006**, *53*, 2582–2588. [CrossRef]

25. Gonzalez-Medina, J.M.; Ruiz, F.G.; Martin, E.G.; Godoy, A.; Gámiz, F. Simulation study of the electron mobility in few-layer MoS_2 metal-insulator-semiconductor field-effect transistors. *Solid-State Electron.* **2015**, *114*, 30–34. [CrossRef]

26. Segura, A.; Marí, B.; Martinez-Pastor, J. Three-dimensional electrons and two-dimensional electric subbands in the transport properties of tin-doped n-type indium selenide: Polar and homopolar phonon scattering. *Phys. Rev. B* **1991**, *43*, 4953–4965. [CrossRef]

27. Chau, R.; Datta, S.; Doczy, M.; Doyle, B.; Kavalieros, J.; Metz, M. High-κ/Metal-Gate Stack and Its MOSFET Characteristics. *IEEE Electron Device Lett.* **2004**, *6*, 408–410. [CrossRef]

28. Cassé, M.; Thevenod, L.; Guillaumot, B.; Tosti, L.; Martin, F.; Mitard, J.; Weber, O.; Andrieu, F.; Ernst, T.; Reimbold, G.; et al. Carrier Transport in HfO_2/Metal Gate MOSFETs: Physical Insight into Critical Parameters. *IEEE Trans. Electron Devices* **2006**, *53*, 759–768. [CrossRef]

micromachines

MDPI

Article

3D Numerical Simulation of a Z Gate Layout MOSFET for Radiation Tolerance

Ying Wang [1,*], Chan Shan [2], Wei Piao [1], Xing-ji Li [3], Jian-qun Yang [3], Fei Cao [1] and Cheng-hao Yu [1]

[1] Key Laboratory of RF Circuits and Systems, Ministry of Education, Hangzhou Dianzi University, Hangzhou 310018, China; luoxin07@hrbeu.edu.cn (W.P.); caofei@hdu.edu.cn (F.C.); yuchenghao@hdu.edu.cn (C.-h.Y.)

[2] College of Information Engineering, Jimei University, Xiamen 361021, China; shanchan@jmu.edu.cn

[3] National Key Laboratory of Materials Behavior and Evaluation Technology in Space Environment, Harbin Institute of Technology, Harbin 150080, China; lxj0218@hit.edu.cn (X.-j.L.); yangjianqun@hit.edu.cn (J.-q.Y.)

* Correspondence: wangying01@hdu.edu.cn

Received: 11 November 2018; Accepted: 11 December 2018; Published: 14 December 2018

Abstract: In this paper, for the first time, an n-channel metal-oxide-semiconductor field-effect transistor (NMOSFET) layout with a Z gate and an improved total ionizing dose (TID) tolerance is proposed. The novel layout can be radiation-hardened with a fixed charge density at the shallow trench isolation (STI) of 3.5×10^{12} cm^{-2}. Moreover, it has the advantages of a small footprint, no limitation in W/L design, and a small gate capacitance compared with the enclosed gate layout. Beside the Z gate layout, a non-radiation-hardened single gate layout and a radiation-hardened enclosed gate layout are simulated using the Sentaurus 3D technology computer-aided design (TCAD) software. First, the transfer characteristics curves (I_d-V_g) curves of the three layouts are compared to verify the radiation tolerance characteristic of the Z gate layout; then, the threshold voltage and the leakage current of the three layouts are extracted to compare their TID responses. Lastly, the threshold voltage shift and the leakage current increment at different radiation doses for the three layouts are presented and analyzed.

Keywords: bulk NMOS devices; radiation hardened by design (RHBD); total ionizing dose (TID); Sentaurus TCAD; layout

1. Introduction

The total ionizing dose (TID) effect is one of the mechanisms that causes radiation-induced anomalies in semiconductor devices. The TID mechanism induces the generation of trapped charges in the dielectrics and interface states along the Si/SiO$_2$ interfaces, causing degradation of a transistor's performance [1–4]. Due to the downscaling, the net-charge trapping in oxides with a thickness of less than 10 nm is modest [5–9]. Since the thickness of the gate oxide of the simulated transistors is 2 nm, in this work, the net-charge trapping in the oxides is negligible. Therefore, the effects on thick oxides, such as the shallow trench isolation (STI), dominate the TID response of metal-oxide-semiconductor field-effect transistors (MOSFETs) [10]. Moreover, the charge trapped in the spacer oxide or at its interface modifies the parasitic series' resistance, reducing the drive current [11].

In a conventional non-radiation-hardened single gate layout, the STI's parasitic conduction path (the red arrow in Figure 1a) induced by the TID effect, which is visible only in an n-MOSFET, occurs along the sidewall oxide between the source and the drain, and leads to an increase in the drain current as the radiation dose increases [1]. A widely studied layout with radiation hardness, called the enclosed gate layout [12–14], which requires tradeoffs in application [14,15], is presented in Figure 1b.

For instance, a very small width over length ratio (W/L) is not realistic for an Enclosed Layout Transistor (ELT), for which the minimum achievable W/L is 2.26 [15], which is a significant concern in analog circuits [16]. Moreover, a larger gate capacitance will cause a longer time delay, which is not favorable for digital circuits. A large footprint is another disadvantage of the enclosed gate layout. In circuit design, the area penalty induced by design has been the main drawback [17].

In order to eliminate the parasitic path and overcome the disadvantages of an enclosed gate layout, for the first time, an n-MOSFET layout with a Z gate is proposed. Moreover, the proposed Z gate layout is applicable to more complicated structures, such as fin-field-effect transistors (FinFETs), tunnel-field-effect transistors (TFETs), and nanowires [18–22]. In this paper, devices with the proposed Z gate layout achieve total-dose hardness by eliminating these edges, but at the expense of fabrication feasibility due to the asymmetric active area design, as shown in Figure 1c. First, the effectiveness of the proposed layout to eliminate the leakage current is demonstrated by I_d-V_g curves. Then, the total shift of the threshold voltage and the variation of the leakage current, before and after the radiation is applied, are calculated for the single gate layout, the enclosed gate layout, and the Z gate layout, respectively. Further, the three simulated layouts are compared with respect to the threshold voltage shift and the leakage current increase as a function of the fixed charge density. Comparing the static characteristics of the different transistor layouts, it is found that the Z gate layout exhibits the best TID response compared with the conventional layouts and ELTs.

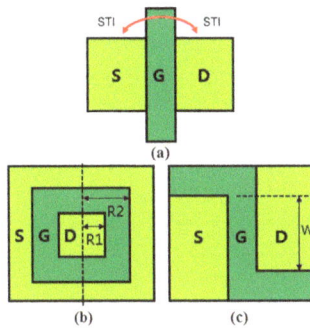

Figure 1. Schematic structures of (**a**) the conventional layout, (**b**) the H gate layout, and (**c**) the proposed layout. STI, shallow trench isolation.

2. Device Structure and Simulation

The Z gate layout achieves the radiation hardness by introducing two short extra gates that separate the active area and the isolation oxides. It should be noted that the precise, effective W/L ratio model of the proposed layout is not available at present; so, the channel width of the Z gate layout in this work is defined as shown in Figure 1c. A report [15] proposed an effective W/L model of an enclosed gate layout, and concluded that the only way to obtain a low aspect ratio is to increase the L value. In the rectangular shape of an enclosed gate layout, the minimum W/L achievable is 2.26, and is almost reached with $L = 7$ um [15], which implies a considerable waste of area and a large capacitance issue. Although a precise W/L model of the Z gate layout is not available at present, the drain current level of the Z gate layout, when compared with the drain current of a single gate layout with the same W, L, and the overdrive voltage (V_{gt}, $V_{gt} = V_{gs} - V_{th}$), is nearly the same. It assumes that a Z gate layout does not need to increase the L value that high to achieve the same effective W/L with a single gate layout, and, thus, has a smaller footprint and gate capacitance.

In the simulation, the main parameters were kept the same for all three layouts. The lateral spacers were formed by a layer of SiO_2 and a thick layer of Si_3N_4, and the STI was inserted using the SiO_2. Because the enclosed gate layout was not able to achieve a small W/L at $L = 0.12$ μm, the values of R_1 and R_2 were 0.15 μm and 0.27 μm, respectively, and the effective W/L was calculated by the formula given in [14], and it was equal to 13.6. The main parameters of the transistors in the simulation are listed in Table 1.

The TID effect on the MOSFET was modeled by adopting the fixed-charge insulator model provided by the sentaurus technology computer-aided design (TCAD) software, which can be used to set a fixed charge density between the STI and the active region [23]. All simulations were performed using a hydrodynamic model with high-field saturation and mobility degradation models that included doping dependence and carrier–carrier scattering. We simulated the effects of the total radiation dose by increasing the fixed charge density on the sidewall oxide [24]. It should be noted that this work is focused only on the effects of fixed charges; so, the interface states were neglected for the reasons below. When a complementary metal-oxide-semiconductor (CMOS) device is exposed to radiation, hole trapping results in fixed charges and interface states in the thick oxides. According to a report on the radiation-induced fixed charge density and interface state density in MOS capacitors [25], the radiation-induced flat band voltage is predominantly shifted by the fixed charges. The effect of the interface states is minor. Therefore, in this simulation, the interface states were neglected, and only the fixed charge density was modified to reflect the total ionizing dose effect [24]. Moreover, the effects of interface traps were left out of the simulations due to a lack of empirical information about several parameters of interface traps, such as trap energy and density and the capture cross-section, which are necessary for accurate simulations [26]. In addition, we can see from the literature [26] that the tendencies of ΔV_{th} and ΔSS extracted from the simulation results are in good agreement with those from the experimental data of 5 Mrad. Through the three-dimensional (3D) simulation results, they confirm that, for sub-100 nm gate-all-around metal-oxide-semiconductor field-effect transistors (GAA MOSFETs), the fixed charges in the gate spacer predominantly determine ΔV_{th} and ΔSS, i.e., the TID effect. Note that interface traps were not taken in the simulation in this paper. Although that may result in some disagreement in the current levels as obtained with the experimental counterparts, this case does not have much impact on our findings, because the focus of this paper is not on the exact values of currents but on the general trends and relative results of Z gate, enclosed gate, and single gate layouts due to the TID effect.

Table 1. The parameters that were used for the device's simulation.

Parameter	Value
Length of channel	0.12 μm
Width of channel	0.21 μm
Thickness of n-type poly gate	100 nm
Thickness of gate oxide	2 nm
Doping of source/drain region	1.0×10^{19} cm^{-3}
Depth of source/drain region	100 nm
Doping of p-type substrate	4.0×10^{17} cm^{-3}

3. Results and Discussion

3.1. The I_d-V_g Simulation Results

In order to verify that the Z gate layout is able to work well in a non-radiation environment, we simulated the I_d-V_g curves of the Z gate layout, the enclosed gate layout, and the single gate layout at the fixed charge density of 3×10^{10} cm^{-2} to model the non-radiation scenario. The following simulation results focus on the analysis of the radiation tolerance characteristics of the proposed layout. The degradation of devices is mainly characterized by the threshold-voltage shift and the off-state leakage current [27]. As we know, I_{DSS} is the maximum current that flows through a FET transistor, which is when the gate voltage (VG) supplied to the FET is 0 V. Additionally, it is only valid when the FET transistor is a junction field-effect transistor (JFET) or depletion MOSFET. However, as the proposed Z gate layout transistor is an enhanced MOSFET, we think that the parameters of I_{DSS} and the $I_{DSS}/Ioff$ ratio are unnecessary to investigate. The threshold-voltage is determined by the linear extrapolation method in the linear region; thus, the simulation was performed with a very small V_{DS}, i.e., 20 mV, 50 mV, and 100 mV [24,27,28]. In this paper, V_{DS} was taken as 20 mV. Moreover, the TID effect will be more serious when V_{DS} reaches V_{DD} [27]. This can be mainly attributed to the fact that more trapped charges at the STI/body interface will sufficiently reduce the potential barrier and result in a larger leakage current at a high drain voltage. In addition, the use of a 20 mV drain bias gives the best results for the Z gate layout in comparison to the alternatives (results not shown). Thus, in this paper, the three layout types are simulated at the drain bias of 20 mV, which sweeps the gate bias from 0 V to 1.5 V.

The simulation results of the I_d-V_g curves of the single gate layout are shown in Figure 2a, where it can be seen that the leakage current significantly increases as the fixed charge density increases, and that the on-current increases slightly as the fixed charge density increases. The simulation results of the I_d-V_g curves of the enclosed gate layout are shown in Figure 2b, where the I_d-V_g curves almost overlap with each other, demonstrating a small impact of the TID effect on the enclosed gate layout.

The simulation results of the I_d-V_g curves of the Z gate layout are shown in Figure 2c, wherein it can be seen that the leakage current increases slightly as the fixed charge density increases. The radiation tolerance characteristic of the Z gate layout was verified by comparison with that of the single gate layout. The curves of the Z gate layout are similar to those of the enclosed gate layout; namely, the leakage current increased very little as the fixed charge density increased, demonstrating that the Z gate layout was radiation tolerant at the fixed charge densities at the STI of 3.5×10^{12} cm^{-2}, the same as the enclosed gate layout.

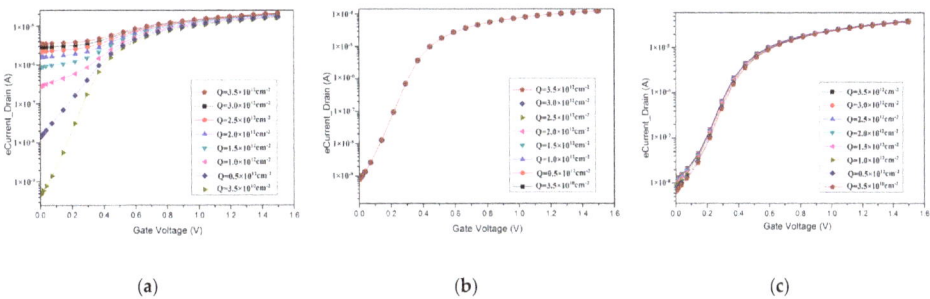

Figure 2. The simulation results of the I_d-V_g curve of (**a**) the single gate layout, (**b**) the enclosed gate layout, and (**c**) the Z gate layout.

3.2. Comparison of Key Transistor Performance Parameters

To compare the TID response of the transistors fairly, the threshold voltage and leakage current parameters were extracted at the fixed charge density of 3×10^{10} cm^{-2} and 3.5×10^{12} cm^{-2} to model the pre- and post-radiation scenarios, respectively. The results of the threshold voltage and leakage current are listed in Tables 2 and 3, respectively.

In Table 2, the threshold voltage of the non-radiation-hardened single gate layout at pre- and post-radiation is 363 mV and 138 mV, respectively, and the total shift is 225.56 mV; for the other two radiation-hardened layouts, the total shift is below 30 mV. Thus, regarding the shift value in descending order, the order of three layout types is the single gate layout, the Z gate layout, and the enclosed gate layout.

Table 2. V_{th} in the pre- and post-radiation scenarios.

Layout	V_{th}-pre (mV)	V_{th}-post (mV)	ΔV_{th} (mV)
single gate	363	138	226
enclosed gate	374	374	<1
Z gate	354	329	25

In Table 3, the leakage current of the single gate layout pre- and post-radiation is 0.458 nA and 3.44 µA, respectively, and the total shift is approximately 3.44 µA. The order of magnitude of the leakage current increase of the other two radiation-hardened layouts was about 1×10^{-9} A compared to the enclosed gate layout. Regarding the increment value in descending order, the order of the three layout types is the single gate layout, the Z gate layout, and the enclosed gate layout.

Table 3. I_{off} in the pre- and post-radiation scenarios.

Layout	I_{off}-pre (A)	I_{off}-post (A)	Increment
single gate	4.58×10^{-10}	3.44×10^{-6}	3.44 µA
enclosed gate	7.91×10^{-10}	7.95×10^{-10}	0.004 nA
Z gate	6.46×10^{-9}	1.17×10^{-8}	5.24 nA

The above-presented comparison of the three layout types regarding the two parameters demonstrates that the enclosed gate layout achieved the best radiation-hardness performance, and the Z gate layout was more effective in mitigating the TID effect on the transistor than the single gate layout.

The threshold voltage shift and the leakage current of the single gate layout, the Z gate layout, and the enclosed gate layout at different charge densities are depicted in Figures 3 and 4, respectively. In Figure 3, it can be seen that the threshold voltage shift of the single gate layout changed non-linearly with the fixed charge density. The shift increased rapidly at a low charge density, and then slowly decreased after reaching the peak value at the fixed charge density of about 2×10^{12} cm^{-2}. As can be clearly seen in the inner figure in Figure 3, the enclosed gate layout's shift was kept very small, and the Z gate layout's shift was similar to that of the single gate layout. It is shown that the shift value of the single gate layout at different charge densities was much larger than that of the other two layouts. The largest shift value of the single gate layout, the Z gate layout, and the enclosed gate layout was 37 mV, 6 mV, and 0.09×10^{-3} mV, respectively. In the inner figure in Figure 3, it can be seen that the enclosed gate layout's shift was smaller than that of the Z gate layout. The enclosed gate layout achieved great performance regarding the radiation hardness; however, the enclosed gate layout comes with disadvantages that cannot be ignored, hindering its application to certain circuits. In such situations, the Z gate layout is a better solution.

The difference between the three layouts regarding the leakage current was even more obvious. As shown in Figure 4, the leakage current of the single gate layout increased rapidly at a low charge density, and then slowly decreased after reaching the peak value at the fixed charge density of about 2×10^{12} cm^{-2}, showing a similar trend as the other two layouts (the trend of the enclosed gate layout is not shown in Figure 4), but at a different order of magnitude. The order of magnitude of the leakage current increase of the single gate layout was about 1×10^{-7} A, and for the other two layouts, it was about 1×10^{-10} A. Thus, it is shown that, compared with the single gate layout, the leakage current increase of the other two layouts was very small, demonstrating the great radiation-hardened characteristic of these two layouts. As can be seen in the inner figure in Figure 4, the leakage current increase of the enclosed gate layout was still relatively small, showing the best radiation tolerance among the three layouts. The largest leakage current increase of the single gate layout, the enclosed gate layout, and the Z gate layout was 0.66 µA, 1.06 nA, and 0.8×10^{-3} nA, respectively. The reduction in the leakage current of the enclosed gate layout and of the Z gate layout compared with the single gate layout was 0.660 μA and 0.659 μA, respectively. According to the results, the Z gate layout performed differently from the enclosed gate layout; however, the Z gate layout was still as effective as the enclosed gate layout regarding the leakage current reduction. Consequently, the Z gate is a better solution at the fixed charge density of 3.5×10^{12} cm^{-2}. In addition, the radiation effects will deteriorate even more as the channel length shrinks due to short-channel effects, which are attributed to the positive charge trapped at the STI/body interface by the radiation [27,28]. This is a serious problem for a highly scaled device operating in an irradiated environment.

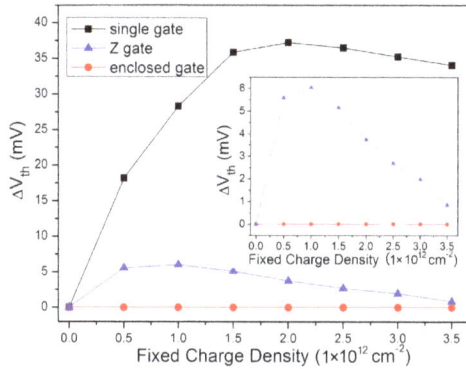

Figure 3. The threshold voltage shift of the three layouts at different charge densities.

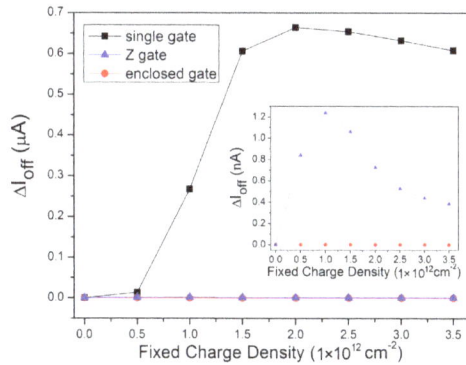

Figure 4. The leakage current increment of the three layouts at different charge densities.

4. Conclusions

A novel n-MOSFET layout with a Z gate was proposed and analyzed using the Sentaurus 3D TCAD software. By comparing the proposed layout with the single gate layout and the enclosed gate layout with respect to the threshold voltage and the leakage current, the radiation-hardened characteristic of the Z gate layout was verified. Besides this, the proposed layout effectively reduces the impact of the TID effect on the transistor's performance compared with the single gate layout; also, the Z gate layout overcomes the drawbacks of the enclosed gate layout, such as a large footprint, a limitation in the W/L's design, and a large capacitance. Thus, the Z gate is a better solution at the fixed charge density of 3.5×10^{12} cm^{-2}.

Author Contributions: Conceptualization, Y.W.; Investigation, C.S.; Software, W.P.; Investigation, X.-j.L., J.Y., F.C., and C.-h.Y.

Funding: This work was funded in part by the National Natural Science Foundation of China (grant number 61774052) and in part by the Excellent Youth Foundation of Zhejiang Province of China (grant number LR17F040001).

Conflicts of Interest: The authors declare no conflict of interest.

References

1. Schwank, J.R.; Shaneyfelt, M.R.; Fleetwood, D.M.; Felix, J.A.; Dodd, P.E.; Paillet, P.; Ferlet-Cavrois, V. Radiation Effects in MOS Oxides. *IEEE Trans. Nucl. Sci.* **2008**, *55*, 1833–1853. [CrossRef]

2. Borghello, G.; Faccio, F.; Lerario, E.; Michelis, S.; Kulis, S.; Fleetwood, D.M.; Bonaldo, S. Dose rate sensitivity of 65 nm MOSFETs exposed to ultra-high doses. *IEEE Trans. Nucl. Sci.* **2018**, *65*, 1482–1487. [CrossRef]

3. Saremi, M.; Privat, A.; Barnaby, H.J.; Clark, L.T. Physically Based Predictive Model for Single Event Transients in CMOS Gates. *IEEE Trans. Electron Devices* **2016**, *63*, 2248–2254. [CrossRef]

4. Narasimham, B.; Bhuva, B.L.; Schrimpf, R.D.; Massengill, L.W.; Gadlage, M.J.; Amusan, O.A.; Benedetto, J.M. Characterization of digital single event transient pulse-widths in 130-nm and 90-nm CMOS technologies. *IEEE Trans. Nucl. Sci.* **2007**, *54*, 2506–2511. [CrossRef]

5. Barnaby, H.J. Total-ionizing-dose effects in modern CMOS technologies. *IEEE Trans. Nucl. Sci.* **2006**, *53*, 3103–3121. [CrossRef]

6. Faccio, F.; Cervelli, G. Radiation-induced edge effects in deep submicron CMOS transistors. *IEEE Trans. Nucl. Sci.* **2005**, *52*, 2413–2420. [CrossRef]

7. Zhang, C.M.; Jazaeri, F.; Borghello, G.; Faccio, F.; Mattiazzo, S.; Baschirotto, A.; Enz, C. Characterization and Modeling of Gigarad-TID-induced Drain Leakage Current of 28-nm Bulk MOSFETs. *IEEE Trans. Nucl. Sci.* **2018**. [CrossRef]

8. Sajid, M.; Chechenin, N.G.; Torres, F.S.; Hanif, M.N.; Gulzari, U.A.; Arslan, S.; Khan, E.U. Analysis of Total Ionizing Dose effects for highly scaled CMOS devices in Low Earth Orbit. *Nucl. Instrum. Methods Phys. Res. Sect. B Beam Interact. Mater. Atoms* **2018**, *428*, 30–37. [CrossRef]

9. Fleetwood, D.M. Evolution of Total Ionizing Dose Effects in MOS Devices with Moore's Law Scaling. *IEEE Trans. Nucl. Sci.* **2018**, *65*, 1465–1481. [CrossRef]

10. Dodd, P.E.; Shaneyfelt, M.R.; Schwank, J.R.; Felix, J.A. Current and future challenges in radiation effects on CMOS electronics. *IEEE Trans. Nucl. Sci.* **2010**, *57*, 1747–1763. [CrossRef]

11. Faccio, F.; Borghello, G.; Lerario, E.; Fleetwood, D.M.; Schrimpf, R.D.; Gong, H.; Zhang, E.X.; Wang, P.; Michelis, S.; Gerardin, S.; et al. Influence of LDD spacers and H+ transport on the total-ionizing-dose response of 65 nm MOSFETs irradiated to ultra-high doses. *IEEE Trans. Nucl. Sci.* **2018**, *65*, 164–174. [CrossRef]

12. Snoeys, W.J.; Faccio, F.; Burns, M.; Campbell, M.; Cantatore, E.; Carrer, N. Layout techniques to enhance the radiation tolerance of standard CMOS technologies demonstrated on a pixel detector readout chip. *Nucl. Instrum. Methods Phys. Res. A* **2000**, *439*, 349–360. [CrossRef]

13. Snoeys, W.J.; Gutierrez, T.A.; Anelli, G. A new NMOS layout structure for radiation tolerance. *IEEE Trans. Nucl. Sci.* **2002**, *49*, 1829–1833. [CrossRef]

14. Gingrich, D.M.; Böttcher, S.; Buchanan, N.J.; Liu, S.; Parsons, J.A.; Sippach, W. Radiation tolerant ASIC for controlling switched-capacitor arrays. *IEEE Trans. Nucl. Sci.* **2004**, *51*, 1324–1332. [CrossRef]

15. Anelli, G.; Campbell, M.; Delmastro, M.; Faccio, F.; Florian, S.; Giraldo, A.; Hejine, E.; Jarron, P.; Kloukinas, K.; Marchioro, A.; et al. Radiation tolerant VLSI circuits in standard deep submicron CMOS technologies for the LHC experiments: Practical design aspects. *IEEE Trans. Nucl. Sci.* **1999**, *46*, 1690–1696. [CrossRef]

16. Giraldo, A.; Paccagnella, A.; Minzoni, A. Aspect ratio calculation in n-channel MOSFET's with a gate-enclosed layout. *Solid-State Electron.* **2000**, *44*, 981–989. [CrossRef]

17. Gaillardin, M.; Martinez, M.; Girard, S.; Goiffon, V.; Paillet, P.; Leray, J.L.; Magnan, P.; Ouerdane, Y.; Boukenter, A.; Marcandella, C.; et al. High Total Ionizing Dose and Temperature Effects on Micro- and Nano-Electronic Devices. *IEEE Trans. Nucl. Sci.* **2015**, *62*, 1226–1232. [CrossRef]

18. Imenabadi, R.M.; Saremi, M.; Vandenberghe, W.G. A Novel PNPN-Like Z-Shaped Tunnel Field-Effect Transistor with Improved Ambipolar Behavior and RF Performance. *IEEE Trans. Electron Devices* **2017**, *64*, 4752–4758. [CrossRef]

19. Sachid, A.B.; Manoj, C.R.; Sharma, D.K.; Rao, V.R. Gate fringe-induced barrier lowering in underlap FinFET structures and its optimization. *IEEE Electron Device Lett.* **2008**, *29*, 128–130. [CrossRef]

20. Boucart, K.; Ionescu, A.M. Double-Gate Tunnel FET with High-K Gate Dielectric. *IEEE Trans. Electron Devices* **2007**, *54*, 1725–1733. [CrossRef]

21. Abadi, R.M.I.; Saremi, M.A. Resonant Tunneling Nanowire Field Effect Transistor with Physical Contractions: A Negative Differential Resistance Device for Low Power Very Large Scale Integration Applications. *J. Electron. Mater.* **2018**, *47*, 1091–1098. [CrossRef]

22. Gnani, E.; Gnudi, A.; Reggiani, S.; Baccarani, G. Theory of the Junctionless Nanowire FET. *IEEE Trans. Electron Devices* **2011**, *58*, 2903–2910. [CrossRef]

23. Esqueda, I.S.; Barnaby, H.J.; Holbert, K.E.; El-Mamouni, F.; Schrimpf, R.D. Modeling of Ionizing Radiation-Induced Degradation in Multiple Gate Field Effect Transistors. *IEEE Trans. Nucl. Sci.* **2011**, *58*, 499–505. [CrossRef]

24. Lee, M.S.; Lee, H.C. Dummy Gate-Assisted n-MOSFET Layout for a Radiation-Tolerant Integrated Circuit. *IEEE Trans. Nucl. Sci.* **2013**, *60*, 3084–3091. [CrossRef]

25. Fernandez-Martinex, P.; Cortes, I.; Hidalgo, S.; Flores, D.; Palomo, F.R. Simulation of total ionizing dose in MOS capacitors. In Proceedings of the 8th Spanish Conference on Electron Devices, Palma de Mallorca, Spain, 8–11 February 2011.

26. Moon, J.-B.; Moon, D.-I.; Choi, Y.-K. Influence of Total Ionizing Dose on Sub-100 nm Gate-All-Around MOSFETs. *IEEE Trans. Nucl. Sci.* **2014**, *61*, 1420–1425. [CrossRef]

27. Wang, J.; Wang, W.; Huang, R.; Pei, Y.; Xue, S.; Wang, X.A.; Wang, Y. Deteriorated radiation effects impact on the characteristics of MOS transistors with multi-finger configuration. *Microelectron. Reliab.* **2010**, *50*, 1094–1097. [CrossRef]

28. Nam, J.; Kang, C.Y.; Kim, K.P.; Yeo, H.; Lee, B.J.; Seo, S.; Yang, J.-W. Influence of Ionizing Radiation on Short-Channel Effects in Low-Doped Multi-Gate MOSFETs. *IEEE Trans. Nucl. Sci.* **2012**, *59*, 3021–3026. [CrossRef]

micromachines

MDPI

Article

Improving ESD Protection Robustness Using SiGe Source/Drain Regions in Tunnel FET

Zhaonian Yang *, Yuan Yang, Ningmei Yu and Juin J. Liou

Shaanxi Key Laboratory of Complex System Control and Intelligent Information Processing, Xi'an University of Technology, Xi'an 710048, China; yangyuan@xaut.edu.cn (Y.Y.); yunm@xaut.edu.cn (N.Y.); eenian2@126.com (J.J.L.)
* Correspondence: yzn@xaut.edu.cn; Tel.: +86-029-823-12431

Received: 14 November 2018; Accepted: 9 December 2018; Published: 12 December 2018

Abstract: Currently, a tunnel field-effect transistor (TFET) is being considered as a suitable electrostatic discharge (ESD) protection device in advanced technology. In addition, silicon-germanium (SiGe) engineering is shown to improve the performance of TFET-based ESD protection devices. In this paper, a new TFET with SiGe source/drain (S/D) regions is proposed, and its ESD characteristics are evaluated using technology computer aided design (TCAD) simulations. Under a transmission line pulsing (TLP) stressing condition, the triggering voltage of the SiGe S/D TFET is reduced by 35% and the failure current is increased by 17% in comparison with the conventional Si S/D TFET. Physical insights relevant to the ESD enhancement of the SiGe S/D TFET are provided and discussed.

Keywords: band-to-band tunneling (BTBT); electrostatic discharge (ESD); tunnel field-effect transistor (TFET); Silicon-Germanium source/drain (SiGe S/D); technology computer aided design (TCAD)

1. Introduction

A traditional metal-oxide-semiconductor field-effect transistor (MOSFET) has a 60 mV/dec subthreshold swing at room temperature, which limits the application of this device in ultra-low power integrated circuits (ICs) [1,2]. The tunnel field-effect transistor (TFET) is a promising candidate for replacing the conventional MOSFET in low power ICs [3–5]. The TFET employs a band-to-band tunneling (BTBT) mechanism and is able to theoretically achieve a subthreshold swing smaller than 60 mV/dec. However, the TFET has a very low driving current compared with the MOSFET, which means it is difficult to realize a high-speed circuit using pure TFETs. Recently, the mixed TFET–MOSFET circuit design methodology was reported, by skillfully designing the circuits such as static random access memory (SRAM), level shifter, and even electrostatic discharge (ESD) protection circuits with two kinds of devices, where both high performance and low standby current can be achieved [6–9]. ESD protection is a very challenging reliability issue of modern integrated circuits (ICs), especially in advanced nanoscale technologies [9–14]. As mentioned in reference [9], TFET can be used to replace the traditional diodes in an ESD protection network to enhance the ESD robustness in nanoscale technology ICs. The ESD behavior of the TFET has been studied using experiments and technology computer aided design (TCAD) simulations [15–18]. However, these results show that the ESD robustness of TFET under positive ESD stress is low.

It has been verified that using a silicon-germanium (SiGe) source in the TFET can increase the driving current compared with the silicon TFET [19–22]. This is because Ge has a narrower band-gap and lower carrier effective mass than Si, and these features increase the tunneling probability. The SiGe engineering has also been introduced in the ESD protection devices to enhance the ESD performance [12,13]. However, as for ESD protection applications, the physical processes mainly occur on the drain side of the TFET. As such, using a SiGe source does not benefit TFET's ESD characteristics [16].

In this paper, we propose a new TFET with SiGe both in the source and drain (S/D) regions for ESD protection. The performance of the proposed device will be investigated using TCAD simulations. The simulation results will show that both the triggering voltage and the failure current of the SiGe S/D TFET are improved over those of the conventional Si TFET. The impact of various technology parameters on the ESD behavior of the SiGe S/D TFET will also be given.

2. Basic Concept of Electrostatic Discharge (ESD) Protection Tunnel Field-Effect Transistor (TFET) and the Protection Network

TFET is essentially a reverse biased gated p-i-n diode. As for ESD protection, TFET, the gate terminal is connected to the source by default. Under the negative ESD stress, namely, ESD current is injected into the source terminal of TFET the with drain terminal grounded. TFET will operate in a positive diode conduction mode and has a high current discharge capability as illustrated in Figure 1a. Whereas under positive ESD stress, the ESD current is injected into the drain terminal with the source terminal grounded. TFET will operate in avalanche breakdown mode to discharge the ESD current as illustrated in Figure 1b. Since avalanche breakdown requires a relatively high electric field, the conduction voltage of TFET under positive ESD stress is high, making it unacceptable in advanced nanoscale technologies. Thus, the research on TFET under ESD stress mainly focuses on the positive discharge mode.

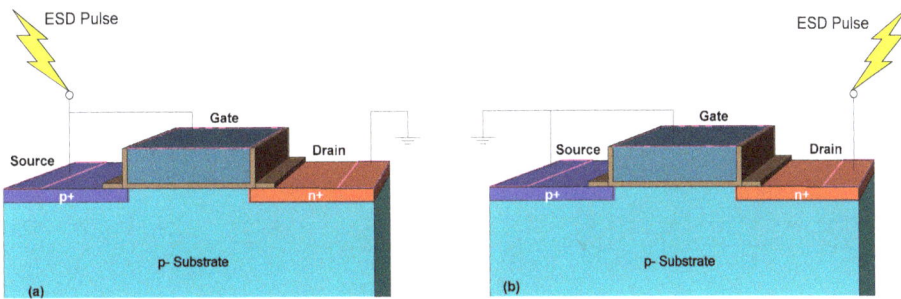

Figure 1. Schematics of tunnel field-effect transistor (TFET) under (**a**) negative and (**b**) positive electrostatic discharge (ESD) stresses.

It should be mentioned that, since TFET has a relatively low positive mode ESD robustness, it cannot be used as a single protection device in an IC, but can be used to implement a protection network as shown in Figure 2, in which TFET is used to replace the traditional diode to enhance the whole chip ESD robustness. As for the pin-to-pin ESD event, their discharge paths exist in the TFET based protection network as shown in Figure 2, whereas in the traditional diode-based protection network only Path2 exists. In Path1 and Path3, TFET1 and TFET4 operate in avalanche breakdown mode with low ESD robustness. Thus, it is necessary to improve the robustness of TFET under positive ESD stress.

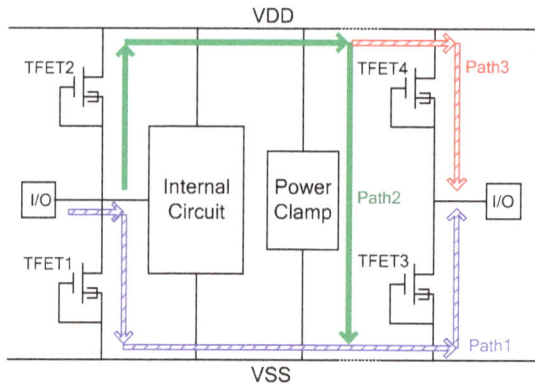

Figure 2. Schematic of the ESD protection network with TFETs.

3. Device Structure and Simulation Setup

As illustrated in Figure 3, the device structure proposed in this work is identical to the conventional silicon point-tunneling TFET except that the source and the drain regions are made of SiGe. The device size is not set to a very small value for better heat dissipation [11,23]. The default device parameters are: Thickness of the gate oxide (HfO$_2$) T$_{ox}$ = 4 nm, thickness of the silicon T$_{Si}$ = 1 µm, width of the device W$_{Si}$ = 1 µm, depth of the junction X$_j$ = 10 nm, length of the gate L$_G$ = 100 nm, and source and the drain side silicide blocking lengths SOP = DOP = 100 nm. Silicide blocking is used in ESD protection devices to reduce the current crowding effect [12,16]. The doping concentrations of the source, drain, and substrate are N$_S$ = 1 × 10^{20} cm^{-3}, N$_D$ = 5 × 10^{19} cm^{-3}, and N$_{Sub}$= 1 × 10^{16} cm^{-3}, respectively. Abrupt doping profile is used in the simulation. In order to avoid possible high defect density at the SiGe/Si interface, the default Ge mole fraction is set at 0.4 [22].

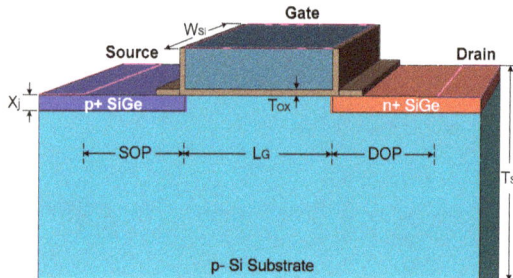

Figure 3. Schematic of the proposed SiGe source/drain (S/D) TFET.

The SiGe S/D TFET can be fabricated using the following process flow. First, the source region is recessed into the p-Si substrate by an etching process. Then the p+ SiGe source region is grown by epitaxy. Similarly, the drain region is recessed into the p-Si substrate by the etching process and the n+ SiGe drain region is grown by epitaxy. Afterward, the gate dielectric and the gate stack are deposited and patterned. Finally, the spacers are formed.

Simulations are carried out in the Sentaurus simulator. The dynamic nonlocal BTBT model is used instead of the local BTBT model. This is because the dynamic model takes into account the spatial variation of the energy bands and therefore can model the BTBT probability more accurately. The fitted coefficients of the SiGe BTBT probability are calculated by linear interpolation between the parameters of pure Si and pure Ge [23]. The lattice temperature is calculated using the thermodynamic model. Van Overstraeten-de Man avalanche generation model, high field saturation, and Philips unified mobility models, band-gap narrowing model and doping dependent Shockley-Read-Hall recombination model are also used.

Transmission line pulsing (TLP) pulses, which mimic the stressing of the human body model (HBM), are used to simulate the quasi-static current-voltage (*I-V*) behavior of the devices during the ESD conditions. The drain terminal of the TFET was stressed with TLP pulses while keeping the gate and the source terminals grounded. The rise time and the pulsewidth are set at 10 ns and 100 ns, respectively. The voltage samples are obtained by averaging the transient data in the range of 60 ns to 90 ns [16].

4. Simulation Results and Discussion

The TLP *I-V* curves of both the SiGe S/D and Si TFETs are shown in Figure 4. The triggering voltage and failure current of the SiGe S/D TFET are 4.1 V and 0.7 mA/μm, respectively, which are 35% lower and 17% higher than those of the Si counterpart. These improved key parameters will make the new TFET easier to fit into the modern ESD design window and offer higher ESD protection capability. It should be noted that the SiGe source has nearly no influence on the ESD characteristics [15], and the improvement is achieved by introducing the SiGe drain in the TFET.

The reduction of the triggering voltage of the TFET is achieved by introducing the SiGe material in the drain region. The Ge material has the following three advantages in triggering the TFET at a lower voltage. First, Ge has a higher BTBT probability than Si due to its narrower bandgap and lower carrier effective mass. The TFET has a BTBT-assisted avalanche generation mechanism, hence a higher BTBT probability gives rise to a more significant avalanche breakdown [16,17]. Second, Ge has a higher impact ionization coefficient than Si under the same electric field [24]. This means that the critical electric field required for avalanche breakdown in the SiGe S/DTFET is lower than that in the Si TFET. Third, the drain/substrate heterojunction offers an enhanced electric field, which helps to reduce the triggering voltage [25]. SiGe and Si have similar electron affinities, thus the bandgap difference approximately equals the valence band offset. Figure 5a shows the energy bands of the SiGe S/D TFET stressed under a TLP current density of 0.5 mA/μm. It can be seen that there is a valence band offset at the drain/substrate interface. This obstructs the holes from moving to the source, causing some holes to accumulate on the drain side, as evidenced by the hole concentration plot shown in Figure 5b, with a significant hole density peak at a distance of 5 nm below the Si/SiO2 interface on the drain side. This leads to an enhancement in the electric field at the drain/substrate interface and consequently a reduction in the trigger voltage.

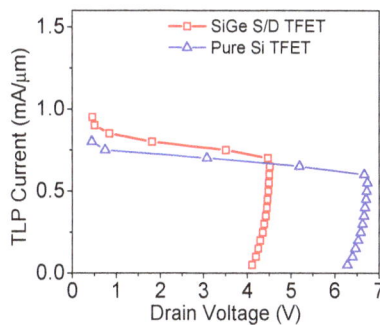

Figure 4. Transmission line pulsing (TLP) *I-V* curves of the SiGe S/D TFET and Si TFET.

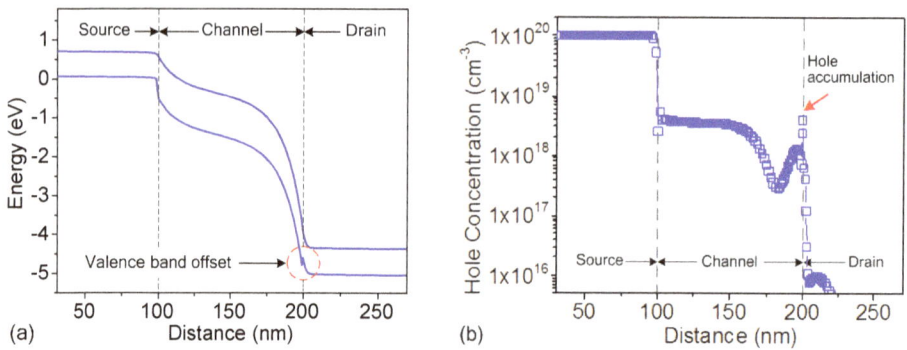

Figure 5. (a) Energy bands and (b) hole concentration simulated at 90 ns under a TLP current density of 0.5 mA/µm, at a distance of 5 nm below the Si/SiO$_2$ interface.

As shown in Figure 2, the failure current of the new TFET is also improved. Under an ESD event, the Joule heat is the main heat component in the device, and it can be expressed as in reference [26],

$$H_{\text{Joule}} = H_p + H_n = \frac{|J_p|^2}{pq\mu_p} + \frac{|J_n|^2}{nq\mu_n} \tag{1}$$

where H is the heat, J is the current density, μ is the mobility, and subscripts n and p denote electrons and holes, respectively. The hole Joule heat is higher than the electron Joule heat because the impact generated holes move from the drain interface to the source through the channel region, whereas the electrons are collected by the drain terminal without traveling. Furthermore, the high electric field and carrier scattering significantly degrade the mobility, especially near the drain and the source interfaces. These, in turn, cause a large amount of hole Joule heat generated at the interface regions as shown in Figure 6. The hole mobility in the SiGe S/D TFET is higher than that in the conventional Si TFET as shown in Figure 7. Thus, the SiGe S/D TFET has an elevated robustness due to the fact that the hole Joule heat is the dominate heat source and hole mobility in SiGe is higher than that in Si.

The thermal conductivity is another important factor influencing the ESD thermal breakdown. SiGe has a lower thermal conductivity compared with Si, which hinders the heat dissipation [27,28]. However, the volume of the SiGe regions are relatively small and the reduction in the triggering voltage implies that less Joule heat is generated.

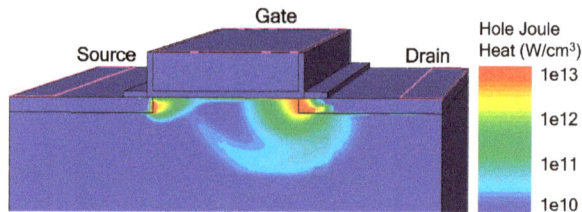

Figure 6. Contour plot of hole Joule heat simulated at 90 ns under a TLP current density of 0.5 mA/µm.

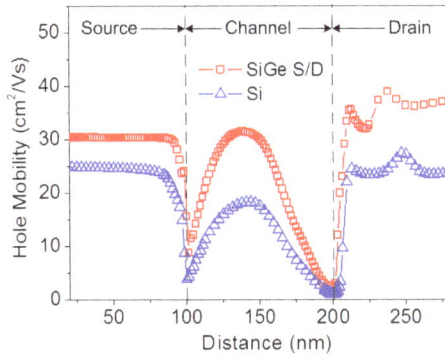

Figure 7. Hole mobilities at a distance of 5 nm below the Si/SiO$_2$ interface.

In the SiGe S/D TFET, the increase in the Ge mole fraction (x) can cause a reduction in the triggering voltage, and a slight increase in the failure current as shown in Figure 8. This trend can be easily understood from the preceding discussions. However, when the Ge mole fraction is higher than 0.4, the defect density at the SiGe/Si interface may degrade the device performance.

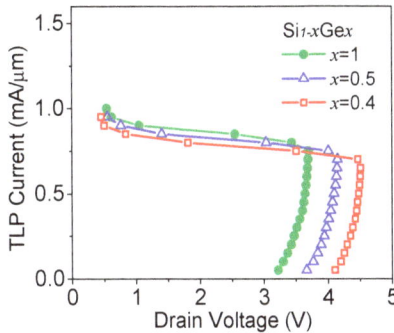

Figure 8. TLP *I-V* curves of SiGe S/D TFET with different Ge mole fractions.

Dimensions have significant influences on the characteristics of ESD protection devices. From Figure 9, it can be seen that with a large DOP and SOP value (see Figure 1), although the triggering voltage is slightly increased, the failure current is significantly increased. This can be attributed to two reasons. The increase in device volume offers a better heat dissipation and thus a reduced temperature in the device. In addition, when DOP and SOP are increased, the series resistance in the discharge path is increased, hence the ballasting effect suppresses the current crowding along the lateral direction [16]. The contour plots of lattice temperature with two DOP and SOP values are shown in Figure 10.

The gate length can also affect the ESD performance. As listed in Table 1, the scaling in the gate length reduces the triggering voltage and the failure current of the TFET. The former can be attributed to the increase in the lateral electric field, which enhances the reverse biased p-n junction tunneling and impact ionization [29]. In addition, the increase in spreading resistance may also play a role [30,31]. However, since the gate is grounded, the electric field near the drain/substrate junction is strongly affected by the gate, and the impact of gate length on the triggering voltage is not very significant [18]. The failure current increases with increasing gate length owning to the larger size and improved conduction uniformity.

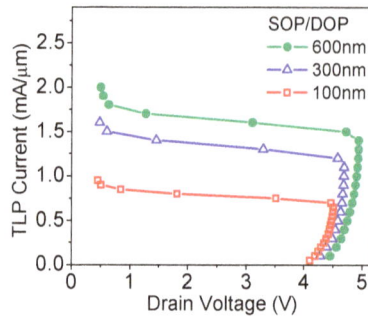

Figure 9. TLP *I-V* curves of SiGe S/D TFET with different SOP and DOP values.

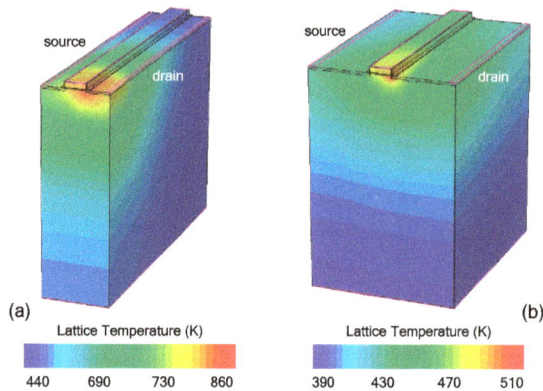

Figure 10. Contour plots of lattice temperatures in TFETs under TLP current density of 0.5 mA/μm with different DOP/SOP values: (**a**) DOP = SOP = 100 nm and (**b**) DOP = SOP = 300 nm.

Table 1. Triggering voltages and failure currents with different gate lengths.

Gate Length	50 nm	100 nm	150 nm	200 nm
Triggering Voltage	4.06 V	4.1 V	4.18 V	4.28 V
Failure Current	0.65 mA/μm	0.7 mA/μm	0.725 mA/μm	0.75 mA/μm

The impact of drain doping level on SiGe S/D TFET's ESD *I-V* characteristic is shown in Figure 11. It can be observed that with the increase in drain doping level, the triggering voltage is reduced. This can be attributed to the enhanced BTBT, and reduction in the critical electric field required for avalanche breakdown. The failure current is slightly increased with the increase in drain doping level, and this is because the reduction in drain voltage results in less Joule heat. It should be mentioned that, since the BTBT and avalanche generations mainly occur on the drain side, the source doping level nearly does not influence the TFET's ESD characteristics [16].

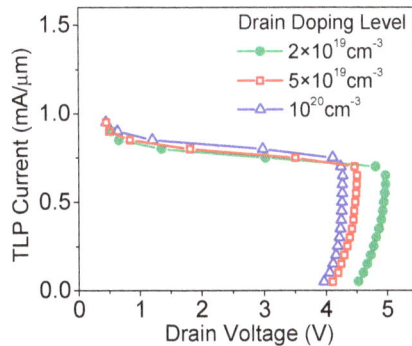

Figure 11. TLP *I-V* curves of SiGe S/D TFET with different Ge mole fractions.

5. Conclusions

In this paper, a new grounded-gate TFET with SiGe source and drain regions was proposed and its ESD characteristics were investigated using TCAD simulations. Compared to the conventional Si TFET, the triggering voltage of the SiGe S/D TFET is reduced because the SiGe regions offer a high BTBT probability, a higher impact ionization coefficient, and a higher electric field due to the SiGe/Si heterostructure. The failure current of the SiGe S/D TFET is also increased due to the combination of a lower triggering voltage and a smaller Joule heat resulting from a higher hole mobility in SiGe. This enhanced ESD performance will be beneficial for constructing robust TFET-based ESD protection networks in the future.

Author Contributions: Z.Y. provided the concept, designed the structures, performed the simulations, and wrote the manuscript. All authors discussed the results, read, and approved the final manuscript.

Funding: This work was funded by the project of National Natural Science Foundation of China (grant number 61804123), and in part by the projects of the China Postdoctoral Science Foundation (grant number 2017M623211), Postdoctoral Science Foundation of Shaanxi Province, China (grant number 2017BSHEDZZ31), and Key Project Foundation of the Education Department of Shaanxi Province, China (grant number 18JS082).

Conflicts of Interest: The authors declare no conflicts of interest.

References

1. Cristoloveanu, S.; Wan, J.; Zaslavsky, A. A review of sharp switching devices for ultra-low power applications. *IEEE J. Electron Devices Soc.* **2016**, *4*, 215–226. [CrossRef]
2. Hueting, R.J.E. The Balancing Act in Ferroelectric Transistors: How Hard Can It Be? *Micromachines* **2018**, *9*, 582. [CrossRef]
3. Seabaugh, A.; Zhang, Q. Low-voltage tunnel transistors for beyond CMOS logic. *Proc. IEEE* **2010**, *98*, 2095–2110. [CrossRef]
4. Ionescu, A.M.; Riel, H. Tunnel field-effect transistors as energy efficient electronic switches. *Nature* **2011**, *479*, 329–337. [CrossRef]
5. Chen, S.; Wang, S.; Liu, H.; Li, W.; Wang, Q.; Wang, X. Symmetric U-Shaped Gate Tunnel Field-Effect Transistor. *IEEE Trans. Electron Devices* **2017**, *64*, 1343–1349. [CrossRef]
6. Chen, Y.-N.; Fan, M.-L.; Hu, P.-H.; Su, P.; Chuang, C.-T. Evaluation of stability, performance of ultra-low voltage MOSFET, TFET, and mixed TFET-MOSFET SRAM cell with write-assist circuits. *IEEE J. Emerg. Sel. Topic Circuits Syst.* **2014**, *4*, 389–399. [CrossRef]
7. Strangio, S.; Palestri, P.; Esseni, D.; Selmi, L.; Crupi, F.; Richter, S.; Zhao, Q.T.; Mantl, S. Impact of TFET unidirectionality and ambipolarity on the performance of 6T SRAM cells. *IEEE J. Electron Devices Soc.* **2015**, *3*, 223–232. [CrossRef]

8. Lanuzza, M.; Strangio, S.; Crupi, F.; Palestri, P.; Esseni, D. Mixed tunnel-FET/MOSFET level shifters: A new proposal to extend the tunnel-FET application domain. *IEEE Trans. Electron Devices* **2015**, *62*, 3973–3979. [CrossRef]

9. Sithanandam, R.; Kumar, M.J. A new on-chip ESD strategy using TFETs-TCAD based device and network simulations. *IEEE J. Electron Devices Soc.* **2018**, *6*, 298–308. [CrossRef]

10. Liou, J.J. Challenges of designing electrostatic discharge (ESD) protection in modern and emerging CMOS technologies. *Proc. ISNE* **2014**, *4*, 1–3.

11. Galy, P. ElectroStatic Discharge (ESD) one real life event: Physical impact and protection challenges in advanced CMOS technologies. In Proceedings of the 2014 International Semiconductor Conference (CAS), Sinaia, Romania, 13–15 October 2014; pp. 31–34.

12. Boschke, R.; Linten, D.; Hellings, G.; Chen, S.-H.; Scholz, M.; Mitard, J.; Witters, L.; Collaert, N.; Thean, A.; Groeseneken, G. ESD characterization of diodes and ggMOS in Germanium FinFET technologies. In Proceedings of the 2015 37th Electrical Overstress/Electrostatic Discharge Symposium (EOS/ESD), Reno, NV, USA, 27 September–2 October 2015; pp. 1–9.

13. Boschke, R.; Chen, S.-H.; Scholz, M.; Hellings, G.; Linten, D.; Witters, L.; Collaert, N.; Groeseneken, G. ESD Ballasting of SiGe FinFET ggNMOS devices. In Proceedings of the 2017 IEEE International Reliability Physics Symposium (IRPS), Monterey, CA, USA, 2–6 April 2017.

14. Lin, C.-Y.; Wu, Y.-H.; Ker, M.-D. Low-leakage and low-trigger-voltage SCR device for ESD protection in 28-nm high-k metal gate CMOS process. *IEEE Electron Device Lett.* **2016**, *37*, 1387–1389. [CrossRef]

15. Galy, P.; Athanasiou, S. Preliminary results on TFET-gated diode in thin silicon film for IO design & ESD protection in 28 nm UTBB FD-SOI CMOS technology. In Proceedings of the 2016 International Conference on IC Design and Technology (ICICDT), Ho Chi Minh City, Vietnam, 27–29 June 2016; pp. 1–4.

16. Kranthi, N.K.; Shrivastava, M. ESD behavior of tunnel FET devices. *IEEE Trans. Electron Devices* **2017**, *64*, 28–36. [CrossRef]

17. Yang, Z.; Zhang, Y.; Yang, Y.; Yu, N. Investigation of the double current path phenomenon in gate-grounded tunnel FET. *IEEE Electron Device Lett.* **2018**, *39*, 103–106. [CrossRef]

18. Yang, Z.; Yu, N.; Liou, J.J. Impact of the gate structure on ESD characteristic of tunnel field-effect transistors. In Proceedings of the 2018 7th International Symposium on Next Generation Electronics (ISNE), Taipei, Taiwan, 7–9 May 2018; pp. 1–4.

19. Mayer, F.; Le Royer, C.; Damlencourt, J.-F.; Romanjek, K.; Andrieu, F.; Tabone, C.; Previtali, B.; Deleonibus, S. Impact of SOI, $Si_{1-x}Ge_xOI$ and GeOI substrates on CMOS compatible tunnel FET performance. In Proceedings of the 2008 IEEE International Electron Devices Meeting, San Francisco, CA, USA, 15–17 December 2008; pp. 1–5.

20. Kim, S.H.; Agarwal, S.; Jacobson, Z.A.; Matheu, P.; Hu, C.; Liu, T.-J.K. Tunnel field effect transistor with raised germanium source. *IEEE Electron Device Lett.* **2010**, *31*, 1107–1109. [CrossRef]

21. Schmidt, M.; Schäfer, A.; Minamisawa, R.A.; Buca, D.; Trellenkamp, S.; Hartmann, J.-M.; Zhao, Q.-T.; Mantl, S. Line and point tunneling in scaled Si/SiGe heterostructure TFETs. *IEEE Electron Device Lett.* **2014**, *35*, 699–701. [CrossRef]

22. Wang, W.; Wang, P.F.; Zhang, C.M.; Lin, X.; Liu, X.Y.; Sun, Q.Q.; Zhou, P.; Zhang, D.W. Design of U-shape channel tunnel FETs with SiGe source regions. *IEEE Trans. Electron Devices* **2014**, *61*, 193–197. [CrossRef]

23. Sithanandam, R.; Kumar, M.J. A novel cascade-free 5-V ESD clamp using I-MOS: Proposal and analysis. *IEEE Trans. Device Mater. Reliab.* **2016**, *16*, 200–207. [CrossRef]

24. Sze, S.M. *Physics of Semiconductor Devices*, 3rd ed.; Wiley: New York, NY, USA, 2006; pp. 39–41.

25. Sarkar, D.; Singh, N.; Banerjee, K. A novel enhanced electric-field impact-ionization MOS transistor. *IEEE Electron Device Lett.* **2010**, *31*, 1175–1177. [CrossRef]

26. Synopsys. *User Manual, Ver. I-2013.03, Synopsys TCAD Sentaurus*; Synopsys: San Jose, CA, USA, 2013.

27. Shrivastava, M.; Agrawal, M.; Mahajan, S.; Gossner, H.; Schulz, T.; Kumar Sharma, D.; Rao, V.P. Physical insight toward heat transport and an improved electrothermal modeling framework for FinFET architectures. *IEEE Trans. Electron Devices* **2012**, *59*, 1353–1363. [CrossRef]

28. Aksamija, Z.; Knezevic, I. Reduced thermal conductivity in SiGe alloy-based superlattices for thermoelectric applications. In Proceedings of the International Silicon-Germanium Technology and Device Meeting, Berkeley, CA, USA, 4–6 June 2012; pp. 1–2.

29. Thijs, S.; Griffoni, A.; Linten, D.; Chen, S.-H.; Hoffmann, T.; Groeseneken, G. On gated diodes for ESD protection in bulk FinFET CMOS technology. In Proceedings of the EOS/ESD Symposium Proceedings, Anaheim, CA, USA, 11–16 September 2011; pp. 1–8.
30. Zhang, X.Y.; Banerjee, K.; Amerasekera, A.; Gupta, V.; Yu, Z.; Dutton, R.W. Process and layout dependent substrate resistance modeling for deep sub-micron ESD protection devices. In Proceedings of the 2000 IEEE International Reliability Physics Symposium Proceedings, San Jose, CA, USA, 10–13 April 2000; pp. 295–303.
31. Griffoni, A.; Thijs, S.; Russ, C.; Trémouilles, D.; Linten, D.; Scholz, M.; Collaert, N.; Witters, L.; Meneghesso, G.; Groeseneken, G. Next generation bulk FinFET devices and their benefits for ESD robustness. In Proceedings of the 2009 31st EOS/ESD Symposium, Anaheim, CA, USA, 30 August–4 September 2009; pp. 1–10.

micromachines

MDPI

Article

Variability Predictions for the Next Technology Generations of *n*-type Si$_x$Ge$_{1-x}$ Nanowire MOSFETs

Jaehyun Lee, Oves Badami, Hamilton Carrillo-Nuñez, Salim Berrada, Cristina Medina-Bailon, Tapas Dutta, Fikru Adamu-Lema, Vihar P. Georgiev * and Asen Asenov

School of Engineering, University of Glasgow, Glasgow G12 8QW, UK; Jaehyun.Lee@glasgow.ac.uk (J.L.); Oves.Badami@glasgow.ac.uk (O.B.); Hamilton.Carrillo-Nunez@glasgow.ac.uk (H.C.-N.); salim.berrada@glasgow.ac.uk (S.B.); Cristina.MedinaBailon@glasgow.ac.uk (C.M.-B.); Tapas.Dutta@glasgow.ac.uk (T.D.); Fikru.Adamu-Lema@glasgow.ac.uk (F.A.-L.); Asen.Asenov@glasgow.ac.uk (A.A.)
* Correspondence: Vihar.Georgiev@glasgow.ac.uk; Tel.: +44-(0)141-330-7659

Received: 21 November 2018; Accepted: 30 November 2018; Published: 5 December 2018

check for updates

Abstract: Using a state-of-the-art quantum transport simulator based on the effective mass approximation, we have thoroughly studied the impact of variability on Si$_x$Ge$_{1-x}$ channel gate-all-around nanowire metal-oxide-semiconductor field-effect transistors (NWFETs) associated with random discrete dopants, line edge roughness, and metal gate granularity. Performance predictions of NWFETs with different cross-sectional shapes such as square, circle, and ellipse are also investigated. For each NWFETs, the effective masses have carefully been extracted from $sp^3d^5s^*$ tight-binding band structures. In total, we have generated 7200 transistor samples and performed approximately 10,000 quantum transport simulations. Our statistical analysis reveals that metal gate granularity is dominant among the variability sources considered in this work. Assuming the parameters of the variability sources are the same, we have found that there is no significant difference of variability between SiGe and Si channel NWFETs.

Keywords: line edge roughness; metal gate granularity; nanowire; non-equilibrium Green's function; random discrete dopants; SiGe; variability

1. Introduction

Semiconductor fabrication has witnessed amazing progress in the last about 50 to 60 years which has enabled the scaling of the physical dimensions of the metal-oxide-semiconductor field-effect transistors (MOSFETs) at an exponential rate. According to the Institute of Electrical and Electronics Engineers (IEEE) International Roadmap for Devices and Systems (IRDS) report, by the year 2024, the gate length (L_G) and diameter of transistors are expected to be 10 and 5 nm, respectively, for high-performance logic applications [1]. However, the scaling has slowed down due to increase in a number of detrimental second order effects like source-to-drain tunneling and drain induced barrier lowering (DIBL) [2,3].

In order to overcome these issues, a device with the gate-all-around (GAA) structure is a promising candidate to replace the Fin field-effect transistor (FinFET), which is being adopted in industries [1,4,5]. Devices with the GAA structure showed better electric transport performance thanks to their superior electrostatic integrity. Maheshwaram et al. reported that, by using the vertical GAA Si nanowire MOSFETs (NWFETs) instead of the FinFET, the ring oscillator delay and the power consumption are improved by 33% and 45%, respectively [6]. In addition, nanowires based on different materials and geometry cross-section can be used as transducers, sensors or photovoltaic devices [7–10].

Studying on the channel material engineering as well as the gate structure is very important to overcome the short channel effects in nanoscale devices. SiGe, III-V, and two-dimensional materials such as graphene and transition metal dichalcogenide are attracting attention as the channel material in future devices thanks to their small transport effective masses (m^*_{trans}) [3,11–14]. It is noteworthy that materials with smaller m^*_{trans} can contribute to increases ON-state current (I_{ON}) but increases OFF-sate current (I_{OFF}) as well in the short channel device due to the source-to-drain tunneling currents. Moreover, transistors with small band-gap materials are suffering from the band-to-band leakage currents [13]. Unfortunately, the overwhelmingly superior material that can replace Si has not been found yet. In this paper, we concentrate on SiGe, which is more compatible with the current complementary metal-oxide-semiconductor (CMOS) technology [15]. In addition, material properties of SiGe can be adjusted by the mole fraction to have the advantages of Si and Ge together.

Previous simulation studies have shown that random discrete dopants (RDD), line edge roughness (LER), and metal gate granularity (MGG) induce significant variability in ultra-scaled InGaAs [16] and Si [17,18] channel nanoscale devices. However, the former used classical transport models, whereas the latter considered a very small number of statistical samples due to the computational cost of quantum transport simulations. To the best of our knowledge, a study comparing the impact of different sources of variability of SiGe channel NWFETs using the quantum transport simulations with a large number of samples is missing.

In this paper, we focus on the investigation of the impact of dominant sources of statistical variability (RDD, LER and MGG) in *n*-type Si_xGe_{1-x} channel GAA NWFETs with different cross-section shapes. In order to capture the source-to-drain tunneling in the nanoscale devices, the quantum transport problem for electrons is solved within the parabolic effective mass (PEM) approximation by means of the non-equilibrium Green's function (NEGF) formalism implemented in the Glasgow Nano-Electronic Simulation Software (NESS) [19]. We also confirm that the calibrated confinement and transport effective masses can reproduce the empirical tight binding (ETB) band structures. For a reliable statistical analysis, an ensemble of 200 transistor samples for each set of variability sources has been adopted. All together, we have performed approximately 10,000 quantum transport simulations with 7200 different transistor samples.

The paper is organized as follows. In Section 2, we discuss the details related to the generation of the statistical variability sources such as RDD, MGG and LER, implementation of the NEGF formalism and the effective mass extraction method from $sp^3d^5s^*$ ETB band structure calculations. This is followed by the discussion of the simulation results in Section 3. Finally, we summarize our results in Section 4.

2. Simulation Framework

2.1. Device Structure with the Variability Sources Included

Figure 1 illustrates the schematic diagram of GAA NWFETs with an elliptic cross-sectional shape. The three dominant variability sources including RDD, LER, and MGG are also highlighted in Figure 1. RDD and LER are generated in the channel region (10.0 nm) and in equal portions of the source and the drain (8.0 nm each), resulting in $L_v = 26.0$ nm. The remaining source and drain regions (20.0 nm each) are assumed to have continuous doping profile in order to ensure good convergence of the electrostatic potential.

For the generation of RDD, a rejection technique has been adopted by considering the atomic arrangement in Si_xGe_{1-x} NW crystal structures [20] with the corresponding lattice parameter. We generate a random number between 0 and 1 in each atom of Si_xGe_{1-x}, and substitute the atom with a dopant atom if the random number is less than the criteria (CR). The criteria can be written as:

$$CR = N_D V_{atom}, \qquad (1)$$

where N_D is the doping density at this site and V_{atom} is the volume of the corresponding atom. Because V_{atom} is a constant value determined by the lattice parameter, as N_D increases, the probability that a dopant atom is located increases. Therefore, the total number of dopant atoms follows the Poisson distribution [21].

LER at the interface between Si_xGe_{1-x} and gate oxide is characterized by an auto-correlation function [22]:

$$C(r) = \Delta_m^2 e^{-\sqrt{2}r/L_m},\tag{2}$$

where Δ_m is the root mean square, L_m is the correlation length, and r is the length between two points. Herein, Δ_m and L_m are 0.2 and 1.0 nm, respectively, which is consistent with experimental data for Si [23]. To be consistent, we have used the same value of these parameters for Si_xGe_{1-x} channel devices.

Regarding MGG, the grains in the TiN metal gate region are generated by using the Voronoi algorithm [24,25]. The value of the work-function for each grain can be either 4.4 or 4.6 eV with the probability of 40% or 60% based on previous experimental results [26]. It was reported that, as the grain size increases, the more significant variability is observed, meaning that the small average grain size causes less variability [24]. Therefore, the average grain size of 3.0 nm used in this paper is small enough to expect a relatively less MGG-induced variability.

Following the IRDS specifications for the node "4/3" [1], the *n*-type Si_xGe_{1-x} ($x = 1.0, 0.8, 0.5,$ and 0.2) channel GAA NWFETs with $L_G = 10.0$ nm and a diameter (or width) of 5.0 nm (see Figure 1) are considered. NWFETs with square, circle, and elliptic cross-sectional shapes are also studied and their corresponding cross-section dimensions are chosen to have the same footprint to keep the technology node. Indeed, NWFETs with elliptic cross-sectional shapes can be referred to the nano-sheet MOSFETs [27]. The transport direction in all the devices is along [100]. The equivalent oxide thickness is 0.8 nm. The source-to-drain bias V_{DS} is set to 0.6 V. All simulations are performed at 300 K.

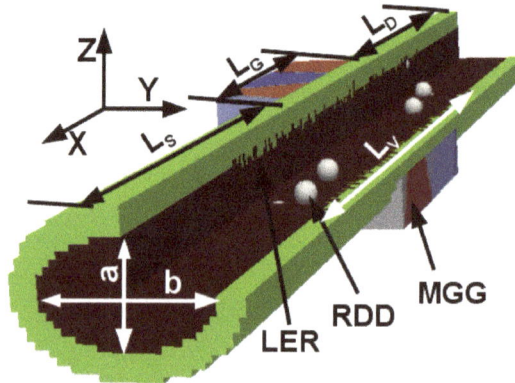

Figure 1. Schematic diagram of the elliptical gate-all-around nanowire metal-oxide-semiconductor field-effect transistors (GAA NWFET) (a = 3 nm and b = 5 nm) highlighting variability sources. For the square and circular nanowires (NWs), a = b = 5 nm. $L_S = L_D = 28$ nm, $L_G = 10$ nm, and $L_V = 26$ nm. The doping concentrations in source/drain and channel regions are 10^{20} (*n*-type) and 10^{15} (*p*-type) cm^{-3}, respectively. RDD–random discrete dopants, LER–line edge roughness and MGG–metal gate granularity.

2.2. Quantum Transport Formalism

The electron quantum transport problem is solved by exploiting the coupled mode NEGF formalism with the PEM Hamiltonian [28]. Assuming steady-state conditions, we briefly summarize the main features of the NEGF approach in matrix notation. Within the PEM approximation, the discretized mode-space Green's function is defined as

$$G_v^r(E) = \left[EI - H_v - \Sigma_{L,v}^r(E) - \Sigma_{R,v}^r(E)\right]^{-1},\tag{3}$$

where I is the identity matrix and H_v represents the mode-space version of the Hamiltonian for the v th conduction band valley. $\Sigma_{L/R}^r$ is the retarded self-energy for the left/right semi-infinite device contact, usually being computed by adopting the recursive algorithm proposed in Ref [29].

The lesser and greater Green's functions are then obtained from

$$G_v^{\lessgtr} = G_v^r\left(\Sigma_{L,v}^{\lessgtr} + \Sigma_{R,v}^{\lessgtr}\right)G_v^{r\dagger}\tag{4}$$

with lesser ($\Sigma^<$) and greater self-energies ($\Sigma^>$). They are related to their corresponding retarded counterpart by

$$\Sigma^r = \frac{1}{2}\left(\Sigma^> - \Sigma^<\right),\tag{5}$$

where the energy variable E has been omitted for brevity. In practice, the real part of the retarded self-energy in Equation (5) is neglected. This approximation shall not introduce significant error in the transport properties [30]. Once the lesser and greater Green's functions are known, physical quantities such as carrier density and current can be computed respectively as,

$$n(x_j, y, z) = -i \times 2 \sum_v \sum_{n,m} \int \frac{dE}{2\pi} G_{nm}^<(x_j, x_j; E)\phi_n(y, z; x_j)\phi_m^\dagger(y, z; x_j),\tag{6}$$

$$I(x_j) = -2 \times \frac{e}{\hbar} \sum_v \sum_{n,m} \int \frac{dE}{2\pi}\left(2\,\mathrm{Re}\left(H_{nm,v}(x_j, x_{j+1})G_{mn}^<(x_{j+1}, x_j; E)\right)\right),\tag{7}$$

in which the factor 2 considers the spin degeneracy. The eigenfunction $\phi_n(y, z; x_j)$ for the mode n is calculated by solving the 2D Schrödinger equation corresponding to the cross-section plane at x_j. In nanostructures, such as the nanowires considered in this paper, only few low energy modes are necessary due to the strong confinement. Therefore, there is a significant gain in the size of the matrices that must be inverted in the recursive algorithm [30] employed in NESS for computing the diagonal and off-diagonal elements of $G^<$ in Equations (6) and (7), respectively. Finally, Equation (6) is self-consistently coupled to Poisson equation. When the convergence criterion for the electrostatic potential is reached, the current is then calculated from Equation (7).

2.3. Extraction of Effective Masses

In order to model the conduction band for the transport simulation, the PEM Hamiltonian is adopted with transport and confinement effective masses extracted from $sp^3d^5s^*$ ETB method with Boykin's parameter set, implemented in Synopsys QuantumATK [31,32]. For Si_xGe_{1-x} materials, virtual crystal approximation is used [33]. Figure 2 shows the conduction band structures of Si and $Si_{0.2}Ge_{0.8}$ NWs as an example. It is highlighted that L-valley is observed in $Si_{0.2}Ge_{0.8}$ NW but not in Si NW. Moreover, it is found that the quantization energy (ΔE_Q), the energy difference of conduction band edges of bulk and NW, of $Si_{0.2}Ge_{0.8}$ NW is larger than that of Si NW.

The transport effective masses are directly calculated from the ETB band structures as follows:

$$m_{trans}^* = \hbar^2\left(\frac{\partial^2 E}{\partial k_x^2}\right)^{-1}.\tag{8}$$

The extraction of confinement effective masses (m^*_{conf}) is more complicated. The least-squares method is used to find the best value of m^*_{conf} to fit ΔE_Q and the energy gap between the first and the second conduction sub-band energies (ΔE_{sub}) as follows:

$$S = \left(\Delta E_Q^{ETB} - \Delta E_Q^{PEM}\right)^2 + \left(\Delta E_{sub}^{ETB} - \Delta E_{sub}^{PEM}\right)^2, \tag{9}$$

where ΔE_Q^{ETB} (ΔE_{sub}^{ETB}) and ΔE_Q^{PEM} (ΔE_{sub}^{ETB}) are ΔE_Q (ΔE_{sub}) obtained from ETB and PEM methods, respectively. It is noteworthy that ΔE_Q^{PEM} and ΔE_{sub}^{ETB} are the function of m^*_{conf}. Herein, minimized the squared residue S indicates m^*_{conf} are well extracted. As a result, the PEM method successfully reproduces the ETB conduction band structures. The extracted m^*_{trans} and m^*_{conf} are summurized in Table 1.

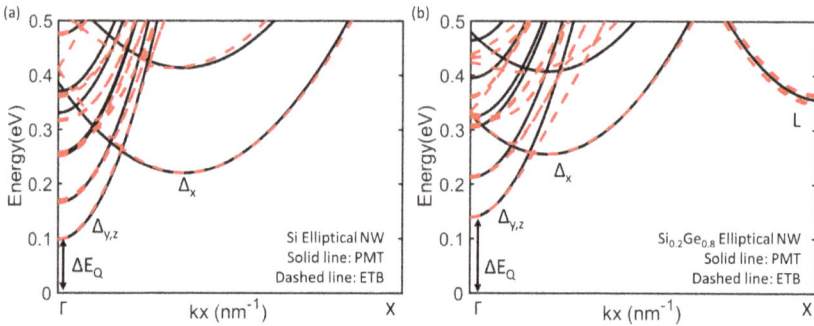

Figure 2. Band structures of (**a**) Si and (**b**) Si$_{0.2}$Ge$_{0.8}$ 5 × 3 nm^2 elliptical NWs. The bulk conduction band edge is set to 0.0 eV. ΔE_Q is also remarked.

Table 1. Calculated effective masses of Si and Si$_x$Ge$_{1-x}$ nanowires (NWs) with various cross-sectional shapes. Herein, unit is m_0, the rest electron mass.

		Degeneracy	Square			Circle			Ellipse		
			m_x	m_y	m_z	m_x	m_y	m_z	m_x	m_y	m_z
Si	Δ_x	2	0.918	0.240	0.240	0.915	0.224	0.224	0.927	0.464	0.146
	Δ_y	2	0.233	0.953	0.237	0.236	0.887	0.215	0.241	0.839	0.220
	Δ_z	2	0.233	0.242	0.875	0.236	0.208	0.896	0.241	0.206	0.886
Si$_{0.8}$Ge$_{0.2}$	Δ_x	2	0.861	0.235	0.235	0.849	0.287	0.287	0.875	0.321	0.198
	Δ_y	2	0.240	0.884	0.221	0.235	1.342	0.262	0.251	0.757	0.224
	Δ_z	2	0.240	0.220	0.885	0.235	0.259	1.366	0.251	0.192	0.905
Si$_{0.5}$Ge$_{0.5}$	Δ_x	2	0.799	0.241	0.241	0.788	0.286	0.286	0.818	0.392	0.179
	Δ_y	2	0.250	0.864	0.224	0.247	1.042	0.272	0.268	0.674	0.210
	Δ_z	2	0.250	0.224	0.816	0.247	0.270	1.015	0.268	0.194	0.809
Si$_{0.2}$Ge$_{0.8}$	Δ_x	2	0.759	0.237	0.237	0.739	0.285	0.285	0.788	0.448	0.174
	Δ_y	2	0.266	0.788	0.217	0.258	0.952	0.272	0.286	0.657	0.206
	Δ_z	2	0.266	0.213	0.798	0.258	0.272	0.958	0.286	0.186	0.828
	L	4	0.350	0.134	0.297	0.500	0.147	0.449	0.600	0.327	0.152

3. Simulation Results and Discussion

Figure 3 shows the statistical transfer characteristics for Si$_{0.2}$Ge$_{0.8}$ channel elliptical GAA NWFETs considering different sets of statistical variability sources. The drain current is normalized by the diameter (width) of 5 nm of NWFETs. A statistical ensemble of 200 devices has been used in this work. Significant statistical variability is observed in terms of I_{ON}, I_{OFF} and threshold voltage

(V_{th}). Figure 3a,b show that the change in RDD-induced variability when adding LER is small, whereas Figure 3c clearly shows that MGG is the dominant source of variability in the devices under consideration although very small average grain size of 3.0 nm is used. It is also found that the median of subthreshold slope (SS) with RDD, LER, and MGG is 62.8 mV/dec, which is comparable to the value of SS (63.0 mV/dec) for the corresponding ideal device. Standard deviation of SS is 0.78 mV/dec suggesting that SS does not change much due to the impact of statistical variability sources.

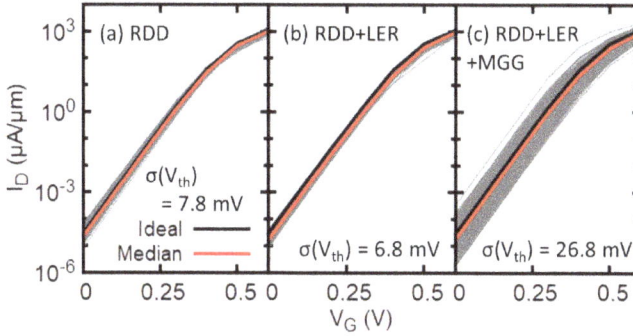

Figure 3. Transfer characteristics of $Si_{0.2}Ge_{0.8}$ elliptical GAA NWFETs associated with (**a**) random discrete dopants (RDD), (**b**) RDD and line edge roughness (LER) and (**c**) RDD, LER and metal gate granularity (MGG. The ideal device refers to a device with continuous and uniform doping profiles in the source and drain and no variability sources. Corresponding standard deviation of V_{th} $\sigma(V_{th})$ is also indicated. $V_{DS} = 0.6$ V.

Figure 4 shows the probability distribution of V_{th} with RDD, LER, and MGG. There is a shift in the median, but the distribution shapes (bell shapes) and standard deviations are similar regardless of the mole fraction. Similar qualitative results for the combination of other architectures and materials are observed.

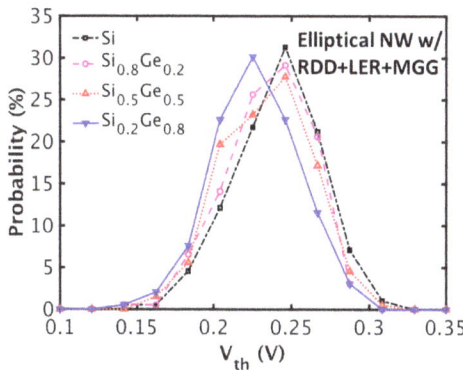

Figure 4. Distributions of threshold voltage (V_{th}) for the elliptical NWFETs with different mole fractions. RDD, LER, and MGG are taken into account.

Medians of I_{ON} and I_{OFF} of all simulated devices considered are summarized in Table 2. Herein, I_{ON} is defined at $V_{DS} = V_{GS} = 0.6$ V and I_{OFF} is defined at $V_{DS} = 0.6$ V and $V_{GS} = 0.0$ V. Variation in I_{OFF} is significant with respect to the Ge mole fraction as compared to I_{ON}, but all I_{OFF} satisfy the IRDS criterion of staying below 100 nA/μm [1].

Table 2. Medians of I_{ON} and I_{OFF} for the Si_xGe_{1-x} nanowire metal-oxide-semiconductor field-effect transistors (NWFETs). Random discrete dopants (RDD), line edge roughness (LER), and metal gate granularity (MGG) are considered.

Si_xGe_{1-x}	I_{ON} (mA/μm)/I_{OFF} (pA/μm)		
	Square	Circular	Elliptical
Si	1.59/397	1.37/98.9	0.771/9.26
$Si_{0.8}Ge_{0.2}$	1.71/427	1.50/127	0.862/11.7
$Si_{0.5}Ge_{0.5}$	1.70/473	1.51/151	0.861/12.7
$Si_{0.2}Ge_{0.8}$	1.84/668	1.63/210	0.958/18.1

Figure 5 summarizes the correlations between important figures-of-merits (FoMs) in terms of scatter plots and correlation coefficients: I_{ON}, I_{OFF}, V_{th} and DIBL. Herein, V_{th} is calculated using the constant current method with the current criteria $I_{th} = 100$ nA/μm. As data in Figure 5 shows, the correlation coefficients ρ for the different Ge mole fraction are comparable and very similar to those for Si. In addition, as expected, the V_{th} and I_{OFF} show negative correlation with ρ almost equal to 1. Negatively correlated are I_{ON} and V_{th} with ρ which still has very high value (around -0.85) but less than the ρ value between the V_{th} and I_{OFF}. I_{OFF} and I_{ON} show positive correlation with a correlation coefficient close to 0.85. As expected, the DIBL parameter is not correlated to any of the other FoMs, as shown by the value of ρ very close to 0. Hence, our results suggest that replacing Si channel by Si_xGe_{1-x} channel will not solve the variability issues in sub-10 nm gate-length NWFETs.

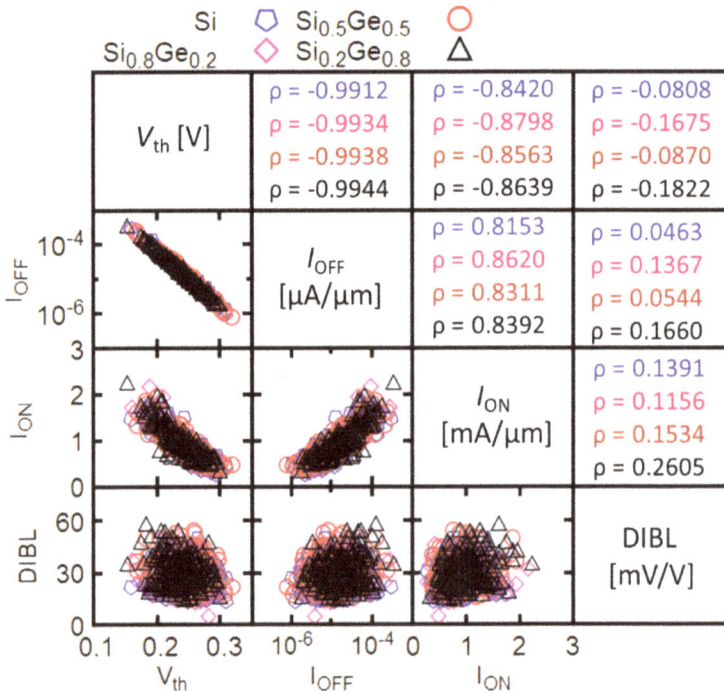

Figure 5. Correlation between important FoMs for the elliptical GAA NWFETs with different Ge mole fraction. The bottom left of the table shows correlation scatter plots and the top right shows correlation coefficients which are also listed in the following order: Si (blue), $Si_{0.8}Ge_{0.2}$ (magenta), $Si_{0.5}Ge_{0.5}$ (red), and $Si_{0.2}Ge_{0.8}$ (black).

Figure 6 shows the variation of V_{th} for elliptical GAA NWFETs with different Ge mole fractions and different sets of variability sources. It is found that, despite the small average grain size of 3 nm, MGG is the dominant source of variability in the considered devices regardless of the Ge mole fraction. Moreover, the median of V_{th} increases as the number of the variability sources included in the simulations increases. We have also found that, as the Ge mole fraction increases, V_{th} decreases. This can be attributed to the increase in the contribution of the L-valley (see Figure 2) [34]. Therefore, the $Si_{0.2}Ge_{0.8}$ channel devices have larger I_{ON} than the Si devices considered in this paper as shown in Table 2.

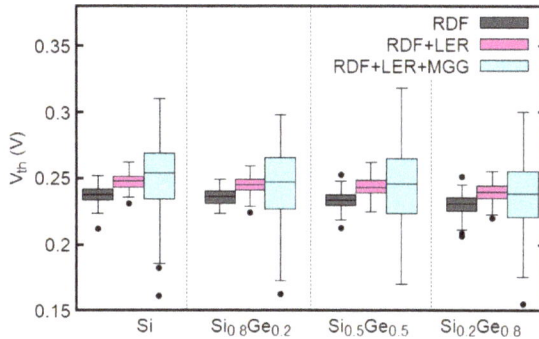

Figure 6. Dependence of V_{th} of the elliptical GAA NWFETs on the variability sources and the Ge mole fraction.

Figure 7a shows the variation of V_{th} for GAA NWFETs with different mole fractions of Ge and different cross-sectional shapes considering the effects of RDD, LER, and MGG. The Ge mole fraction and the shape of the cross-section do not have significant effect on V_{th} variability. Regardless of the cross-sectional shapes, V_{th} is smaller for the larger mole fraction of Ge, which is in a good agreement with the results in Figure 6. Additionally, it is found that the median of V_{th} decreases when the cross-sectional shape is changed from ellipse to circle and to square, in this order. This trend is consistent with the dependence of V_{th} on the inverse of the cross-sectional area, which increases in the aforementioned order. Therefore, the elliptical devices have smaller I_{ON} than the other devices (see Table 2).

The variation of DIBL is plotted in Figure 7b. DIBL calculated from the ideal device is underestimated with respect to its median when considering variability sources (see Table 3). It is interesting to note that $Si_{0.2}Ge_{0.8}$ channel devices with larger I_{ON} (see Table 2) also show larger DIBL than others, regardless of the cross-sectional shape. Furthermore, the median and the variation of DIBL of the elliptical devices are smaller than that of square and circular devices.

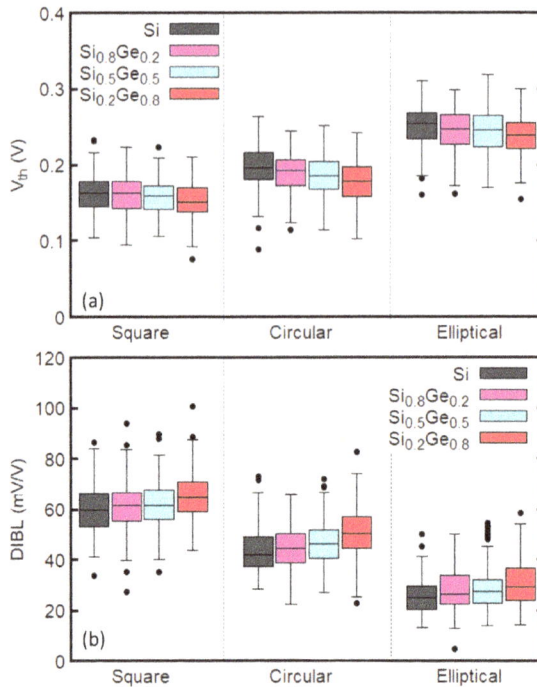

Figure 7. Dependence of (a) V_{th} and (b) drain induced barrier lowering (DIBL) on the Ge mole fraction and cross-sectional shape. RDD, LER, and MGG are considered

Table 3. The comparison of drain induced barrier lowering (DIBL) in $Si_{0.2}Ge_{0.8}$ channel devices obtained from the ideal devices and statistical simulations.

Cross-Sectional Shape (RDD + LER + MGG)	Ideal Device	Median
Square	62.4 mV/V	64.7 mV/V
Circle	42.8 mV/V	50.2 mV/V
Ellipse	20.3 mV/V	29.2 mV/V

4. Conclusions

We have performed a comprehensive variability analysis of n-type Si_xGe_{1-x} ($x = 1.0, 0.8, 0.5$, and 0.2) channel GAA NWFETs using 7200 samples. The electron transport has been modeled by means of the coupled-mode space NEGF formalism implemented in NESS. Our results show that the Ge mole fraction and cross-sectional shapes do not affect significantly the variability in GAA NWFETs, and MGG is the dominant source of variability as when compared to RDD and LER. It is noticeable that the small average grain size of 3 nm is considered in this paper, which is expected to cause relatively less MGG-induced variability. We have also found that $Si_{0.2}Ge_{0.8}$ channel devices have not only smaller V_{th} but also larger DIBL compared to the devices with lower Ge mole fractions indicating that they suffer the most from short channel effects. In addition, elliptical GAA NWFETs have smaller DIBL compared to square and circular devices, while providing smaller I_{ON}.

Author Contributions: Writing—Original Draft Preparation: J.L. and O.B.; Methodology: (Variability) J.L., (NEGF) H.C.-N. and S.B., (Effective Mass Extraction) O.B. and J.L.; Writing—Review and Editing: C.M.-B., T.D., F.A.-L., and V.P.G.; Supervision: V.P.G. and A.A.

Funding: This project has received funding from the European Union's Horizon 2020 Research and Innovation Programme under Grant No. 688101 and Engineering and Physical Sciences Research Council (EPSRC) United Kingdom Research and Innovation (UKRI) Innovation Fellowship scheme (EP/S001131/1).

Conflicts of Interest: The authors declare no conflict of interest.

References

1. IEEE International Roadmap for Devices and Systems (IRDS). 2016. Available online: https://irds.ieee.org/reports (accessed on 25 September 2018).
2. Kawaura, H.; Sakamoto, T.; Baba, T. Observation of source-to-drain direct tunneling current in 8 nm gate electrically variable shallow junction metal–oxide–semiconductor field-effect transistors. *Appl. Phys. Lett.* **2000**, *76*, 3810–3812. [CrossRef]
3. Grillet, C.; Logoteta, D.; Cresti, A.; Pala, M.G. Assessment of the Electrical Performance of Short Channel InAs and Strained Si Nanowire FETs. *IEEE Trans. Electron Devices* **2017**, *64*, 2425–2431. [CrossRef]
4. Badami, O.; Caruso, E.; Lizzit, D.; Osgnach, P.; Esseni, D.; Palestri, P.; Selmi, L. An Improved Surface Roughness Scattering Model for Bulk, Thin-Body, and Quantum-Well MOSFETs. *IEEE Trans. Electron Devices* **2016**, *63*, 2306–2312. [CrossRef]
5. Al-Ameri, T.; Georgiev, V.P.; Adamu-Lema, F.; Asenov, A. Simulation study of vertically stacked lateral Si nanowires transistors for 5 nm CMOS applications. *IEEE J. Electron Devices Soc.* **2017** *5*, 466–472. [CrossRef]
6. Maheshwaram, S.; Manhas, S.K.; Kaushal, G.; Anand, B.; Singh, N. Vertical Silicon Nanowire Gate-All-Around Field Effect Transistor Based Nanoscale CMOS. *IEEE Electron Device Lett.* **2011**, *32*, 1011–1013. [CrossRef]
7. Zheng, G.; Lu, W.; Jin, S.; Lieber, C.M. Synthesis and Fabrication of High-Performance n-Type Silicon Nanowire Transistors. *Adv. Mater.* **2004**, *16*, 830–834. [CrossRef]
8. Tian, B.; Cohen-Karni, T.; Qing, Q.; Duan, X.; Xie, P.; Lieber, C. Three-Dimensional, Flexible Nanoscale Field-Effect Transistors as Localized Bioprobes. *Science* **2010**, *329*, 1890–1893. [CrossRef]
9. Kim, S.K.; Day, R.W.; Cahoon, J.F.; Kempa, T.J.; Song, K.D.; Park, H.G.; Lieber, C.M. Tuning Light Absorption in Core/Shell Silicon Nanowire Photovoltaic Devices through Morphological Design. *Nano Lett.* **2012**, *12*, 4971–4976. [CrossRef]
10. Kim, J.; Lee, H.C.; Kim, K.H.; Hwang, M.S.; Park, J.S.; Lee, J.M.; So, J.P.; Choi, J.H.; Kwon, S.H.; Barrelet, C.J.; et al. Photon-triggered nanowire transistors. *Nat. Nanotechnol.* **2017**, *12*, 963–968. [CrossRef]
11. Vasen, T.; Ramvall, P.; Afzalian, A.; Thelander, C.; Dick, K.A.; Holland, M.; Doornbos, G.; Wang, S.W.; Oxland, R.; Vellianitis, G.; et al. InAs nanowire GAA n-MOSFETs with 12–15 nm diameter. In Proceedings of the 2016 IEEE Symposium on VLSI Technology, Honolulu, HI, USA, 14–16 June 2016; pp. 1–2. [CrossRef]
12. Selvakumar, C.S.; Hecht, B. SiGe-channel n-MOSFET by germanium implantation. *IEEE Electron Device Lett.* **1991**, *12*, 444–446. [CrossRef]
13. Lee, J.; Shin, M. Performance Assessment of III-V Channel Ultra-Thin-Body Schottky-Barrier MOSFETs. *IEEE Electron Device Lett.* **2014**, *35*, 726–728. [CrossRef]
14. Radisavljevic, B.; Radenovic, A.; Brivio, J.; Giacometti, V.; Kis, A. Single-layer MoS2 transistors. *Nat. Nanotechnol.* **2011**, *6*, 147–150. . [CrossRef] [PubMed]
15. Alher, M.; Mosleh, A.; Cousar, L.; Dou, W.; Grant, P.C.; Ghetmiri, S.A.; AlKabi, S.; Du, W.; Benamara, M.; Li, B.; et al. CMOS Compatible Growth of High Quality Ge, SiGe and SiGeSn for Photonic Device Applications. *ECS Trans.* **2015**, *69*, 269–278. [CrossRef]
16. Seoane, N.; Indalecio, G.; Comesaña, E.; Aldegunde, M.; García-Loureiro, A.J.; Kalna, K. Random Dopant, Line-Edge Roughness, and Gate Workfunction Variability in a Nano InGaAs FinFET. *IEEE Trans. Electron Devices* **2014**, *61*, 466–472. [CrossRef]
17. Valin, R.; Martinez, A.; Barker, J. Non-equilibrium Green's functions study of discrete dopants variability on an ultra-scaled FinFET. *J. Appl. Phys.* **2015**, *117*, 164505. [CrossRef]

18. Georgiev, V.P.; Towie, E.A.; Asenov, A. Impact of precisely positioned dopants on the performance of an ultimate silicon nanowire transistor: A full three-dimensional NEGF simulation study. *IEEE Trans. Electron Devices* **2013**, *60*, 965–971. [CrossRef]

19. Berrada, S.; Carrillo-Nuñez, H.; Lee, J.; Medina-Bailon, C.; Dutta, T.; Duan, M.; Adamu-Lema, F.; Georgiev, V.; Asenov, A. NESS: New Flexible Nano-Transistor Simulation Environment. In Proceedings of the International Conference on Simulation of Semiconductor Processes and Devices 2018, Austin, TX, USA, 24–26 September 2018.

20. Frank, D.J.; Taur, Y.; Ieong, M.; Wong, H.-S. Monte Carlo modeling of threshold variation due to dopant fluctuations. In Proceedings of the 1999 Symposium on VLSI Technology Digest of Technical Papers, Kyoto, Japan, 14–16 June 1999; pp. 169–170. [CrossRef]

21. Asenov, A. Random dopant induced threshold voltage lowering and fluctuations in sub 50 nm MOSFETs: A statistical 3D 'atomistic' simulation study. *Nanotechnology* **1999**, *10*, 153–158. [CrossRef]

22. Kim, S.; Luisier, M.; Paul, A.; Boykin, T.B.; Klimeck, G. Full Three-Dimensional Quantum Transport Simulation of Atomistic Interface Roughness in Silicon Nanowire FETs. *IEEE Trans. Electron Devices* **2011**, *58*, 1371–1380. [CrossRef]

23. Goodnick, S.M. Surface roughness at the Si(100)-SiO$_2$ interface. *Phys. Rev. B* **1985**, *32*, 8171–8186. [CrossRef]

24. Wang, X.; Brown, A.R.; Idris, N.; Markov, S.; Roy, G.; Asenov, A. Statistical Threshold-Voltage Variability in Scaled Decananometer Bulk HKMG MOSFETs: A Full-Scale 3-D Simulation Scaling Study. *IEEE Trans. Electron Devices* **2011**, *58*, 2293–2301. [CrossRef]

25. Vardhan, P.H.; Mittal, S.; Ganguly, S.; Ganguly, U. Analytical Estimation of Threshold Voltage Variability by Metal Gate Granularity in FinFET. *IEEE Trans. Electron Devices* **2017**, *64*, 3071–3076. [CrossRef]

26. Dadgour, H.; Endo, K.; De, V.; Banerjee, K. Modeling and analysis of grain-orientation effects in emerging metal-gate devices and implications for SRAM reliability. In Proceedings of the 2008 IEEE International Electron Devices Meeting, San Francisco, CA, USA, 15–17 December 2008; pp. 1–4. [CrossRef]

27. Loubet, N.; Hook, T.; Montanini, P.; Yeung, C.; Kanakasabapathy, S.; Guillom, M.; Yamashita, T.; Zhang, J.; Miao, X.; Wang, J.; et al. Stacked nanosheet gate-all-around transistor to enable scaling beyond FinFET. In Proceedings of the 2017 IEEE Symposium on VLSI Technology, Kyoto, Japan, 5–8 June 2017; pp. T230–T231.

28. Luisier, M.; Schenk, A.; Fichtner, W. Quantum transport in two- and three-dimensional nanoscale transistors: Coupled mode effects in the nonequilibrium Green's function formalism. *J. Appl. Phys.* **2006**, *100*, 043713. [CrossRef]

29. Lopez-Sancho, M.P.; Lopez-Sancho, J.M.; Sancho, J.M.L.; Rubio, J. Highly convergent schemes for the calculation of bulk and surface Green functions. *J. Phys. F Met. Phys.* **1985**, *15*, 851. [CrossRef]

30. Svizhenko, A.; Anantram, M.P. Role of scattering in nanotransistors. *IEEE Trans. Electron Devices* **2003**, *50*, 1459–1466. [CrossRef]

31. Atomistix Toolkit Version 2017.2, Synopsys QuantumWise A/S. Available online: https://www.quantumwise.com (accessed on 25 September 2018).

32. Timothy, T.; Klimeck, G.; Oyafuso, F. Valence band effective-mass expressions in the sp3d5s* empirical tight-binding model applied to a Si and Ge parametrization. *Phys. Rev. B* **2004**, *69*, 115201. [CrossRef]

33. Paul, A.; Mehrotra, S.; Luisier, M.; Klimeck, G. Performance Prediction of Ultrascaled SiGe/Si Core/Shell Electron and Hole Nanowire MOSFETs. *IEEE Electron Device Lett.* **2010**, *31*, 278–280. [CrossRef]

34. Fischetti, M.V.; Laux, S.E. Band structure, deformation potentials, and carrier mobility in strained Si, Ge, and SiGe alloys. *J. Appl. Phys.* **1996**, *80*, 2234–2252. [CrossRef]

micromachines

MDPI

Review

Modeling of Gate Stack Patterning for Advanced Technology Nodes: A Review

Xaver Klemenschits *, Siegfried Selberherr and Lado Filipovic

Institute for Microelectronics, Technische Universität Wien, Vienna 1040, Austria; selberherr@iue.tuwien.ac.at (S.S.); filipovic@iue.tuwien.ac.at (L.F.)
* Correspondence: klemenschits@iue.tuwien.ac.at; Tel.: +43-1-58801-36026

Received: 7 November 2018; Accepted: 25 November 2018; Published: 29 November 2018

Abstract: Semiconductor device dimensions have been decreasing steadily over the past several decades, generating the need to overcome fundamental limitations of both the materials they are made of and the fabrication techniques used to build them. Modern metal gates are no longer a simple polysilicon layer, but rather consist of a stack of several different materials, often requiring multiple processing steps each, to obtain the characteristics needed for stable operation. In order to better understand the underlying mechanics and predict the potential of new methods and materials, technology computer aided design has become increasingly important. This review will discuss the fundamental methods, used to describe expected topology changes, and their respective benefits and limitations. In particular, common techniques used for effective modeling of the transport of molecular entities using numerical particle ray tracing in the feature scale region will be reviewed, taking into account the limitations they impose on chemical modeling. The modeling of surface chemistries and recent advances therein, which have enabled the identification of dominant etch mechanisms and the development of sophisticated chemical models, is further presented. Finally, recent advances in the modeling of gate stack pattering using advanced geometries in the feature scale are discussed, taking note of the underlying methods and their limitations, which still need to be overcome and are actively investigated.

Keywords: technology computer-aided design (TCAD); metal oxide semiconductor field effect transistor (MOSFET); topography simulation; metal gate stack; level set; high-k; fin field effect transistor (FinFET)

---

1. Introduction

Ongoing miniaturization of metal oxide semiconductor field effect transistors (MOSFETs) is essential for the continued advances in computing performance, reduction of chip area, and lowered power dissipation in modern integrated circuits in accordance with Moore's Law [1]. For decades, the design of MOSFETs did not change drastically [2], while its size was scaled thanks to more advanced lithography techniques and improved fabrication processes. Additionally, thinner insulating layers and smaller dimensions allowed for faster switching, thereby increasing speed and improving performance. However, smaller sizes presented new challenges. For example, the insulating silicon dioxide (SiO_2) layer between the conducting channel and the gate became so thin that quantum tunneling resulted in gate leakage currents too high to sustain stable MOSFET operation [3]. However, the insulating SiO_2 layer is required to be as thin as possible in order to reach the high gate to channel capacitance required for effective switching characteristics, while a physically thicker layer helps to reduce tunneling. Effective switching and reduced tunneling were achieved by replacing SiO_2 with a material with a higher dielectric constant (high-k material), to balance the increased distance between the gate and the channel. The most prominent of those materials used today are Hafnium (Hf)-based insulators, usually HfO_2. The combination of a high dielectric constant and a wide band gap [4], needed to create a potential barrier to the silicon channel and thus to act as an insulator, make HfO_2 ideal for this purpose. Therefore, it is the most commonly used material for gate insulation ever since its introduction in the 45 nm technology node [5]. However, new materials, such as Al_2O_3, are currently being investigated as possible alternatives [6].

The gate contact material must be chosen carefully, since its work function controls the threshold voltage, above which the channel will be inverted. Fine control over this important parameter of a MOSFET was achieved with a polycrystalline silicon (Poly-Si) gate, doped depending on the type of transistor desired. The dopant concentration inside the gate material influences the work function and thus allows the channel band to be shifted either towards or away from the Fermi energy level, decreasing or increasing the threshold voltage, respectively. However, doping of the gate contact results in the unwanted penetration of dopants into the dielectric and the channel, leading to numerous unwanted side effects [7]. Furthermore, other unfavorable characteristics of Poly-Si, such as Fermi pinning and gate depletion [8], created the need for different materials to be considered for the gate contact and led to the re-introduction of metals into the gate stack. However, not aluminum, but rather titanium nitride (TiN) is nowadays the most commonly used material, since its work function is close to the middle of the silicon band gap, meaning that it can be used for both p-type and n-type transistors and does not require doping [9]. An additional benefit of TiN over Poly-Si is that it has a lower electrical resistance and can act as an oxygen diffusion barrier, increasing the stability of the dielectric. Furthermore, its integration into fabrication is simple since it has already been used in the complementary metal oxide semiconductor (CMOS) fabrication process as a dielectric in interconnects and for diffusion barriers. However, the threshold voltage in a metal gate stack can only be tuned by doping the channel itself, degrading some of its characteristics. Therefore, there has been a considerable interest in finding advanced gate metals of desirable work functions [10]. Among others, TiC [11], TiAlC [12], and Ru [13] are heavily investigated as potential future materials.

The introduction of new materials inevitably led to the need for more complex fabrication techniques and intricate patterning steps to achieve smaller feature sizes as laid out in the International Roadmap to Semiconductors (ITRS) [14] and the International Roadmap for Devices and Systems (IRDS) [15]. Several complex deposition, etch, and cleaning steps are necessary in order to manufacture highly controlled gate profiles without damaging neighboring materials. The introduction of three-dimensional structures such as fin field-effect transistors (FinFETs) has added further complexities to the gate patterning process [16], as straight etch profiles must be obtained despite the exposure of sections of the underlying material. Therefore, a combination of highly directional as well as selective patterning techniques and intermediate cleaning steps must be applied in order to achieve the required accuracies [6].

In order to develop new techniques for reliable patterning of more and more complex gate structures and to improve the understanding of the underlying mechanisms, modeling becomes increasingly important. On one hand, a comparison of simulation results with experimental data can give insight into the physical properties of different processes, as disagreements between the two indicates the presence of additional phenomena, which must be considered. On the other hand, reliable and predictive models for processing steps allow for quick testing of new designs without the need for expensive experiments. Especially in the case of gate stack etching sequences, such predictive models can reduce development costs greatly as the fabrication of prototypes is expensive and time-consuming. However, complex gate structures require the careful combination of the different etch techniques described above, including their influence on the subsequent etch steps, resulting in the need for sophisticated modeling of the underlying physical phenomena. In order to enable such complex simulations, even fundamental computational techniques must be considered carefully to achieve physically meaningful results. Therefore, this review will cover the fundamental techniques of process simulation, such as methods for describing moving surfaces and particle transport inside a plasma chamber, and move on to modern patterning techniques and sophisticated models used to describe them. The particular aim of this review is to summarize recent achievements in the simulation of the etching of advanced node multi-layered gate stacks.

2. Methods

The underlying numerical methods driving simulators have important implications on the modeling capabilities. It may be highly inefficient or even impossible to model certain physical processes using a specific method, but straightforward using another. Therefore, the choice of appropriate numerical methods depends strongly on the desired modeling capabilities. In order to judge the applicability of a certain method to a problem, a deep understanding of the relevant physics, as well as of the method itself is necessary. The fundamental methods used to describe the wafer surface and the flow of atoms, ions, and molecules within the feature scale are discussed in this section, highlighting their respective consequences to the final modeling capabilities of a simulator.

2.1. Surface Representations

The simulation of microelectronic fabrication techniques requires accurate descriptions of the topography of different materials and their interfaces, since certain processing steps, such as deposition and etching, can result in complex surface deformations. Therefore, the ability of the surface description to represent such changes over time in a robust way is essential. The manner in which the surface will move is usually calculated for every surface element and applied for a discrete time step [17], to be calculated again for the resulting new topography [18]. The manner in which this surface evolution is applied depends strongly on the surface representation used. The surface representation can be explicit or implicit, both of which are addressed in this section. Implicit surface descriptions have become standard in modern technology computer-aided design (TCAD) simulators due to their robustness and computational efficiency.

2.1.1. Explicit Surfaces

Many applications, such as graphics rendering, rely on explicit surface representations which define surface elements by interconnected points on the surface [19]. This representation has several desirable properties, such as no principal limitation on feature size or resolution and minimal memory requirements, since the number of surface elements scales directly with the total area [20]. Furthermore, it can be visualized easily, since the absolute coordinates of all elements are known by design. Therefore, all volume and surface elements have fixed sizes, which is useful when modeling stress, which can develop when an element grows in size while surrounded by other elements. One such process is oxidation, where the substrate below a mask grows and moves the mask, leading to stress within the materials [21].

The movement of a surface is realized by shifting the defined nodes in a desired direction and connecting the points again to obtain new surface elements [22]. This can lead to a non-physical intersection of surface elements as shown in Figure 1, since there is no strict definition of which side of the surface represents the material and which one is the ambient space [19]. Figure 1 shows the merger of two surfaces, which creates a non-physical geometry in the center, due to the overlap of two materials. This is a common concern with explicit surface definitions. Testing for such self-intersections is a computationally expensive process and thus not favorable when describing moving surfaces. The separation of surfaces, or indeed any movement of a surface, can lead to similar problems as different nodes must be identified and connected correctly to achieve accurate descriptions of the interfaces [23]. Furthermore, topography changes can lead to a wide separation of neighboring points and therefore large surface element areas, reducing the accuracy of the surface representation. Due to these potential problems, the surface must be remeshed regularly to obtain proper and efficient representations after each time step. This includes recalculating nodes and surface elements in order to satisfy certain minimum mesh quality criteria, such as equal area triangles or equal edge lengths [24].

This additional step can be computationally expensive for large surfaces and is therefore not desirable in complex, three-dimensional simulations.

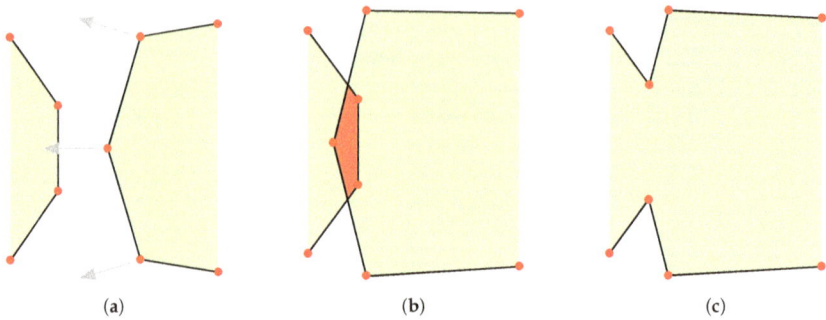

(a)   (b)   (c)

Figure 1. Two explicit surfaces merging by movement of nodes: (**a**) initial geometry with included area (green); (**b**) broken final geometry due to a surface overlap (red); (**c**) correct surface after merging of the two surfaces. In order to reach the correct surface representation, additional meshing steps are necessary, decreasing performance.

2.1.2. Implicit Surfaces

Implicit surfaces are isosurfaces described by a function $\phi(\vec{x})$, defined at every point in space. It is not solved for one of its variables, but rather used to find the set of points which let the function go to a specific scalar value, usually zero [23,25]. Therefore, all points on the surface, $\vec{x} \in S$, must satisfy $\phi(\vec{x}) = 0$, which is why these points are called the zero level set. Since it is not feasible to represent all possible surfaces algebraically, $\phi(\vec{x})$ is constructed using signed distance transforms. These construct $\phi(\vec{x})$ from the distance d between any point $\vec{x}$ and the surface S bounding the volume M:

$$\phi(\vec{x}) = \begin{cases} -d, & \text{for } \vec{x} \in M, \\ 0, & \text{for } \vec{x} \in S, \\ d, & \text{for } \vec{x} \notin M. \end{cases} \tag{1}$$

Therefore, every point in the simulation domain is known to be inside or outside M, by examining the sign of $\phi(\vec{x})$, without the need for further analysis. Robust and fast algorithms for signed distance transforms exist [26–28], allowing for simple integration into process simulators. These algorithms construct $\phi(\vec{x})$ from an explicit representation, such as a triangulated surface, by traversing the simulation domain and finding the smallest distance between $\vec{x}$ and the surface iteratively. The fast marching method is optimized for level sets and thus it is the most efficient method in converting between explicit surfaces and level sets [29].

The time evolution of a surface is usually captured in a scalar field denoting the surface normal speed $v(\vec{x})$ [30]. For simple surfaces, such as planes, the velocity field can be subtracted from the signed distance function to move the surface with velocity $v(\vec{x})$:

$$\frac{\partial \phi(\vec{x}, t)}{\partial t} = -v(\vec{x}). \tag{2}$$

However, more complex surfaces with non constant gradients must be moved differently in order to retain their shape, which is achieved using the gradient of the signed distance function to normalize $v(\vec{x})$, leading to the level set equation [31]:

$$\frac{\partial \phi(\vec{x}, t)}{\partial t} + v(\vec{x})|\nabla \phi(\vec{x}, t)| = 0. \tag{3}$$

Since Equation (3) is a form of the Hamilton–Jacobi equation, often encountered in mathematics, many algorithms are available to solve it using finite difference schemes [31,32]. In order to use $\phi(\vec{x})$ in numerical simulations, the values of the function are usually stored at points defined on a regular grid to achieve an approximate representation, as shown in Figure 2b. The regular spacing of grid points enables the use of well-known finite difference algorithms to solve the differential equations needed for the calculation of surface normal vectors, surface curvature, or the time evolution as described above [33].

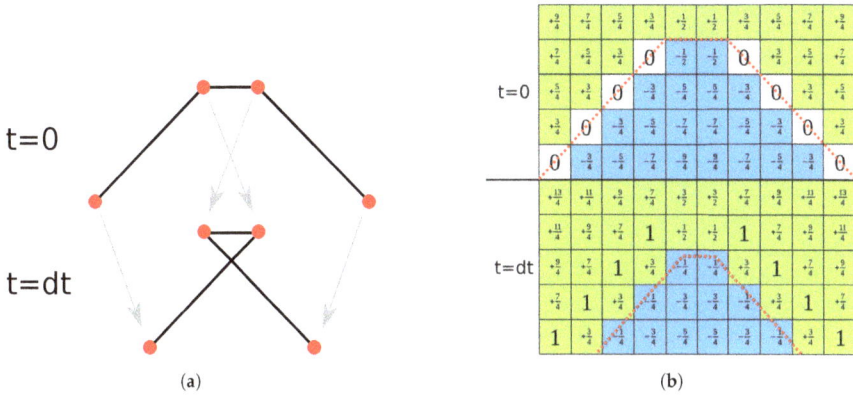

(a)   (b)

Figure 2. Schematic comparison between explicit and implicit surfaces being moved, highlighting self-intersection in explicit surfaces. In contrast, the level set method applied in (**b**) shows surface movement without self-intersection, albeit losing some features such as the expected sharp peak in the center of the geometry. (**a**) Nodes of an explicit surface moved by a velocity field, creating a self-intersection. Additional steps are required to form a correct representation of the surface. (**b**) Implicit surface being moved by adding unity to all values stored on a regular grid. The numbers represent the level set value stored at the center of each cell. Negative regions inside the surface are highlighted in blue, regions outside with positive values in green. Dashed, red lines indicate the explicit location of the surface.

Since it is not the exact location of the surface, which is stored but the distance to it at regular intervals in the entire simulation domain, the position of the grid points does not change with the moving surface, but only their respective level set values. This means that self-intersection and similar problems occurring in explicit surfaces are not encountered using implicit level set methods. Figure 2 shows these clear differences by comparing the movement of a surface in these representations. Figure 2b also highlights another characteristic of the level set method, which is the loss of sharp features on the surface [34]. As can be seen clearly, the peak expected after the evolution of the surface is flattened due to the size and limited resolution of the grid. Increasing the number of grid points will dampen this effect, however, increasing the computational cost greatly, since the number of grid points scales with the domain volume. Nevertheless, these negative effects are not expected to reduce the quality significantly because the modeled processes do not tend to create radically outstanding features, but rather smooth profiles varying over a number of grid points. Nevertheless, great care must be taken when choosing the number of grid points to balance computational cost with simulation accuracy. Additionally, the spacing of grid points influences the output of numerical schemes for calculating the curvatures or normal vectors at grid points. Therefore, the accuracy is influenced by the chosen grid resolution in several ways.

Furthermore, the merger or separation of surfaces does not require additional consideration as there is no ambiguity about whether a point lies inside or outside the surface. Hence, two surfaces growing towards each other must merge, when there are no more oppositely signed points between them, since there can be no part of the encapsulated volume between them. However, this can lead to surfaces merging too quickly, when there are no oppositely signed points between the fronts, leading the surface evolution to jump up to one grid spacing just before merging. The same effect can be observed for separating surfaces or thin layers being removed entirely, which is shown in Figure 3. Therefore, the grid spacing also sets a minimum layer thickness in all directions.

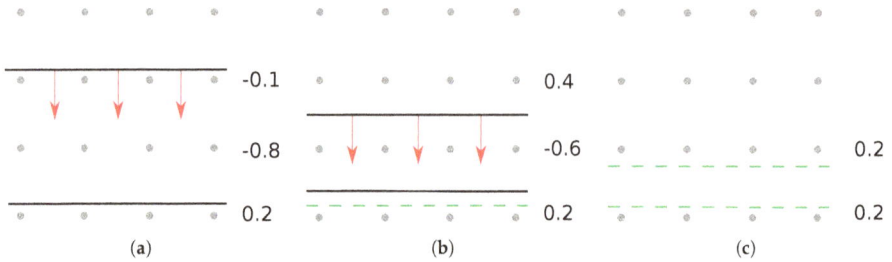

Figure 3. Top surface of a thin layer is moved until the layer disappears entirely. Grey points indicate the grid, black lines the surface, red arrows the movement per time step, and dashed green lines the correct position of the surface. The level set values of each row are shown to its right. (**a**) Initial layer, only two grid spacings wide, whose top surface is moved downwards. (**b**) As the layer is thinned to only one grid point, the level set values are not normalized anymore, resulting in symmetric shrinking. (**c**) Once the last row of grid points is outside of the layer, it ceases to exist, although it should still be almost one grid spacing wide.

Especially in modern gate stack etching sequences, the accurate description of thin layers is of critical importance to the combination of different chemical processes, which deposit thin layers of different materials while etching the structure. These thin layers have a considerable impact on the subsequent etch steps as they have very different chemical properties to the substrate. Despite being very thin, they can protect the underlying material from etching, since they might etch very slowly. If, however, they are ignored or disappear too early in the simulation, the underlying material is exposed, resulting in an inaccurate modeling of the physical process. The problem of quick merging and the disappearance of layers, as well as the symmetric shrinking is usually due to few grid points and therefore a lack of information about the surface position. This can be overcome by describing thin layers by not only the material they consist of, but also including the sum of all materials beneath them [35]. The materials below are also stored as a separate level set so that the original thin layer is not lost and can be extracted again. The advantages of this material representation over defining single materials separately is highlighted in Figure 4, which strongly increases accuracy, especially when considering the thin layers. In order to recover the single materials, the lower ones must be subtracted from the top material, which can be performed efficiently using the level set method, since any boolean operation can be carried out element-wise at each grid point [36]. The intersection between two level sets, for example, can be achieved by comparing the two values at each grid point and choosing the greater of the two. This results in a stable conversion to single layers, although thin materials might not be represented correctly in a separate level set, due to the effects described above.

The memory requirements for storing a level set surface are high compared to explicit surfaces, as they scale with volume rather than with surface area [31]. Figure 2b highlights that only the points around the surface are needed to describe the set of zeros defining it, as the surrounding values increase linearly in an ideal level set. Therefore, only a few layers around the boundary, a so-called narrow band [37], influence the surface description. If the surface evolves towards the edge of the narrow band, a new band must be initialized with the surface at its center. Since re-initialization is computationally expensive, there is an optimal width of the narrow band, which uses the smallest number of grid points for calculations and avoids re-initialization for as long as possible. The optimal width found in the original publication [37], was between 6 and 12 layers of active grid points. An extension of the narrow band approach is the sparse field algorithm, which significantly reduces computational cost of re-initialization by approximating distances from the surface in a stable way [33]. Using the sparse field algorithm, re-initialization can be performed at every time step, which allows for the use of only a single layer of grid points, thus achieving optimal storage efficiency [38]. Neighboring grid points for the calculation of surface normals and curvatures can be calculated for each time step using the same efficient distancing algorithm.

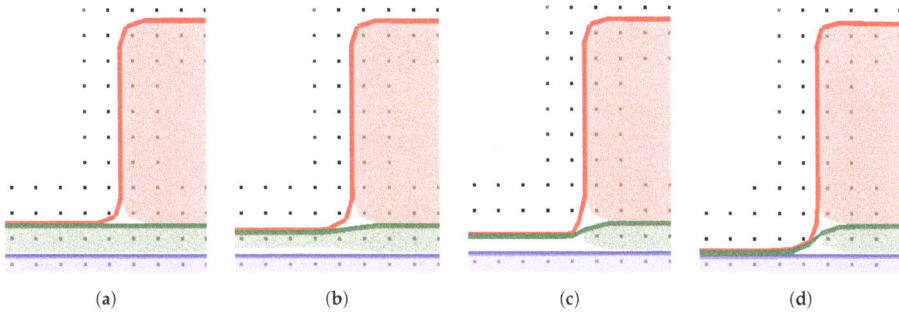

Figure 4. Schematic comparison of the difference in surface representations when layers are wrapped around lower ones (volume inside solid lines) or if they only encapsulate a single material (colored areas) in uniform etching of a thin layer (green) under a mask (red). Different problems when representing only a thin layer with level sets are shown. (**a**) Initial layout with only minor discrepancies between the two representations. (**b**) Symmetric shrinking of a single layer as the level set value in the center decreases: the bottom of the thin layer (green) lifts up as there is only one grid point defining the distance, reducing the layer symmetrically as shown in Figure 3b. (**c**) Complete removal of the thin layer as no grid point is inside the surface anymore. (**d**) Final layout with receded surfaces: the thin layer is still intact for wrapped level sets, but is completely removed for individual materials. (used from [35] under CC BY 4.0 / cropped from original).

2.1.3. Cell Based Methods

Another common approach to describing surfaces is considering them only as interfaces between different materials. By describing only the volume occupied by a material, the interfaces are simply described by its boundaries. In cell-based methods, this is usually realized on a regular grid, where every grid point represents a unit cube, or voxel, at its location, storing relevant information [39]. Usually, a number denoting the material and a filling fraction are stored, allowing for the calculation of the exact location of the boundary [40].

At high resolutions, it is also possible to describe geometries using only a single material per cell, which allows for simpler modeling. If no filling fraction is stored, but only the material of the cell (i.e., the filling fraction is binary), this method can also be considered a voxel based explicit surface representation [41]. The geometry can be extracted easily, as each voxel can be included explicitly, although this leads to stepped surfaces. If the cell size is close to the size of physical atoms, cell based methods can come close to atomistic modeling, enabling more accurate descriptions of physical processes, such as diffusion or ion implantation [42].

As shown in Figure 5, this approach is similar to the level set method because it represents the boundaries implicitly on a regular grid. However, this leads to the same shortcomings, such as resolution limitations and scaling problems. These can be overcome with similar techniques as described above, such as the narrow band approach. Cell-based methods share their robustness for complex topographies with other implicit methods. Moving a cell based surface is more trivial than one defined by level sets, as conservation of mass can be used to add or remove volume from a voxel, greatly simplifying surface velocity calculations.

However, numerical schemes associated with cell based methods are not as efficient as those used in level set representations [43] and if filling fractions are non-binary, conversion to explicit surfaces is complex [44]. Since there is no information about where the material lies inside a cell, the explicit boundary must be reconstructed from the surrounding cells. This can be quite complex, especially when considering several different materials or thin layers of materials, where there is little material within a cell while it stretches across the entire cell width. Reconstructing an explicit surface might thus be ambiguous and not reliable for complex structures. Therefore, most simulating frameworks use level set representations for moving surfaces as they usually are more robust to complex deformations and computationally more efficient when modeling large structures. Cell based methods are better suited for describing smaller geometries, incorporating mixed materials and volume characteristics, such as implanted ions.

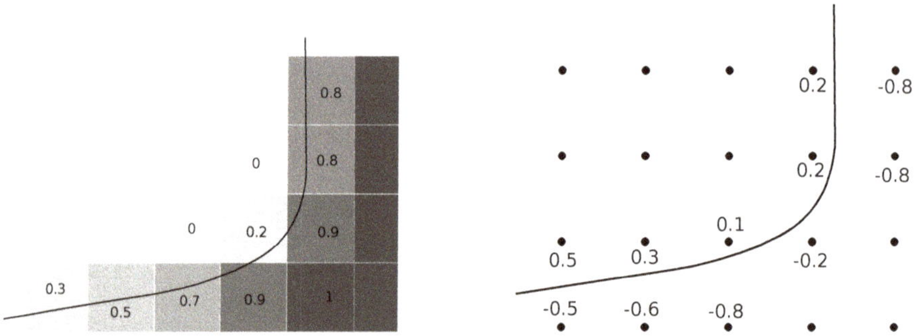

Figure 5. Comparison of cell-based and level set representations in narrow-band implementations. While the former is intrinsically associated with volume, level set representations describe an interface or boundary. (**a**) Cell-based representation of an explicit surface (black line) with numbers indicating the filling fraction of each cell. The darker a cell, the higher the filling fraction. (**b**) Level set representation of an explicit surface (black line) with level set values, related to the normal distance from the surface, for each grid point.

2.2. Surface Velocity Calculation

As outlined in Section 2.1, the surface evolution is governed by a scalar field of velocities $v(\vec{x})$, describing how much each discretized element of the surface should move in the next time step. Therefore, the most crucial part in process simulation is the calculation of those velocities using models, which match the described physical process as closely as possible, whether empirically or physically. Despite recent advances in atomistic modeling [45], the structures considered in process simulations are usually too large to take into account individual atoms and, therefore, the surface is approximated as a continuum, using the surface representations described in Section 2.1. This results in the loss of microscopic information, such as surface roughness, while considerably decreasing computational complexity [46]. Velocities must be calculated for every discretized surface element, in order to move the surface correctly. Therefore, a velocity value for each triangle of an explicit surface, for each grid point of a level set, and for each cell of a cell-based representation must be defined in order to advance the surface.

A simple way to generate these velocities, is to extract geometric parameters, such as etch depth, from experimental data and changing the surface to replicate the result of the fabrication process.

This approach is called process emulation, as no physical behavior is modeled, but rather simple geometric rules are applied to mimic the result of a fabrication process [47]. Constant deposition, for example, can be approximated by expanding the surface by the same amount in each direction, meaning the growth rate is the same everywhere on the surface. More complex processes can also be emulated by applying more sophisticated geometric rules [48]. Since no physical processes have to be modeled, this approach is computationally efficient. Therefore, this method is useful for creating large structures quickly for device characterization or for feasibility studies, due to its high efficiency [49–51]. However, it is not very accurate, especially when describing complex processing steps.

Since process emulation does not take into account any physical properties of the surface or the etch chemistry, it cannot be used for any physical analysis. In order to identify dominant etch or deposition mechanics, or even predict the properties of new fabrication processes, a sophisticated physical description of the involved physics and chemical reactions is necessary. This approach is called process simulation and is focused on in the following sections, which cover the modeling of the transport of atoms, ions, and molecules through the feature scale region, as well as the modeling of surface reactions leading to etching or deposition. From these models, the surface velocity field $v(\vec{x})$ can be generated, leading to a physically accurate deformation of the simulated wafer surface.

2.3. Transport of Molecular Entities in Plasma Environments

In order to simulate how much material is removed or deposited on a surface during a fabrication process, the rates at which different atoms, ions or molecules impinge on the surface must be found. Collectively, atoms, ions and molecules are hereafter referred to as molecular entities. These rates can depend strongly on different geometrical effects and transport phenomena inside the reactor [52]. The way in which molecular entities traverse the reactor depends on their specific properties, as well as on the thermodynamics of the chosen process. In order to describe this transport, the reactor space is usually divided into reactor-scale and feature-scale regions separated by a plane $\mathcal{P}$, as shown in Figure 6. This simplifies the description of neutral atoms and molecules because their motion in the reactor-scale region is governed by the Maxwell–Boltzmann distribution, since this region is large compared to their mean free path [53]. In contrast, the feature-scale region is small in relation to their mean free path, meaning collisions with the surface are much more common than those with other parts of the gas phase. Therefore, ballistic transport is commonly used to describe the propagation of molecular entities through the feature scale region [54], which can result in shadowing and reflection effects. This transport can then be simulated in a straightforward manner using ray tracing methods. Knudsen diffusion has also been used successfully to describe the transport of atoms, ions and molecules in simple geometries, such as straight trenches [55,56], eliminating the need for complex modeling of the molecular entities' trajectories. However, a process description close to the physical reality can only be obtained by considering particle transport directly.

Physically, each infinitesimal element dA on $\mathcal{P}$ can be considered as an individual source of molecular entities with certain properties and emitting fluxes, as highlighted in Figure 6. Usually, the angular dependence of the neutral molecular entities' distribution $\Gamma_{neutral}$ is assumed to follow a cosine due to the angular projection of dA on the emission angle with $\mathcal{P}$ [57]. The distribution of accelerated ions Γ_{ion}, present in ion-enhanced plasma etching processes, are usually described by more focused power cosine or normal distributions [46]. This leads to highly directional properties which reflect those found in experiments. In order to describe the motion of ions as straight trajectories in the feature scale region, the electromagnetic field distortion by the surface must be small enough to not influence the paths of ions drastically, which is a reasonable assumption given the short path lengths of ions in the feature scale region and the strong directional electric fields used to guide the ions.

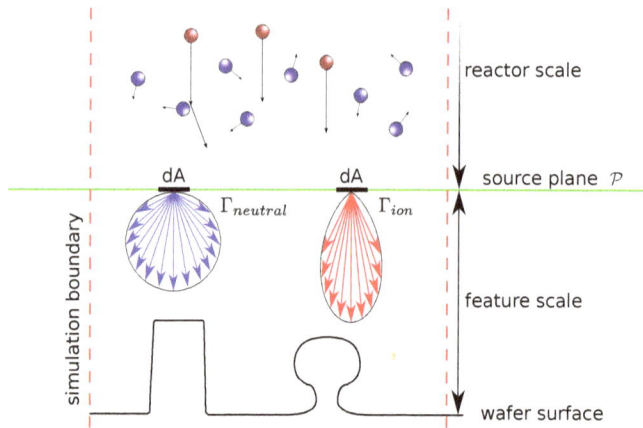

Figure 6. Schematic representation of the traversal of neutral molecular entities and ions through the reactor and feature scale regions. While the motion through the reactor scale is dominated by random collisions with other molecular entities, the path through the feature scale region is dominated by ballistic transport. The directional distribution of neutral atoms and molecules $\Gamma_{neutral}$ (blue) and ions Γ_{ion} (red) entering the feature scale region is shown as blue and red arrows, respectively. The molecular entities will then traverse the feature scale region in straight lines, only colliding with the surface.

When the physics described above are simulated, the reactor and feature scale are usually treated separately, as the only input needed for feature scale simulations are the source distributions $\Gamma_{neutral}$ and Γ_{ion}. These can be obtained from experiments or from chemical kinetic simulations. Simulating the feature scale requires some additional consideration, due to the limited size of the simulation domain and other computational limits. For example, due to the large number of molecular entities in a physical process, it is not feasible to simulate all of them. They are usually simulated with Monte Carlo techniques using particles, where each particle represents multiple molecular entities. Computational methods addressing this problem are described in the following sections.

Another important factor for simulations is the appropriate choice of boundary conditions due to the limited size of the simulation domain, when compared to the size of an actual wafer. If the simulated wafer contains only a single structure which fills the simulation domain and is planar otherwise, particles which leave the simulation domain can be ignored, since they cannot return back into it. However, if the same structure is repeated across the wafer, periodic boundary conditions are more appropriate since particles which are reflected to leave the domain can be mapped back into it, as if they originated from the neighboring structure, increasing simulation accuracy. This enables the consideration of parts of the wafer, which cannot be simulated directly due to the limited size of the simulation domain. The same applies to reflective boundary conditions, where the neighboring structures are mirrored to the considered one.

2.3.1. Top-Down Flux Calculation

The fluxes at which different particles impinge on each part of the surface can be found by launching a large number of particles from the source plane $\mathcal{P}$ and using ray tracing to find the point of impact of each particle on the surface [58]. After all particles have been traced, the number of impacts is counted for each discretized surface element, which may differ depending on the surface representation used. Each simulation particle may represent a single molecular entity or several of the same species, depending on the number of particles used to simulate the transport. Simulating the maximum number of particles, each describes only a single molecular entity. However, this is not practical due to the large number of actual molecular entities usually involved in a physical fabrication process. Therefore, fewer particles are used, each representing a number of molecular entities. The particle flux at each discretized surface element is then found by the number of incident particles times the number of molecular entities represented per particle.

Monte Carlo methods are employed to generate particles according to the probability distributions describing the neutral flux $\Gamma_{neutral}$ and ion flux Γ_{ion} [59], while ray tracing methods, as used in computer graphics [60], enable the simulation of a large number of particles. These are launched from several points on the source plane $\mathcal{P}$, usually spaced at regular intervals, forming a grid of particle sources. A large enough quantity of particles thus results in a good description of the effect on the surface of the original source distribution. Figure 7 schematically shows how particles, launched from different locations of the source plane in pseudo-random directions, might interact with the surface, with some particles experiencing multiple reflections. The starting direction and the probability of reflection and re-emission are both determined probabilistically, meaning numerous pseudo-random numbers must be generated, increasing the performance requirements [61].

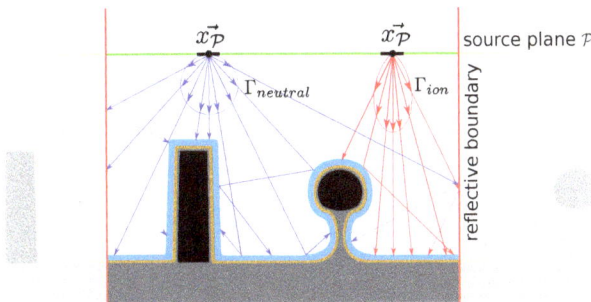

Figure 7. Schematic depiction of discrete particles being traced from the source plane to the surface using rays. Each particle either describes neutral atoms or molecules (blue) or ions (red), governed by the source distributions $\Gamma_{neutral}$ and Γ_{ion}, respectively. They define the relative probability of particle direction, energy and other properties. Specular reflections are shown for ions, and diffuse reflections for neutral species.

Balancing computational efficiency with simulation accuracy is one of the main concerns in this method, as the modeling complexity can theoretically be extended to model every single molecular entity without systematic limits, due to the physical nature of this approach. However, the computational cost of tracing a large number of particles is high as several intersection tests with the surface have to be performed for each particle [46]. The minimum number of particles depends on the exact implementation and whether smoothing is used in order to avoid abrupt changes in the fluxes along the surface. In general, it must be ensured that each discretized surface element is intersected several times, if one is to achieve physically meaningful results. This means that several particles must reach each triangle in explicit surfaces, grid point in level sets, or cells in cell-based methods. The complex geometries used in modern gate stacks require a large number of particles in order to create a physically meaningful flux profile everywhere on the surface.

Although implicit surface representations are often encountered in process simulations, as well as visualization tasks, some form of explicit surface is usually required in an intermediate step during ray tracing. An explicit surface representation is often necessary, since intersection tests used in ray tracing algorithms are much more efficient on explicit surfaces than on implicit ones. Once the fluxes have been found and the velocity field $v(\vec{x})$ is generated, an implicit representation can be used again to move the surface. Conversion between the two representations can be quite time-consuming, creating a bottle-neck for simulation efficiency. However, perfectly closed explicit surfaces are not strictly required for ray tracing as small self-intersects and other minor flaws in the geometry do not have a great effect on the final result. Intermediate, explicit surfaces can be created by triangulated, cell-based approximations of the surface as produced by marching cubes algorithms [62] or more crudely by approximating the surface using discs [57] or spheres [63]. In this approach, each implicit grid point is approximated explicitly by a disc or sphere with a radius of at least one grid point separation, resulting in a closed surface due to the overlap of discs or spheres, as shown in Figure 8. Spheres can be placed directly on the grid points and do not require any translation to the surface normal, making them more efficient, albeit less accurate. This allows for quick conversion between the surface representations, while still enabling the use of advantageous explicit ray tracing methods.

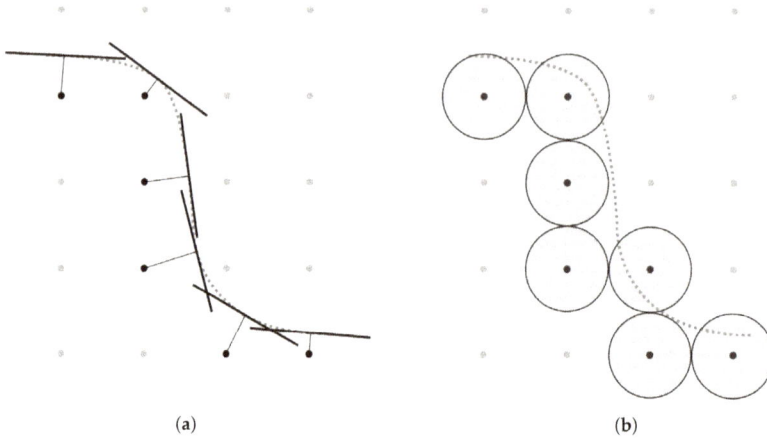

(a) (b)

Figure 8. Two ways to approximate an implicit surface efficiently by explicit shapes on active grid points (black), in order to simplify intersection tests for ray tracing. The line segments and circles are replaced by disks and spheres in three dimensions. (**a**) tangential line segments used to form an explicit approximation of the surface, as described in [57]; (**b**) surface approximated by explicit circles centered at active grid points, as described in [63].

Due to the physical nature of the top-down method, even complex reflective properties, such as specular reflection, can be modeled straightforwardly, as the incoming angle of a ray is found easily from the intersection test and extracting the surface normal and curvature is intuitive when using the level set method [25]. If diffuse reflections are to be considered, some form of random number generation must also be applied several times per ray to find the reflected direction, which increases the simulation time. Additionally, other effects, such as different material properties due to variations in the crystal orientation [64], can be included for a more physical description. This approach also allows for particle–particle collisions to be considered, if the simulated geometries are too large for the assumption of ballistic transport to hold. This can be the case for large aspect ratio geometries, where particles may travel far in one direction without a surface intersection. Therefore, due to the physical approach of this method, it is the most accurate one as it does not limit the number of effects which can be included in modeling the physical processes. However, it usually requires more computational effort than alternatives.

2.3.2. Bottom-Up Flux Calculation

As mentioned in Section 2.3.1, the source plane $\mathcal{P}$ is usually described numerically as a regular grid of particle sources. Instead of tracing many rays from each source to the wafer surface, it is also possible to do the reverse. In the bottom-up method, a single discretized surface element is considered, and all the particle sources visible to it are summed [65,66]. This is achieved by iterating over all discretized particle sources on the source plane $\mathcal{P}$. For each source, it is verified whether the source, located at $\vec{x_P}$, is visible from the considered discretized surface element at $\vec{x}$ (Figure 9). The particle flux incident on this point on the surface is then found by considering the particle source distribution Γ_{src}. Summing the contributions of all particle sources gives the total particle flux incident on the discretized surface element at $\vec{x}$:

$$F_0(\vec{x}) = \sum_{\vec{x_P}} \Gamma_{src}(\vec{x_P}, \vec{x}) Y(\vec{x_P}, \vec{x}). \tag{4}$$

Here, the angular dependence of the source is captured by $\vec{x_P}$ and $\vec{x}$, as their relative positions give the relevant angle of emission and impact, respectively. The visibility function, $Y(\vec{x_P}, \vec{x})$, describes whether a particle source at $\vec{x_P}$ is visible to a surface element at $\vec{x}$, and is unity if the point is visible and zero otherwise. Visible points are indicated by a green arc in Figure 9, so the fluxes of all discretized particle sources within this arc are included in the total flux.

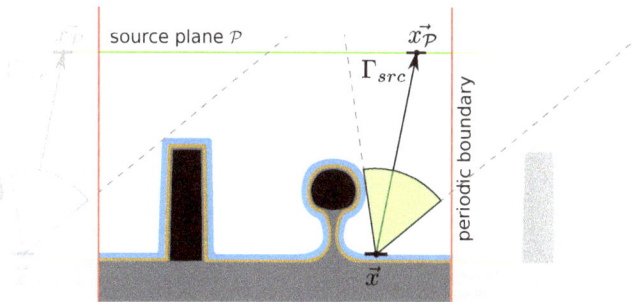

Figure 9. Schematic representation of the bottom-up flux calculation for modern gate structures, using periodic boundary conditions, meaning the entire simulation domain is repeated at the boundaries. The black arrow indicates the direction used to find the direct flux incident on $\vec{x}$ from a single particle source at $\vec{x_P}$ with a source distribution of Γ_{src}. The flux of all visible source plane elements, indicated by the green arc, is summed to give the total direct flux on $\vec{x}$.

Equation (4) only gives the particle flux incident on the surface directly from the source plane and does not include any reflection or re-emission effects from other locations on the surface. It is possible to formulate an analytical solution to include reflection and re-emission and numerically solve for the total flux [66]. However, this approach can be memory intensive for large geometries due to the large matrices built to describe the correlations between large numbers of discretized surface elements and particle sources on the source plane [67]. For highly symmetric geometries, such as high aspect ratio holes or trenches, the calculations can be simplified by considering their symmetries and calculating the fluxes only for non-degenerate parts of the surface. These fluxes are then extended to the entire structure, resulting in a full description. However, this approach only works for symmetric geometries and fails once even small irregularities, such as surface roughness, break the symmetry, which is unavoidable in most practical simulations.

Therefore, iterative approaches are commonly used, which first calculate the direct flux F_0 and then the fluxes to be reflected F_{refl} or re-emitted F_{reem} from each discretized surface element. All of the discretized surface elements can then be described as a particle source with distributions Γ_{refl} and Γ_{reem}, which is shown schematically in Figure 10. These source distributions also include the description of the reflected and re-emitted particles in terms of angular dependence and other surface properties. The contributions from a particle source at $\vec{x}'$ visible to $\vec{x}$ can then be found using Γ_{refl} and Γ_{reem}. All of these contributions are summed to give the total incident flux, similar to the direct flux calculation, which leads to an expression for the total reflected and re-emitted flux incident on $\vec{x}$. The total flux after the first iteration F_1 is thus:

$$F_1(\vec{x}) = \sum_{\vec{x}'} Y(\vec{x}', \vec{x}) \left[\Gamma_{refl}(\vec{x}', \vec{x}, F_{refl}, E) + \Gamma_{reem}(\vec{x}', \vec{x}, F_{reem}, E) \right]. \tag{5}$$

This process can be repeated n times, until all particles have been adsorbed or a satisfying accuracy has been reached. The necessary number of iterations depends strongly on the properties of Γ_{refl} and Γ_{reem} as they dictate how much of the incoming flux is reflected or re-emitted again, respectively. The number of necessary iterations can also be set using a minimum change in arrived fluxes, which should be achieved at each iteration. If the fluxes change less than this margin, no further iterations are needed. The final result for the flux incident at $\vec{x}$ is given by

$$F(\vec{x}) = \sum_0^n F_n(\vec{x}). \tag{6}$$

After each step, only the flux to be used in the subsequent iteration is saved. Therefore, specular reflections cannot be modeled accurately using the bottom-up method, but only by assigning an average incoming and thus outgoing direction [46]. Therefore, energetic ion reflections, which are mostly specular, cannot be modeled accurately using this method, as they are approximated by a common reflection distribution, Γ_{refl}.

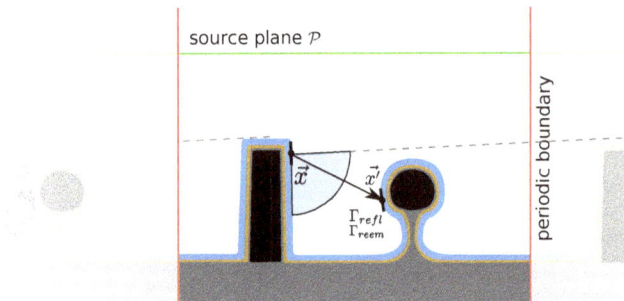

Figure 10. Calculation of reflected or reemitted fluxes using a bottom up technique with periodic simulation boundaries. The black arrow indicates the direction used to find the reflected and re-emitted flux incident on $\vec{x}$ from a particle source at $\vec{x}'$. The source distributions Γ_{refl} and Γ_{reem} define the flux emitted towards $\vec{x}$. The total indirect flux at point $\vec{x}$ is found by summing the flux from all visible surface points, highlighted by the blue arc.

2.4. Chemical Modeling

Similarly to the transport of molecular entities, discussed in Section 2.2, it is not feasible to simulate all chemical reactions taking place inside a reactor. In order to simplify the modeling complexity, usually only surface reactions are considered [68], while reactions in the gas phase are assumed to reach a steady state due to the short reaction times compared to the time it takes to traverse the reactor. In order to find the effective particle flow, captured by the source flux distribution Γ_{src}, it is essential for reactor-scale simulations to identify dominant reactions in the gas phase and properly approximate the particle flow to the surface. This flow can either be found experimentally or simulated for specific reactor geometries [69], which can indeed be challenging for complex processes. Due to the large number of different chemical elements used for each etching step during gate stack patterning and the high temperatures in the reactor, simulating the expected source flux distribution is cumbersome and time-consuming. Even if dominant reactions can be identified, simulating them in the energetic environment of a plasma reactor can present a challenge. Therefore, experimental data of these processes are vital for accurate simulations.

Simulating surface reactions can still be highly challenging as volatile atoms and molecules are usually involved in etch processes, creating a wide variety of possible end products. Especially modern plasma etch processes pose a challenge as many different chemical species are present, leading to countless reactions and thus great computational effort required to find reasonable results. Since modern gate stack etch sequences consist of a combination of such processes, which can influence each other strongly, physically meaningful models must consider highly complex chemical phenomena. Even if all possible reactions could be modeled efficiently, the wide variety of reactor and surface geometries, as well as high sensitivity to minor changes in chemical composition of reactants, makes it impractical to gather meaningful data from experiments in order to test computational models. Therefore, semi-empirical models are still the most robust options in simulating modern fabrication processes.

Each of the different physical processes involved are usually described by coefficients in a general surface rate model, which can then be used to find the overall surface normal velocity $v(\vec{x})$ [70]. These coefficients must be fitted to a particular technology by comparing them to fabricated structures [71]. For an arbitrary plasma etch simulation, all possible physical processes have to be taken into account, including chemical etching, ion-enhanced etching, sputtering, and deposition. Depending on the actual properties of each of these processes, different coefficients are used to describe them. To simplify the modeling and to allow for a description in discrete time steps, necessary for process simulations, the effect of every physical process on the surface can be computed by considering the relative concentrations of materials on the surface, found using flux calculations described in Section 2.2.

For further simplification, the chemicals involved are grouped into a smaller number of types, representing their effect on the surface [72]. The rates used for simulation usually describe neutral etchant particles, passivating particles, passivation etchant particles, and ions, as illustrated in Figure 11. Re-emitted etch products and sputtered material are usually included during ray tracing and thus affect surface fluxes directly, without the need for any further considerations.

The rates of particle types impinging on the surface can be summed to give coverages of different particle types, ϕ_x, where x is a particle type. Therefore, coverages of etchant ϕ_e, polymer ϕ_p, polymer etchant ϕ_{pe}, and ions ϕ_i describe the amount of material covering this surface. Stochastically, this can also be seen as the probability of a given particle being at that location on the surface. Although ions would not deposit and cover the surface, ϕ_i is used to capture the number of ions which impinged on the surface, described as a coverage here for simplicity. The coverages can then be used in different models to find the surface normal velocity $v(\vec{x})$ described in Section 2.1, hence describing etching or deposition. Assuming steady-state conditions for the different surface coverages, a system of linear equations describing all physical deposition and etching processes is set up [73]:

$$\frac{d\phi_e}{dt} = J_e S_e (1 - \phi_e - \phi_p) - k_{ie} J_i Y_{ie} \phi_e - k_{ev} J_{ev} \phi_e \approx 0, \tag{7}$$

$$\frac{d\phi_p}{dt} = J_p S_p - J_i Y_p \phi_p \phi_{pe} - \Delta_p \approx 0, \tag{8}$$

$$\frac{d\phi_{pe}}{dt} = J_e S_{pe} (1 - \phi_{pe}) - J_i Y_p \phi_{pe} \approx 0. \tag{9}$$

Each term in Euqaitons (7)–(9) describes a physical process, which changes the surface, and includes the necessary coefficients, where J_x denotes the respective arriving fluxes on the surface element, S_x the respective sticking probabilities, Y_x the yields (e.g., etching or sputtering yield), and k_x are the stoichiometric factors, which describe how much of one material, compared to its reactant, is needed to form the reaction product. Sticking probabilities and coverages are bound to the range $[0, 1]$, where 1 stands for a fully covered surface, or fully balanced polymer by etchant in the case of ϕ_{pe}, since this coverage is normalized to ϕ_p. Δ_p describes the amount of material needed to advance the surface through deposition. The first terms in Euqaitons (7)–(9) describe the incoming flux adsorbed onto the surface, where the sticking coefficient describes the probability of adsorption. The second terms in the above equations are proportional to the ion flux and describe the loss of particles through ion enhanced etching, which may remove all types of particles from the surface. However, the removal of material through evaporation or chemical etching is only considered for the etchant species, since it reacts to form compounds, which, by definition, have a much lower binding energy to the surface than other materials present on the surface [74].

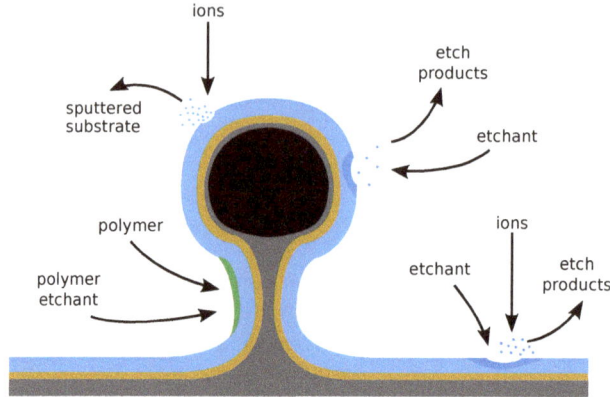

Figure 11. Four physical processes, which are considered when describing a modern plasma etch process used for gate stack pattering. Passivating species form polymer layers on sidewalls, as there are fewer energetic ion impacts to remove them from these surfaces. Ion sputtering removes material from the substrate by physical sputtering without the involvement of a chemical etchant. Purely chemical etching removes material by forming volatile etch products, which then desorp from the surface. Ion-enhanced etching speeds up this process by breaking existing bonds, enhancing the formation of volatile etch products.

The change in surface coverages must include all relevant mechanisms which add or remove particle types from the surface. Since mass cannot be lost, the number of particles arriving and departing from the surface must balance [75]. Therefore, the surface coverages reach a steady state with respect to etching and deposition time scales almost instantaneously.

Considering the polymer coverage, as shown in Equation (8), the first term is directly proportional to the deposition rate DR_p of the polymer,

$$DR_p = \frac{1}{\rho_p} \left(J_p S_p \right),$$ (10)

while the second term is proportional to the etch rate ER_p of the polymer,

$$ER_p = \frac{1}{\rho_p} \left(J_i Y_p \phi_p \phi_{pe} \right).$$ (11)

If the etch and deposition rates balance perfectly, no polymer will be deposited since $\Delta_p = 0$ and there is no material left for deposition. If more particles impinge than are removed through ion-enhanced etching, the additional material will deposit onto the surface, advancing the surface at $\vec{x}$ by a velocity

$$v(\vec{x}) = \frac{\Delta_p}{\rho_p} = \frac{1}{\rho_p} \left(J_i Y_p \phi_{pe} - J_p S_p \right).$$ (12)

In Equation (12), ρ_p represents the density of the polymer, and ϕ_p is ignored as it must be unity if the entire surface is covered by the polymer. Since ϕ_{pe} can be found easily from Equation (9), the surface normal velocity field $v(\vec{x})$ necessary to advance a surface, as described in Section 2.1.2, can be found straightforwardly.

The same field can be constructed using Equation (12), if etching dominates, provided the underlying material is the polymer. If all the polymer has been removed and the substrate is etched, different effects must be taken into account, which depend on the specific etch chemistries used. Assuming that the substrate can be removed by chemical etching, ion-enhanced etching, and physical ion sputtering, as is the case for many plasma chemistries, the normal velocity of the surface at $\vec{x}$ is expressed as:

$$v(\vec{x}) = \frac{1}{\rho_{sub}} \left[\underbrace{J_{ev}\phi_e}_{\text{chemical etching}} + \underbrace{J_i Y_{ie}\phi_e}_{\text{ion-enhanced etching}} + \underbrace{J_i Y_s(1 - \phi_e)}_{\text{ion sputtering}} \right]. \tag{13}$$

The first two terms in Equation (13) describe loss mechanisms, which also remove etchant species from the surface coverage and therefore also appear in Equation (7), where the amount of etchant needed to remove a unit of the substrate is captured in the stoichiometric factors, k_x. The etchant coverage, ϕ_e, depends on the other coverages in this case, as all of them are active and can be found by solving Equations (7)–(9). Therefore, surface normal speeds for polymer deposition and etching, Equation (12), as well as substrate etching, Equation (13), can be found using only the incoming particle fluxes, J_i, J_e, and J_p.

Additionally, each etching mechanism can be described more accurately by considering certain dependencies, such as ion-enhanced etching or ion sputtering, which depend strongly on the energy and incoming angle of ions [76]. The choice of coefficients for each process is the most crucial step and usually requires data from experiments, other reactor-scale or ab initio simulations, or a combination of both. A model encapsulating a large number of etch mechanisms could, in theory, describe numerous different processes. The choice of sticking probabilities, etch yields, and stoichiometric factors constitutes the only differentiating property for a variety of process models. They are therefore fitting parameters, which have a basis in the physical and chemical surface reactions taking place. Model calibration is therefore one of the most important and time-consuming parts in process simulation.

Any physical effect modifying the number of molecular entities arriving on the surface could be included, adding more coefficients and thus increasing the complexity, although simple models are frequently sufficient, even when describing complex processes [77]. It is therefore crucial to identify the dominant physical processes for each chemistry in order to make sure they are adequately represented in the model, as some effects might dominate the behavior of one chemistry, while they can be neglected in others. The model described above can be considered as an illustration of how such a model might be set up and does not describe all potentially relevant physical effects exhaustively. At the same time, some chemistries might already be described well in a simpler model, such as pure chlorine etching, where the effect of ions might be negligible compared to the chemical etching properties. The dominant mechanisms during each patterning step of gate stacks, the required modeling techniques, and their dependence on the basic concepts introduced earlier are discussed in the next section.

3. Simulation Software

Several proprietary and open-source simulation frameworks with the capability to describe the complex processes occurring in gate stack patterning, are readily available. Some well-known frameworks and the numerical methods on which they are based, are discussed here briefly:

Sentaurus Topography [78] is a commercial simulator developed by *Synopsys* (Mountain View, CA, USA), which uses level set based surface descriptions for topography changes, cell based representations for chemical surface reactions, and provides Monte Carlo methods for particle transport [79]. Dunn et al. [80] successfully used this tool to simulate the fabrication of FinFET structures of the 7 nm node. The Florida Object Oriented Process Simulator (FLOOPS) [81] provides similar capabilities. It was incorporated into Sentaurus, but a version is also available as an open-source project.

Victory Process [82] is a proprietary process simulator distributed by *Silvaco* (Santa Clara, CA, USA). It allows level set surface descriptions, as well as explicit surfaces to be used. Nanda et al. [83] were able to simulate the fabrication of strained FinFETs using this framework. Victory Cell [84] is a related tool, which uses cell based and explicit surface representations in order to improve the description of ion implantation and diffusion. It was used by Maiti et al. [85] to simulate the fabrication of stressed FinFETs.

ViennaTS [86] is an open-source feature scale process simulation tool developed at the *Institute for Microelectronics, TU Wien*. Surfaces are represented using level sets and top-down, as well as bottom-up methods, are implemented. The software provides predefined etch and deposition models, including several for the simulation of advanced node etching processes [77].

The Monte Carlo Feature Profile Model (MCFPM) [87] is one of several software components developed at the *Computational Plasma Science and Engineering Group, University of Michigan*. Cell based methods are used to describe different materials and top-down approaches are used to describe particle transport. Combined with other software components, it was used by Huard et al. [88] to simulate the fabrication of advanced-node FinFETs.

Phietch [89] and K-Speed[90] are also widely used simulation frameworks, while SEMulator3D [91] provides a framework for process emulation. University and open source tools, such as ViennaTS and MCFPM, usually provide the underlying methods, algorithms and implemented models of the framework. Commercial tools usually do not disclose these. Therefore, the methods discussed in this review are primarily based on studies performed with open academic tools.

4. Plasma Chemistries for Gate Stack Etching

The etching sequence of a gate stack used for advanced technology nodes of 14 nm and below consists of several highly different processes with unique properties and etch mechanisms due to the different materials used in the gate stacks [92]. It includes highly anisotropic dry etch processes, as well as highly selective or isotropic ones, depending on the different materials included in the gate stack [93]. Wet etch processes have generally fallen out of favor in advanced node gate stack patterning due to additional cleaning steps required to remove residues left on the wafer after wet etching. Furthermore, their isotropic etch properties are not ideal for the high vertical etching accuracy needed for modern three-dimensional structures [94]. Therefore, widely adopted process flows in industry today rely on dry plasma etch processes [95].

Even though two processes may have similar etch properties, the underlying mechanics might differ completely, leading to diverse effects in complex geometries or chemically different environments. In the following, the applicability and reliability of the earlier introduced concepts to these real processes will be discussed in reference to advanced node metal gate stacks. These metal gate stack geometries, shown in Figure 12, usually consist of a thin layer of high-k dielectric, such as hafnium dioxide, a contact metal, such as titanium nitride, and poly-Si [94,96]. These layers cover the conducting silicon channel, on top of an insulating silicon dioxide substrate. In order to achieve better switching characteristics, the contact area between the gate and the conducting channel should be maximized An established approach to achieving these high contact areas, while reducing the footprint on the wafer, is using three-dimensional structures, such as the ones shown in Figure 12: a trigate, where the channel is as thin and high as possible [97], and an Ω-gate [98], where the channel is almost completely surrounded by the gate. The variety of the incorporated materials leads to the need for a sophisticated and carefully tuned etch sequence, removing each layer without damaging masked regions or the layers below. Gate all-around (GAA) structures and stacked nanowire gates, which achieve a gate contact around the full circumference of the channel, are researched heavily as a promising improvement on the finFET and omega gate structures [99]. Many different approaches exist, some incorporating a variety of new materials. Therefore, the specific materials and the fabrication techniques, which gate all-around structures might incorporate, are not discussed here. However, many of the current etch techniques will likely also be applicable to the fabrication of GAA structures [100].

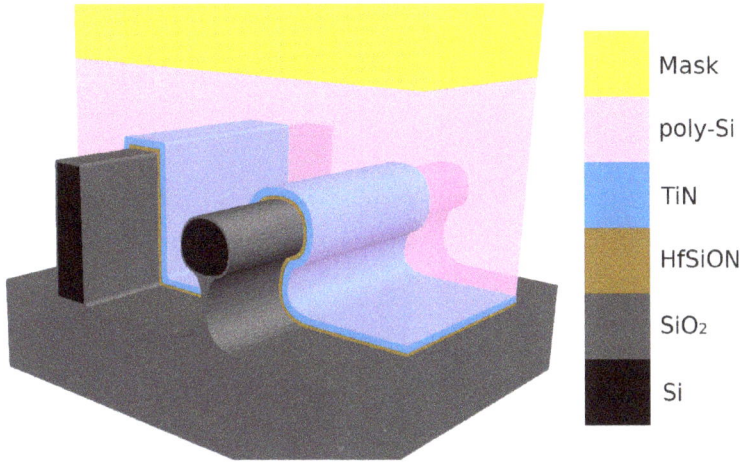

Figure 12. Schematic depiction of multi-layered geometries of modern three-dimensional gate structures after gate etching. A trigate (left) will have three sides of the Si-channel accessible to the gate, while an Ω-gate (right) comes close to an all-around gate structure.

4.1. Silicon Etching

As silicon is the most important material in semiconductor fabrication, many different etching chemistries have been investigated in the past several decades [101,102]. In the following, the three most common chemistries for directional dry etching of silicon are described and their etching mechanics as well as possible use in a gate stack patterning sequence is discussed.

4.1.1. CF Type Chemistries

Fluorocarbon (CF) chemistries have been used to etch Si and SiO_2 for decades, due to the ability of fine tuning of different materials' etch rates, thereby improving selectivity [103]. The use of several additive gases allows for the etch processes characteristics to change drastically [104], which enables the adjustment to better fit a variety of substrates and geometries. In this manner, a high etch selectivity can be reached for certain materials, in addition to the highly anisotropic properties of plasma etch processes [105]. Fluorocarbon plasmas can etch chemically, via ion-enhanced etching or physical sputtering, which is shown in Figure 13. Silicon is removed chemically by reacting with fluoride leading to the etch products evaporating back to the gas phase [106]. The rate of chemical etching depends on the temperature and is reduced by carbon atoms present on the surface. Ion-enhanced etching proceeds through the bombardment of radical CF^+ ions reacting with the substrate, which are either sputtered from the surface or evaporate due to their now smaller binding energy to the surface [107]. Physical sputtering appears only above a threshold ion energy, which is related to the binding energy of the substrate. Deposition takes place through the polymerization of neutrals, covering the surface by forming SiC bonds, or through direct ion deposition, where energetic ions are directly absorbed into the substrate [108]. Thus, most active mechanisms in this chemistry can be described by the illustrative model given in Section 2.4. Only direct ion implantation cannot be included when using level sets, as there is no volume information. However, it can be included in a model when using explicit volume definitions or voxel elements.

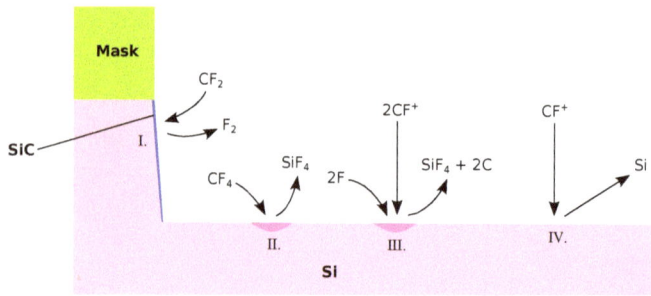

Figure 13. Active etching and deposition mechanics in CF type chemistries used to etch poly-Si: I. chemical deposition of carbon forming an SiC passivation layer, II. chemical etching, III. ion-enhanced etching, and IV. ion sputtering through high energy ions.

4.1.2. SF Type Chemistries

Sulfur Hexafluoride (SF_6) is a good alternative to CF chemistries because of their high etch rates and due to the fine control over the etch properties using additional gases fed into the reactor [109]. Pure SF_6 chemistries etch isotropically, while the addition of oxygen (O) forms a thin silicon oxide passivation layer, which inhibits lateral etching [110]. Oxygen also binds sulfur, prohibiting the recombination with fluoride, more of which is then available for etching, resulting in higher vertical etch rates [111]. However, if the O concentration is too high, it competes for surface adsorption with F, reducing the etch rate [112]. Etching proceeds only on lateral surfaces due to the ion bombardment preventing the buildup of a passivation layer. Fluoride atoms can then attach to silicon atoms on the surface, forming SiF_4, which is then removed chemically or through ion-enhanced processes [113]. If only oxygen is used as an additional gas, the physical processes can be described straightforwardly. However, the introduction of additional gases can create more complex properties. For example, the addition of hydrogen bromide (HBr) and oxygen results in better sidewall passivation and less lateral etching, due to the formation of a SiO_xBr_y passivation layer, which reflects high energy ions, further enhancing the vertical etch rate [114]. This additional interplay of different chemical compounds has to be considered carefully when developing a model and cannot be represented with such a simple description as used in Section 2.4. Passivating species other than oxygen, such as CH_2F_2, have been used successfully to etch silicon [115]. The formation of sidewall passivation in such a chemistry is not due to the deposition from the gas phase, but rather due to sputtering of CF etch products from vertical etching and line of sight deposition [116], shown in Figure 14. Modeling this process requires an additional ray tracing step which could be realized by launching rays from a surface element, if a certain combination of etchant and ion coverage is reached. The additional ray tracing required decreases computational efficiency, but is indispensable in order to model the shadowing effects expected in complex geometries. Approximating the build-up of the passivating layer with deposition from the gas phase is only sufficient in simple geometries [77]. Hence, the modeling of most modern gate geometries require additional ray tracing for an accurate description of the deposition mechanism.

4.1.3. HBr Type Chemistries

When etching a layer of the gate stack, care must be taken in order not to damage the layer below [117]. Poly-Si etching in a metal gate stack is therefore often concluded with a more selective HBr chemistry, despite its lower etch rate [118]. Even before metal gate stack designs, HBr/O_2 chemistries were used due to the high selectivity against the gate oxide, used as a dielectric [119,120]. This chemistry is quite simple as ion-enhanced etching typically dominates, meaning other etching mechanisms can be ignored, while deposition proceeds mainly chemically [114]. As shown in Figure 15, thick SiO_xBr_y layers form on sidewalls and the gate oxide, protecting it from energetic ions. Over the course of the etch process, bromine is removed from the passivation layer and replaced by oxygen, resulting in a denser silicon oxide layer on top of the passivation layers formed earlier in the process compared to an amorphous, bromine rich layer at the side walls formed later [121]. Modeling this desorption of bromine from the surface is not easily achievable, as the densening depends on the fraction of bromine in the passivation layer, which is a volume property, making direct simulation impractical when using implicit surfaces. Since a cleaning step often follows, it might be sufficient to only model the final oxide layer using direct deposition [122].

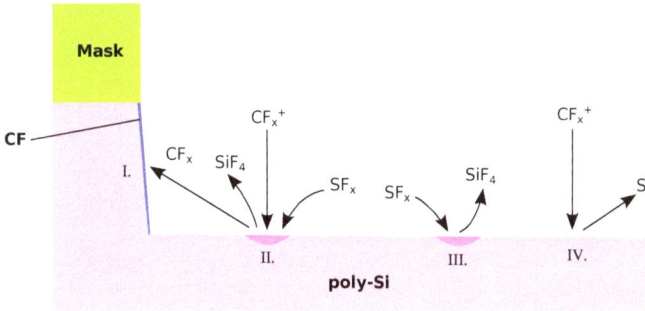

Figure 14. Sulfur with Fluoride (SF) type etching and deposition mechanics with additional CH_2F_2 feed gas. I. Line of sight deposition of a CF passivation layer; II. ion-enhanced etching; III. chemical etching; and IV. physical ion sputtering.

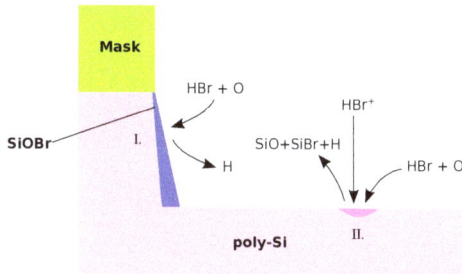

Figure 15. Dominant etch mechanics during silicon etching using hydrogen bromide. I. Sidewall passivation proceeds though chemical deposition from the gas phase, while II. vertical etching is dominated by ion-enhanced etching.

4.2. TiN Etching

Titanium has been used in microelectronics as a mask, as well as an interconnect layer, for decades. Therefore, appropriate etch chemistries were investigated long before its introduction in metal gate stacks [123,124]. Due to the high temperatures required for higher etch rates in fluorinated plasmas, chlorine or bromine based chemistries are applicable more universally [125]. TiN is removed predominantly via ion-enhanced etching [126], forming unstable TiCl/TiBr etch products, which quickly react to other, more stable compounds such as TiO [127] in the presence of oxygen. Fast oxidation of the Ti surface can form an etch stop layer due to the much lower etch rate of TiO_2 [128]. It is therefore important to reduce the amount of oxygen when etching titanium compounds [129], despite the increased etch rate of passivation layers. However, small amounts of oxygen result in a higher fraction of ionized etchant in the gas phase, increasing the etch rates and selectivity below a concentration of 1%. Bromine etches titanium significantly slower than chlorine due to its lower volatility, while HBr can be added to Cl chemistries to achieve better control over certain etch properties [126]. Different additives, such as boron (B) can also be included, which forms non-volatile BO_xN_y polymers chemically, protecting TiN sidewalls and further reducing lateral etching [130]. Due to the relative simplicity of the etching and deposition mechanics, simulation of TiN etching is possible with the model described earlier. However, care must be taken when experimenting with feed gases other than the ones described, as they might change the underlying mechanics drastically. The models would need to be adjusted accordingly and are not predictable across different feed gases, due to the potential changes in the underlying mechanics.

4.3. HfO₂ Etching

Since the dielectric layer is usually very thin, high etch rates are not as important as high selectivity to the silicon substrate and the other materials in the gate stack. Fluorine based chemistries have been shown to reach satisfactory selectivity and etch rates [131], although CF based chemistries form thick fluorocarbon layers, requiring an additional cleaning step after etching, which is not the case for SF chemistries [132]. The thickness of these layers could be reduced by adding hydrogen, which removes carbon from the surface via loosely bound hydrocarbon etch products [133], although even these fewer fluorocarbon residues on the substrate still present a problem [134].

Such problems are not encountered in HBr or Cl based chemistries, due to the surface bonding energies of their etch products [135]. However, due to the small thickness of the HfO_2 layer, a high dielectric to substrate selectivity is necessary in order to meet the requirements for transistors, while these chemistries only offer selectivities of around 15, even when carbon is added as an inhibitor [136]. BCl was initially added to pure Cl chemistries to achieve higher selectivities [137], while pure BCl_3 chemistries were soon chosen due to their near-infinite selectivity to the substrate. This is due to the favored reaction of the present boron with oxygen and silicon over hafnium, forming SiB compounds on the surface of the substrate, but no boron layers on Hf, which is shown in Figure 16. Therefore, the surface of the dielectric is free for chlorine to attach and form volatile etch products, removed via ion-enhanced etching, while the silicon and silicon dioxide surfaces are covered in an SiB layer, prohibiting the further adsorption of etching species. This layer is only removed via high-energy ions, so near-infinite etch selectivities towards Si and SiO_2 can be observed at low bias powers [138]. Although BCl_3 etching is mainly enabled by ion-enhanced etching, the strong dependence on ion energy with respect to selectivity, requires careful fitting of model parameters to the observed rates.

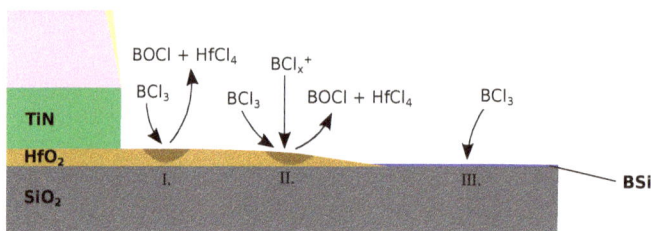

Figure 16. Etch mechanisms leading to infinite etch selectivity of HfO_2 over SiO_2 through I. chemical and II. ion-enhanced etching of HfO_2, with III. additional chemical deposition of BSi on the SiO_2 substrate.

4.4. Full Etch Sequence

Since several of the etch steps described above involve the deposition of passivation layers to protect side walls, the subsequent etch steps can be affected by these residues. Therefore, simulating the combination of all these processes can provide an insight into the full process, including the effects of each etch step on the subsequent ones. As we [77] have documented using the simulator ViennaTS, the final profile of a 14 nm fully depleted silicon-on-insulator (FDSOI) gate stack patterning process is indeed strongly influenced by earlier etch steps, which involve similar chemistries to the ones described above. This influence is shown in Figure 17, which highlights the role of deposited passivation layers for subsequent etch steps in protecting the otherwise exposed poly-Si sidewalls. While the poly-Si is protected by the earlier deposited SiBr and CF layers, the HfO_2 etch step creates an under-etch in the TiN layer since it is not protected. After the final etch step, only remnants of the protecting layers remain, which are removed in a subsequent cleaning step.

Figure 17 also shows why the ability to represent very thin layers is important when modeling gate stacks, since several materials with highly different etch properties must be considered in these simulations. Even the thin layer of CF, deposited in the first etch step, can protect the poly-Si in the last step through its much lower etch rates. Thus, the correct representation of thin layers by the choice of adequate surface description techniques is crucial in this application.

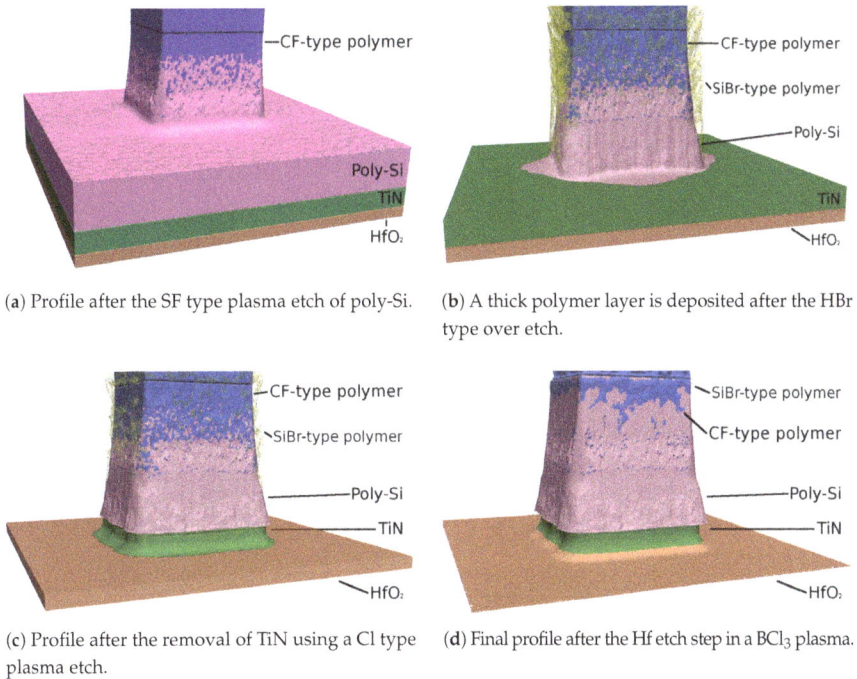

(**a**) Profile after the SF type plasma etch of poly-Si.

(**b**) A thick polymer layer is deposited after the HBr type over etch.

(**c**) Profile after the removal of TiN using a Cl type plasma etch.

(**d**) Final profile after the Hf etch step in a BCl_3 plasma.

Figure 17. Simulated gate stack geometry after each etch step, showing the influence of passivation layers on subsequent etch steps. The poly-Si sidewalls remain protected during the more isotropic later etch steps, which leads to the under-etch in the TiN and Hf layers.

Due to the small layer thickness, which might be comparable to a few atomic sizes, the thin layers do not have sharp material boundaries, but rather consist of a mix of materials. Therefore, the simulated thin layers should be seen as a rough guide to the amount of material present at the interface, rather than abrupt changes in composition. In order to circumvent these sharp boundary representations, rectilinear explicit meshes, as described in Section 2.1.3, can be used to represent single blocks of compounds at the interface. Huard et al. [88] successfully used the MCFPM based on this approach, as described in Section 3, to simulate the gate etching of finFET structures. They compared the profiles created by continuous etching and self-limited atomic layer etching of silicon using ion-enhanced chlorine chemistries. Diffusion and ion-implantation are taken into account, enabled by the cell based surface description. The difference in the resulting layers to level set based simulations can be seen clearly in Figure 18, which shows the thin passivation layers as a mixture of different materials, rather than a single strictly defined layer. However, due to the fixed size of each cell, sharp material boundaries can only be avoided if the cell size is comparable to atomic sizes. Modeling at these scales also enables the inclusion of additional interactions between cells, creating a better physical description. However, the only way to truly simulate the physical behavior is atomistic modeling, which takes into account single atoms and bonds.

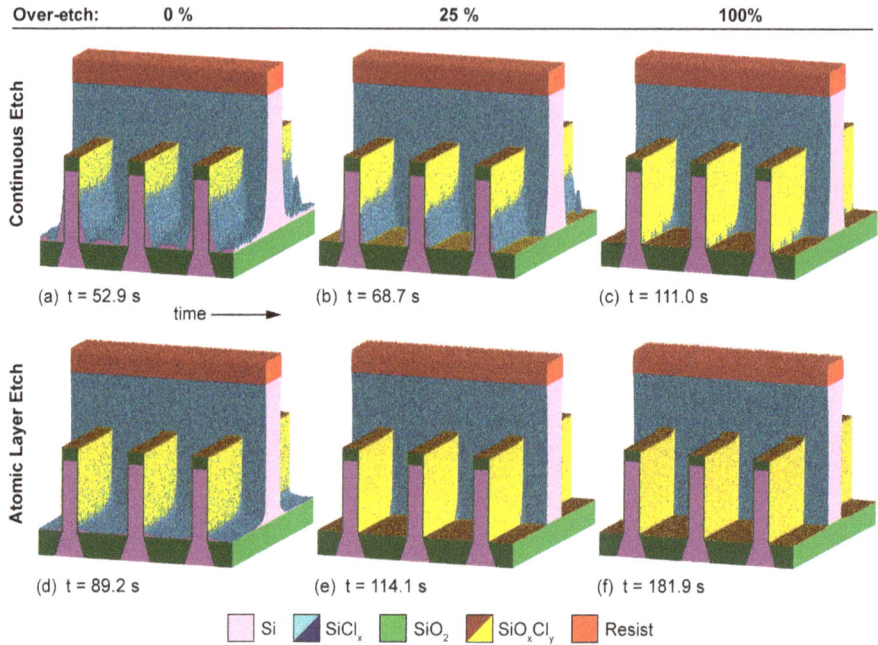

Figure 18. Profiles resulting from continuous Cl etching (**a–c**) and self-limited atomic layer etching using the same chemistry (**d–f**); used with publisher's permission from [88].

5. Conclusions

Modern gate stack patterning processes require sophisticated simulation techniques, starting from the fundamental computational methods used, due to the numerous new materials introduced into the gate stack. In order for complex deformations, as expected in modern dry etch processes, to be represented efficiently, several surface representations are considered for different purposes. While moving surfaces are described more robustly using implicit surface representations, such as level sets or cell based methods, ray tracing techniques used for flux calculations are better suited for explicit surface representations. However, the level set method is still the most widely used technique to describe interfaces in process simulations, due to its robustness during complex deformations. Furthermore, approximate explicit surface descriptions using disks or spheres eliminate the need for computationally expensive conversions to triangulated surfaces for rate calculations with ray tracing techniques.

Robust surface descriptions and physically meaningful rate calculation methods made recent advances in the modeling of gate stack etching processes possible. The correct representation of thin layers, often formed in such processes, contribute to the robust description of sequential etching steps, while ray tracing techniques allow for accurate and physically more meaningful modeling of the transport of molecular entities to the etched surface.

Furthermore, advances in the identification of the dominant etching and deposition mechanics allow for more physical representations of complex etch chemistries in reactor- and feature-scale simulations. The main factors defining the properties of different dry etch techniques and chemistries can thus be identified and incorporated into universal models, allowing for the physically meaningful description of many modern etch processes. Even if the dominant mechanics are identified, some physical processes might not be described adequately in certain surface representations, such as ion-implantation using level sets. Due to the lack of volume information in level set surfaces, ion-implantation cannot be modeled straightforwardly. Explicit or cell based methods are more accurate in describing volume dependent processes, but require more sophisticated methods to advance the surface.

One of the biggest challenges faced today is the synthesis of predictive models over a wide range of reactor geometries, etch chemistries, and feature scale profiles. Due to the complexity of the chemical processes taking place inside the reactor, minor changes in process parameters can result in drastic changes in the etch profile. While process simulation already provides an understanding of the dominant chemical processes by comparison with experiments, the prediction of results requires highly sophisticated surface descriptions, as well as chemical models. Advances in computer performance and numerical models will enable atomistic approaches, where each atom or molecule is described during etch processes. This level of sophistication will likely enable the development of truly predictive models, reducing the need for expensive experiments in order to create advanced node structures.

Author Contributions: Conceptualization, X.K., S.S. and L.F.; Methodology, X.K., S.S. and L.F.; Software, X.K. and L.F.; Validation, X.K. and L.F.; Formal Analysis, X.K. and L.F.; Investigation, X.K.; Resources, S.S. and L.F.; Data Curation, X.K.; Writing-Original Draft Preparation, X.K.; Writing-Review and Editing, S.S. and L.F.; Visualization, X.K.; Supervision, S.S. and L.F.; Project administration, S.S. and L.F.; Funding acquisition, S.S. and L.F.;

Acknowledgments: The research leading to these results has received funding from the European Union's Horizon 2020 research and innovation programme under Grant No. 688101 SUPERAID7.

Conflicts of Interest: The authors declare no conflict of interest.

References

1. Moore, G.E. Progress in digital integrated electronics. [Technical Digest—International Electron Devices Meeting, IEDM, 1975, Vol. 21, pp. 11–13.] *IEEE Solid-State Circuits Soc. Newsl.* **2006**, *11*, 36–37. [CrossRef]
2. Theis, T.N.; Wong, H.S.P. The end of Moore's Law: A new beginning for information technology. *Comput. Sci. Eng.* **2016**, *19*, 41–50. [CrossRef]
3. Lee, B.H.; Song, S.C.; Choi, R.; Kirsch, P. Metal electrode/high-k dielectric gate-stack technology for power management. *IEEE Trans. Electron Devices* **2008**, *55*, 8–20. [CrossRef]
4. Robertson, J. Band offsets of wide-band-gap oxides and implications for future electronic devices. *J. Vac. Sci. Technol. B Microelectron. Nanometer Struct.* **2000**, *18*, 1785–1791. [CrossRef]
5. Mistry, K.; Allen, C.; Auth, C.; Beattie, B.; Bergstrom, D.; Bost, M.; Brazier, M.; Buehler, M.; Cappellani, A.; Chau, R.; et al. A 45nm logic technology with high-k+metal gate transistors, strained silicon, 9 Cu interconnect layers, 193nm dry patterning, and 100% Pb-free packaging. In Proceedings of the 2007 IEEE International Electron Devices Meeting, Washington, DC, USA, 10–12 December 2007; pp. 247–250. [CrossRef]
6. Lee, J.H.; Kim, D.G.; Lee, H.J.; Hwang, C.S. Fabrication of a nano-scaled tri-gate field effect transistor using the step-down patterning and dummy gate processes. *Microelectron. Eng.* **2017**, *173*, 33–41. [CrossRef]
7. Pfiester, J.R.; Baker, F.K.; Mele, T.C.; Tseng, H.H.; Tobin, P.J.; Hayden, J.D.; Miller, J.W.; Gunderson, C.D.; Parrillo, L.C. The effects of boron penetration on p+ poly silicon gated PMOS devices. *IEEE Trans. Electron Devices* **1990**, *37*, 1842–1851. [CrossRef]
8. Hobbs, C.C.; Fonseca, L.R.; Knizhnik, A.; Dhandapani, V.; Samavedam, S.B.; Taylor, W.J.; Grant, J.M.; Dip, L.R.G.; Triyoso, D.H.; Hegde, R.I.; et al. Fermi-level pinning at the polysilicon/metal oxide interface—Part I. *IEEE Trans. Electron Devices* **2004**, *51*, 971–977. [CrossRef]
9. Lin, Y.S.; Huang, K.W.; Lin, H.C.; Chen, M.J. Effective work function modulation of the bilayer metal gate stacks by the Hf-doped thin TiN interlayer prepared by the in-situ atomic layer doping technique. *Solid State Commun.* **2017**, *258*, 49–53. [CrossRef]

10. Huang, K.W.; Cheng, P.H.; Lin, Y.S.; Wang, C.I.; Lin, H.C.; Chen, M.J. Tuning of the work function of bilayer metal gate by in-situ atomic layer lamellar doping of AlN in TiN interlayer. *J. Appl. Phys.* **2017**, *122*, 095103. [CrossRef]

11. Kim, C.K.; Ahn, H.J.; Moon, J.M.; Lee, S.; Moon, D.I.; Park, J.S.; Cho, B.J.; Choi, Y.K.; Lee, S.H. Temperature control for the gate workfunction engineering of TiC film by atomic layer deposition. *Solid-State Electron.* **2015**, *114*, 90–93. [CrossRef]

12. Xiang, J.; Ding, Y.; Du, L.; Li, J.; Wang, W.; Zhao, C. Growth mechanism of atomic-layer-deposited TiAlC metal gate based on TiCl 4 and TMA precursors. *Chin. Phys. B* **2016**, *25*, 037308. [CrossRef]

13. Hayes, M.H.; Dezelah, C.L.; Conley, J.F. Properties of Annealed Atomic-Layer-Deposited Ruthenium from Ru(DMBD)(CO) 3 and Oxygen. *ECS Trans.* **2018**, *85*, 743–749. [CrossRef]

14. *The International Technology Roadmap for Semiconductors 2.0*; The Optical Society: Washington, DC, USA, 2015.

15. *The International Roadmap for Devices and Systems*; The Optical Society: Washington, DC, USA, 2017.

16. Auth, C.; Allen, C.; Blattner, A.; Bergstrom, D.; Brazier, M.; Bost, M.; Buehler, M.; Chikarmane, V.; Ghani, T.; Glassman, T.; et al. A 22nm high performance and low-power CMOS technology featuring fully-depleted tri-gate transistors, self-aligned contacts and high density MIM capacitors. In Proceedings of the 2012 Symposium on VLSI Technology (VLSIT), Honolulu, HI, USA, 12–14 June 2012; pp. 131–132. [CrossRef]

17. Memos, G.; Lidorikis, E.; Kokkoris, G. The interplay between surface charging and microscale roughness during plasma etching of polymeric substrates. *J. Appl. Phys.* **2018**, *123*, 073303. [CrossRef]

18. Bär, E.; Lorenz, J. 3-D simulation of LPCVD using segment-based topography discretization. *IEEE Trans. Semicond. Manuf.* **1996**, *9*, 67–73. [CrossRef]

19. Thurgate, T. Segment-based etch algorithm and modeling. *IEEE Trans. Comput.-Aided Des. Integr. Circuits Syst.* **1991**, *10*, 1101–1109. [CrossRef]

20. Pauly, M.; Gross, M.; Kobbelt, L.P. Efficient simplification of point-sampled surfaces. In Proceedings of the IEEE Visualization VIS 2002, Boston, MA, USA, 27 October–1 Novmber 2002; pp. 163–170. [CrossRef]

21. Law, M. Grid adaption near moving boundaries in two dimensions for IC process simulation. *IEEE Trans. Comput.-Aided Des. Integr. Circuits Syst.* **1995**, *14*, 1223–1230. [CrossRef]

22. Law, M.E.; Cea, S.M. Continuum based modeling of silicon integrated circuit processing: An object oriented approach. *Comput. Mater. Sci.* **1998**, *12*, 289–308. [CrossRef]

23. Bloomenthal, J.; Wyvill, B.; Bajaj, C. *Introduction to Implicit Surfaces*; Morgan Kaufmann Publishers: San Francisco, CA, USA, 1997.

24. Alliez, P.; de Verdire, E.; Devillers, O.; Isenburg, M. Isotropic surface remeshing. In Proceedings of the 2003 Shape Modeling International, Seoul, Korea, 12–15 May 2003; IEEE Computer Society: Washington, DC, USA, 2003; Volume 2003, pp. 49–58. [CrossRef]

25. Sethian, J. *Level Set Methods and Fast Marching Methods: Evolving Interfaces in Computational Geometry, Fluid Mechanics, Computer Vision, And Materials Science*; Cambridge University Press: Cambridge, UK, 1999.

26. Chacon, A.; Vladimirsky, A. Fast two-scale methods for eikonal equations. *SIAM J. Sci. Comput.* **2012**, *34*, A547–A578. [CrossRef]

27. Adalsteinsson, D.; Sethian, J.A. The fast construction of extension velocities in level set methods. *J. Comput. Phys.* **1999**, *148*, 2–22. [CrossRef]

28. Mauch, S. *A Fast Algorithm for Computing the Closest Point and Distance Transform*; Unpublished Technical Report; California Institute of Technology: Pasadena, CA, USA, 2000; pp. 1–17.

29. Zhao, H.K. Fast sweeping method for eikonal equations. *Math. Comput.* **2005**, *74*, 603–627. [CrossRef]

30. Sethian, J.A. A fast marching level set method for monotonically advancing fronts. *Proc. Natl. Acad. Sci. USA* **1996**, *93*, 1591–1595. [CrossRef] [PubMed]

31. Osher, S.; Sethian, J.A. Fronts propagating with curvature-dependent speed: Algorithms based on hamilton-jacobi formulations. *J. Comput. Phys.* **1988**, *79*, 12–49. [CrossRef]

32. Osher, S.; Shu, C.W. High-order essentially nonoscillatory schemes for Hamilton–Jacobi equations. *SIAM J. Numer. Anal.* **1991**, *28*, 907–922. [CrossRef]

33. Whitaker, R.T. A level-set approach to 3D reconstruction from range data. *Int. J. Comput. Vis.* **1998**, *29*, 203–231. [CrossRef]

34. Hsiau, Z.K.; Kan, E.C.; McVittie, J.P.; Dutton, R.W. Robust, stable, and accurate boundary movement for physical etching and deposition simulation. *IEEE Trans. Electron Devices* **1997**, *44*, 1375–1385. [CrossRef]

35. Manstetten, P.L. Efficient Flux Calculations for Topography Simulation. Doctoral Thesis, Institute for Microelectronics, TU Wien, Austria, 2018.

36. Pasko, A.; Adzhiev, V.; Sourin, A.; Savchenko, V. Function representation in geometric modeling: Concepts, implementation and applications. *Vis. Comput.* **1995**, *11*, 429–446. [CrossRef]

37. Adalsteinsson, D.; Sethian, J.A. A fast level set method for propagating interfaces. *J. Comput. Phys.* **1995**, *118*, 269–277. [CrossRef]

38. Ertl, O.; Selberherr, S. A fast level set framework for large three-dimensional topography simulations. *Comput. Phys. Commun.* **2009**, *180*, 1242–1250. [CrossRef]

39. Strasser, E.; Selberherr, S. Algorithms and models for cellular based topography simulation. *IEEE Trans. Comput.-Aided Des. Integr. Circuits Syst.* **1995**, *14*, 1104–1114. [CrossRef]

40. Fujinaga, M.; Kotani, N. 3-D topography simulator (3-D MULSS) based on a physical description of material topography. *IEEE Trans. Electron Devices* **1997**, *44*, 226–238. [CrossRef]

41. Zhang, Y.; Huard, C.; Sriraman, S.; Belen, J.; Paterson, A.; Kushner, M.J. Investigation of feature orientation and consequences of ion tilting during plasma etching with a three-dimensional feature profile simulator. *J. Vac. Sci. Technol. A Vac. Surf. Films* **2017**, *35*, 021303. [CrossRef]

42. Huard, C.M. Nano-Scale Feature Profile Modeling of Plasma Material Processing. Ph.D. Thesis, University of Michigan, Ann Arbor, MI, USA, 2018.

43. Toh, K.; Neureuther, A.; Scheckler, E. Algorithms for simulation of three-dimensional etching. *IEEE Trans. Comput.-Aided Des. Integr. Circuits Syst.* **1994**, *13*, 616–624. [CrossRef]

44. Zhou, Z.F.; Huang, Q.A.; Li, W.H.; Lu, W. A novel 3-D dynamic cellular automata model for photoresist-etching process simulation. *IEEE Trans. Comput.-Aided Des. Integr. Circuits Syst.* **2007**, *26*, 100–114. [CrossRef]

45. Gosalvez, M.A.; Xing, Y.; Sato, K.; Nieminen, R.M. Atomistic methods for the simulation of evolving surfaces. *J. Micromech. Microeng.* **2008**, *18*. [CrossRef]

46. Ertl, O. Numerical Methods for Topography Simulation. Doctoral Thesis, Institute for Microelectronics, TU Wien, Austria, 2010.

47. Schröpfer, G.; King, D.; Kennedy, C.; Mcnie, M. Advanced process emulation and circuit simulation for co-design of MEMS and CMOS devices. In Proceedings of the Design, Test, Integration and Packaging of MEMS and MOEMS, Montreux, Switzerland, 1–3 June 2005.

48. Schropfer, G.; McNie, M.; da Silva, M.; Davies, R.; Rickard, A.; Musalem, F.X. Designing manufacturable MEMS in CMOS-compatible processes: Methodology and case studies. In *MEMS, MOEMS, and Micromachining, Proceedings of the International Society for Optics and Photonics, Strasbourg, France, 26–30 April 2004*; SPIE Europe Ltd.: Cardiff, Wales, 2004; Volume 5455, pp. 116–128. [CrossRef]

49. Vianne, B.; Morin, P.; Beylier, C.; Giraudin, J.C.; Desmoulins, S.; Gonella, R.; Juncker, A.; Fried, D. Investigations on contact punch-through in 28 nm FDSOI through virtual fabrication. In Proceedings of the 2017 IEEE SOI-3D-Subthreshold Microelectronics Technology Unified Conference (S3S), Burlingame, CA, USA, 16–19 October 2017; Volume 2018, pp. 1–2. [CrossRef]

50. Franke, J.H.; Gallagher, M.; Murdoch, G.; Halder, S.; Juncker, A.; Clark, W. EPE analysis of sub-N10 BEOL flow with and without fully self-aligned via using Coventor SEMulator3D. In *Metrology, Inspection, and Process Control for Microlithography XXXI*; Sanchez, M.I., Ukraintsev, V.A., Eds.; SPIE: San Jose, CA, USA, 2017; Volume 10145, p. 1014529.

51. Murdoch, G.; Bommels, J.; Wilson, C.J.; Gavan, K.B.; Le, Q.T.; Tokei, Z.; Clark, W. Feasibility study of fully self aligned vias for 5 nm node BEOL. In Proceedings of the 2017 IEEE International Interconnect Technology Conference (IITC), Sinchu, Taiwan, 16–18 May 2017; pp. 1–4. [CrossRef]

52. Chopra, M.; Helpert, S.; Verma, R.; Zhang, Z.; Zhu, X.; Bonnecaze, R. A model-based, bayesian approach to the CF_4/Ar etch of SiO_2. In *Design-Process-Technology Co-optimization for Manufacturability XII*; SPIE: San Jose, CA, USA, 2018; Volume 10588, p. 105880G. [CrossRef]

53. Rodgers, S.T.; Jensen, K.F. Multiscale modeling of chemical vapor deposition. *J. Appl. Phys.* **1998**, *83*, 524–530. [CrossRef]

54. Cale, T.S. A unified line-of-sight model of deposition in rectangular trenches. *J. Vac. Sci. Technol. B Microelectron. Nanometer Struct.* **1990**, *8*, 1242. [CrossRef]

55. Chatterjee, S. Prediction of step coverage during blanket CVD tungsten deposition in cylindrical pores. *J. Electrochem. Soc.* **1990**, *137*, 328. [CrossRef]

56. Raupp, G.B.; Cale, T.S. Step coverage prediction in low-pressure chemical vapor deposition. *Chem. Mater.* **1989**, *1*, 207–214. [CrossRef]

57. Ertl, O.; Selberherr, S. Three-dimensional topography simulation using advanced level set and ray tracing methods. In Proceedings of the 2008 International Conference on Simulation of Semiconductor Processes and Devices, Hakone, Japan, 9–11 September 2008; pp. 325–328. [CrossRef]

58. Al-Mohssen, H.A.; Hadjiconstantinou, N.G. Arbitrary-pressure chemical vapor deposition modeling using direct simulation Monte Carlo with nonlinear surface chemistry. *J. Comput. Phys.* **2004**, *198*, 617–627. [CrossRef]

59. Cook, R.L. Shade trees. In Proceedings of the 11th Annual Conference on Computer Graphics and Interactive Techniques—SIGGRAPH '84, Minneapolis, MN, USA, 23–27 July 1984; ACM Press: New York, NY, USA, 1984; pp. 223–231. [CrossRef]

60. Singh, J.; Narayanan, P. Real-time ray tracing of implicit surfaces on the GPU. *IEEE Trans. Vis. Comput. Graph.* **2010**, *16*, 261–272. [CrossRef] [PubMed]

61. Marsaglia, G. Choosing a point from the surface of a sphere. *Ann. Math. Stat.* **1972**, *43*, 645–646. [CrossRef]

62. Parker, S.; Parker, M.; Livnat, Y.; Sloan, P.; Hansen, C.; Shirley, P. Interactive ray tracing for large volume visualization. In Proceedings of the SIGGRAPH 2005, Los Angeles, CA, USA, 31 July–4 August 2005; [CrossRef]

63. Yu, J.C.; Zhou, Z.F.; Su, J.L.; Xia, C.F.; Zhang, X.W.; Wu, Z.Z.; Huang, Q.A. Three-dimensional simulation of DRIE process based on the narrow band level set and monte carlo method. *Micromachines* **2018**, *9*, 74. [CrossRef] [PubMed]

64. Kim, S.-H.; Lee, S.-H.; Lim, H.-T.; Kim, Y.-K.; Lee, S.-K. [110] silicon etching for high aspect ratio comb structures. In Proceedings of the IEEE 6th International Conference on Emerging Technologies and Factory Automation Proceedings, EFTA '97, Los Angeles, CA, USA, 9–12 September 1997; pp. 248–252. [CrossRef]

65. Liao, H.; Cale, T.S. Three-dimensional simulation of an isolation trench refill process. *Thin Solid Films* **1993**, *236*, 352–358. [CrossRef]

66. Adalsteinsson, D.; Sethian, J.A. A level set approach to a unified model for etching, deposition, and lithography. *J. Comput. Phys.* **1997**, *138*, 193–223. [CrossRef]

67. Heitzinger, C.; Sheikholeslami, A.; Badrieh, F.; Puchner, H.; Selberherr, S. Feature-scale process simulation and accurate capacitance extraction for the backend of a 100-nm aluminium/TEOS process. *IEEE Trans. Electron Devices* **2004**, *51*, 1129–1134. [CrossRef]

68. Cheimarios, N.; Kokkoris, G.; Boudouvis, A.G. Multiscale modeling in chemical vapor deposition processes: Coupling reactor scale with feature scale computations. *Chem. Eng. Sci.* **2010**, *65*, 5018–5028. [CrossRef]

69. Setyawan, H.; Shimada, M.; Ohtsuka, K.; Okuyama, K. Visualization and numerical simulation of fine particle transport in a low-pressure parallel plate chemical vapor deposition reactor. *Chem. Eng. Sci.* **2002**, *57*, 497–506. [CrossRef]

70. Cooperberg, D.J.; Vahedi, V.; Gottscho, R.A. Semiempirical profile simulation of aluminum etching in a Cl2/BCl3 plasma. *J. Vac. Sci. Technol. A Vac. Surf. Films* **2002**, *20*, 1536–1556. [CrossRef]

71. Belen, R.J.; Gomez, S.; Kiehlbauch, M.; Cooperberg, D.; Aydil, E.S. Feature-scale model of Si etching in SF$_6$ plasma and comparison with experiments. *J. Vac. Sci. Technol. A Vac. Surf. Films* **2005**, *23*, 99–113. [CrossRef]

72. Abraham-Shrauner, B. Analytic models for plasma-assisted etching of semiconductor trenches. *J. Vac. Sci. Technol. B Microelectron. Nanometer Struct.* **1994**, *12*, 2347. [CrossRef]

73. Magna, A.L.; Garozzo, G. Factors affecting profile evolution in plasma etching of SiO$_2$. *J. Electrochem. Soc.* **2003**, *150*, F178. [CrossRef]

74. van Delft, F.C. Mechanistic framework for dry etching, beam assisted etching and tribochemical etching. *Microelectron. Eng.* **1996**, *30*, 361–364. [CrossRef]

75. Barker, R.A. Surface studies of and a mass balance model for Ar+ ion-assisted Cl2 etching of Si. *J. Vac. Sci. Technol. B Microelectron. Nanometer Struct.* **1983**, *1*, 37. [CrossRef]

76. Tuda, M.; Nishikawa, K.; Ono, K. Numerical study of the etch anisotropy in low-pressure, high-density plasma etching. *J. Appl. Phys.* **1997**, *81*, 960–967. [CrossRef]

77. Klemenschits, X.; Selberherr, S.; Filipovic, L. Unified feature scale model for etching in SF$_6$ and Cl plasma chemistries. In Proceedings of the 2018 Joint International EUROSOI Workshop and International Conference on Ultimate Integration on Silicon (EUROSOI-ULIS), Granada, Spain, 19–21 March 2018; pp. 1–4. [CrossRef]

78. Sentaurus. Available online: https://www.synopsys.com/silicon/tcad/process-simulation/sentaurus-process.html (accessed on 19 November 2018).

79. Wu, C.E.; Yang, W.; Luan, L.; Song, H. Photoresist 3D profile related etch process simulation and its application to full chip etch compact modeling. In *Optical Microlithography XXVIII*; Lai, K., Erdmann, A., Eds.; SPIE: San Jose, CA, USA, 2015; Volume 9426, p. 94261Q.

80. Dunn, D.; Sporre, J.R.; Deshpande, V.; Oulmane, M.; Gull, R.; Ventzek, P.; Ranjan, A. Guiding gate-etch process development using 3D surface reaction modeling for 7 nm and beyond. In *Proceedings Advanced Etch Technology for Nanopatterning VI*; Engelmann, S.U., Wise, R.S., Eds.; SPIE: San Jose, CA, USA, 2017; Volume 10149, p. 101490Q.

81. Florida Object Oriented Process Simulator. Available online: www.flooxs.ece.ufl.edu/ (accessed on 19 November 2018).

82. Victory Process. Available online: https://www.silvaco.com/products/tcad/process_simulation/victory_process/victory_process.html (accessed on 19 November 2018).

83. Nanda, R.K.; Dash, T.P.; Das, S.; Maiti, C.K. Beyond silicon: Strained-SiGe channel FinFETs. In Proceedings of the 2015 IEEE International Conference on Man and Machine Interfacing (MAMI), Bhubaneswar, India, 17–19 December 2015; Volume i, pp. 1–4. [CrossRef]

84. Victory Cell. Available online: https://www.silvaco.com/content/kbase/VictoryCell_jan09.pdf (accessed on 19 November 2018).

85. Maiti, C.K.; Dash, T.P.; Dey, S. Performance enhancement of FinFETs at low temperature. In Proceedings of the 2017 IEEE Devices for Integrated Circuit (DevIC), Kalyani, India, 23–24 March 2017; pp. 35–39. [CrossRef]

86. ViennaTS. Available online: https://github.com/viennats/viennats-dev (accessed on 19 November 2018).

87. Monte Carlo Feature Profile Model. Available online: http://uigelz.eecs.umich.edu/ (accessed on 19 November 2018).

88. Huard, C.M.; Zhang, Y.; Sriraman, S.; Paterson, A.; Kanarik, K.J.; Kushner, M.J. Atomic layer etching of 3D structures in silicon: Self-limiting and nonideal reactions. *J. Vac. Sci. Technol. A Vac. Surf. Films* **2017**, *35*, 031306. [CrossRef]

89. Phietch. Available online: http://www.phietch.com/ (accessed on 19 November 2018).

90. K-Speed. Available online: http://www.tbnsolution.com/ (accessed on 19 November 2018).

91. SEMulator3D. Available online: https://www.coventor.com/semiconductor-solutions/semulator3d/ (accessed on 19 November 2018).

92. Robertson, J.; Wallace, R.M. High-k materials and metal gates for CMOS applications. *Mater. Sci. Eng. R Rep.* **2015**, *88*, 1–41. [CrossRef]

93. Posseme, N. *Plasma Etching Processes for CMOS Devices Realization*; Elsevier: Oxford, UK, 2017.

94. Bengoetxea, O.R. Development and Characterization of Plasma Etching Processes for the Dimensional Control and LWR Issues during High-k Metal Gate Stack Patterning for 14FDSOI Technologies. Ph.D. Thesis, Université Grenoble Alpes, Grenoble Alpes, France, 2016.

95. Ros, O.; Pargon, E.; Fouchier, M.; Gouraud, P.; Barnola, S. Gate patterning strategies to reduce the gate shifting phenomenon for 14 nm fully depleted silicon-on-insulator technology. *J. Vac. Sci. Technol. A Vac. Surf. Films* **2017**, *35*, 021306. [CrossRef]

96. Natarajan, S.; Agostinelli, M.; Akbar, S.; Bost, M.; Bowonder, A.; Chikarmane, V.; Chouksey, S.; Dasgupta, A.; Fischer, K.; Fu, Q.; et al. A 14 nm logic technology featuring 2nd-generation FinFET interconnects, self-aligned double patterning and a $0.0588 \mu m^2$ SRAM cell size. In Proceedings of the IEDM, IEEE Technical Digest—International Electron Devices Meeting, San Francisco, CA, USA, 15–17 December 2014; pp. 3.7.1–3.7.3, [CrossRef]

97. Auth, C.; Aliyarukunju, A.; Asoro, M.; Bergstrom, D.; Bhagwat, V.; Birdsall, J.; Bisnik, N.; Buehler, M.; Chikarmane, V.; Ding, G.; et al. A 10nm high performance and low-power CMOS technology featuring 3rd generation FinFET transistors, self-aligned Quad patterning, contact over active gate and cobalt local interconnects. In Proceedings of the 2017 IEEE International Electron Devices Meeting (IEDM), San Francisco, CA, USA, 2–6 December 2017; pp. 29.1.1–29.1.4. [CrossRef]

98. Barraud, S.; Coquand, R.; Casse, M.; Koyama, M.; Hartmann, J.M.; Maffini-Alvaro, V.; Comboroure, C.; Vizioz, C.; Aussenac, F.; Faynot, O.; Poiroux, T. Performance of omega-shaped-gate silicon nanowire MOSFET with diameter down to 8 nm. *IEEE Electron Device Lett.* **2012**, *33*, 1526–1528. [CrossRef]

99. Chang, Y.t.; Peng, K.p.; Li, P.w.; Lin, H.c. Fabrication and characterization of novel gate-all-around polycrystalline silicon junctionless field-effect transistors with ultrathin horizontal tube-shape channel. *Jpn. J. Appl. Phys.* **2018**, *57*, 04FP06. [CrossRef]

100. Barraud, S.; Lapras, V.; Samson, M.; Gaben, L.; Grenouillet, L.; Maffini-Alvaro, V.; Morand, Y.; Daranlot, J.; Rambal, N.; Previtalli, B.; et al. Vertically stacked-NanoWires MOSFETs in a replacement metal gate process with inner spacer and SiGe source/drain. In Proceedings of the 2016 IEEE International Electron Devices Meeting (IEDM), San Francisco, CA, USA, 3–7 December 2016; pp. 17.6.1–17.6.4. [CrossRef]

101. Donnelly, V.M.; Kornblit, A. Plasma etching: Yesterday, today, and tomorrow. *J. Vac. Sci. Technol. A Vac. Surf. Films* **2013**, *31*, 050825. [CrossRef]

102. Wu, B.; Kumar, A.; Pamarthy, S. High aspect ratio silicon etch: A review. *J. Appl. Phys.* **2010**, *108*, 051101. [CrossRef]

103. Heinecke, R.A.H. Control of relative etch rates of SiO_2 and Si in plasma etching. *Solid State Electron.* **1975**, *18*, 1146–1147. [CrossRef]

104. Coburn, J.W.; Kay, E. Some chemical aspects of the fluorocarbon plasma etching of silicon and its compounds. *IBM J. Res. Dev.* **1979**, *23*, 33–41. [CrossRef]

105. Ephrath, L.M. Selective etching of silicon dioxide using reactive ion etching with CF_4-H_2. *J. Electrochem. Soc.* **1979**, *126*, 1419. [CrossRef]

106. Mauer, J.L.; Logan, J.S.; Zielinski, L.B.; Schwartz, G.C. Mechanism of silicon etching by a CF_4 plasma. *J. Vac. Sci. Technol.* **1978**, *15*, 1734–1738. [CrossRef]

107. Tu, Y.Y.; Chuang, T.J.; Winters, H.F. Chemical sputtering of fluorinated silicon. *Phys. Rev. B* **1981**, *23*, 823–835. [CrossRef]

108. Gogolides, E.; Vauvert, P.; Kokkoris, G.; Turban, G.; Boudouvis, A.G. Etching of SiO_2 and Si in fluorocarbon plasmas: A detailed surface model accounting for etching and deposition. *J. Appl. Phys.* **2000**, *88*, 5570–5584. [CrossRef]

109. Thompson, B.E. Polysilicon etching in SF_6 RF discharges. *J. Electrochem. Soc.* **1986**, *133*, 1887. [CrossRef]

110. D'Emic, C.P. Deep trench plasma etching of single crystal silicon using SF_6/O_2 gas mixtures. *J. Vac. Sci. Technol. B Microelectron. Nanometer Struct.* **1992**, *10*, 1105. [CrossRef]

111. D'Agostino, R.; Flamm, D.L. Plasma etching of Si and SiO_2 in $SF_6 - O_2$ mixtures. *J. Appl. Phys.* **1981**, *52*, 162–167. [CrossRef]

112. Anderson, H.M.; Merson, J.A.; Light, R.W. A kinetic model for plasma etching silicon in a SF_6/O_2 RF discharge. *IEEE Trans. Plasma Sci.* **1986**, *14*, 156–164. [CrossRef]

113. Belen, R.J.; Gomez, S.; Cooperberg, D.; Kiehlbauch, M.; Aydil, E.S. Feature-scale model of Si etching in SF_6/O_2 plasma and comparison with experiments. *J. Vac. Sci. Technol. A Vac. Surf. Films* **2005**, *23*, 1430–1439. [CrossRef]

114. Belen, R.J.; Gomez, S.; Kiehlbauch, M.; Aydil, E.S. Feature scale model of Si etching in SF_6/O_2/HBr plasma and comparison with experiments. *J. Vac. Sci. Technol. A Vac. Surf. Films* **2006**, *24*, 350–361. [CrossRef]

115. Shamiryan, D.; Redolfi, A.; Boullart, W. Dry etching process for bulk finFET manufacturing. *Microelectron. Eng.* **2009**, *86*, 96–98. [CrossRef]

116. Luere, O.; Pargon, E.; Vallier, L.; Pelissier, B.; Joubert, O. Etch mechanisms of silicon gate structures patterned in SF_6/CH_2F_2/Ar inductively coupled plasmas. *J. Vac. Sci. Technol. B Microelectron. Nanometer Struct.* **2011**, *29*, 011028. [CrossRef]

117. Lemme, M.C.; Mollenhauer, T.; Gottlob, H.; Henschel, W.; Efavi, J.; Welch, C.; Kurz, H. Highly selective HBr etch process for fabrication of triple-gate nano-scale SOI-MOSFETs. *Microelectron. Eng.* **2004**, *73-74*, 346–350. [CrossRef]

118. Vinet, M.; Poiroux, T.; Widiez, J.; Lolivier, J.; Previtali, B.; Vizioz, C.; Guillaumot, B.; Le Tiec, Y.; Besson, P.; Biasse, B.; et al. Bonded planar double-metal-gate NMOS transistors down to 10 nm. *IEEE Electron Device Lett.* **2005**, *26*, 317–319. [CrossRef]

119. Kim, D.k.; Kim, Y.K.; Lee, H. A study of the role of HBr and oxygen on the etch selectivity and the post-etch profile in a polysilicon/oxide etch using HBr/O_2 based high density plasma for advanced DRAMs. *Mater. Sci. Semicond. Process.* **2007**, *10*, 41–48. [CrossRef]

120. Ohchi, T.; Kobayashi, S.; Fukasawa, M.; Kugimiya, K.; Kinoshita, T.; Takizawa, T.; Hamaguchi, S.; Kamide, Y.; Tatsumi, T. Reducing damage to Si substrates during gate etching processes. *Jpn. J. Appl. Phys.* **2008**, *47*, 5324–5326. [CrossRef]

121. Desvoivres, L.; Vallier, L.; Joubert, O. X-ray photoelectron spectroscopy investigation of sidewall passivation films formed during gate etch processes. *J. Vac. Sci. Technol. B Microelectron. Nanometer Struct.* **2001**, *19*, 420. [CrossRef]

122. Tuda, M.; Shintani, K.; Ootera, H. Profile evolution during polysilicon gate etching with low-pressure high-density Cl_2/HBr/O_2 plasma chemistries. *J. Vac. Sci. Technol. A Vac. Surf. Films* **2001**, *19*, 711–717. [CrossRef]

123. Darnon, M.; Chevolleau, T.; Eon, D.; Vallier, L.; Torres, J.; Joubert, O. Etching characteristics of TiN used as hard mask in dielectric etch process. *J. Vac. Sci. Technol. B Microelectron. Nanometer Struct.* **2006**, *24*, 2262. [CrossRef]

124. Tabara, S. WSi_2/polysilicon gate etching using TiN hard mask in conjunction with photoresist. *Jpn. J. Appl. Phys.* **1997**, *36*, 2508–2513. [CrossRef]

125. Blumenstock, K. Anisotropic reactive ion etching of titanium. *J. Vac. Sci. Technol. B Microelectron. Nanometer Struct.* **1989**, *7*, 627. [CrossRef]

126. Hwang, W.S.; Chen, J.; Yoo, W.J.; Bliznetsov, V. Investigation of etching properties of metal nitride/high-k gate stacks using inductively coupled plasma. *J. Vac. Sci. Technol. A Vac. Surf. Films* **2005**, *23*, 964–970. [CrossRef]

127. Chiu, H.K.; Lin, T.L.; Hu, Y.; Leou, K.C.; Lin, H.C.; Tsai, M.S.; Huang, T.Y. Characterization of titanium nitride etch rate and selectivity to silicon dioxide in a Cl_2 helicon-wave plasma. *J. Vac. Sci. Technol. A Vac. Surf. Films* **2001**, *19*, 455–459. [CrossRef]

128. Muthukrishnan, N.M.; Amberiadis, K. Characterization of titanium etching in Cl_2/N_2 plasmas. *J. Electrochem. Soc.* **1997**, *144*, 1780–1784. [CrossRef]

129. Le Gouil, A.; Joubert, O.; Cunge, G.; Chevolleau, T.; Vallier, L.; Chenevier, B.; Matko, I. Poly-Si/TiN/HfO_2 gate stack etching in high-density plasmas. *J. Vac. Sci. Technol. B Microelectron. Nanometer Struct.* **2007**, *25*, 767. [CrossRef]

130. Tonotani, J.; Iwamoto, T.; Sato, F.; Hattori, K.; Ohmi, S.; Iwai, H. Dry etching characteristics of TiN film using Ar/CHF_3, Ar/Cl_2, and Ar/BCl_3 gas chemistries in an inductively coupled plasma. *J. Vac. Sci. Technol. B Microelectron. Nanometer Struct.* **2003**, *21*, 2163. [CrossRef]

131. Norasetthekul, S.; Park, P.Y.; Baik, K.H.; Lee, K.P.; Shin, J.H.; Jeong, B.S.; Shishodia, V.; Norton, D.P.; Pearton, S.J. Etch characteristics of HfO_2 films on Si substrates. *Appl. Surf. Sci.* **2002**, *187*, 75–81. [CrossRef]

132. Min, K.S.; Park, B.J.; Kim, S.W.; Kang, S.K.; Yeom, G.Y.; Heo, S.H.; Hwang, H.S.; Kang, C.Y. Selective etching of HfO_2 by using inductively-coupled Ar/C_4F_8 plasmas and the removal of etch residue on Si by using an O_2 plasma treatment. *J. Korean Phys. Soc.* **2008**, *53*, 1675–1679. [CrossRef]

133. Takahashi, K.; Ono, K. Selective etching of high-k HfO_2 films over Si in hydrogen-added fluorocarbon (CF_4/Ar/H_2 and C_4F_8/Ar/H_2) plasmas. *J. Vac. Sci. Technol. A Vac. Surf. Films* **2006**, *24*, 437–443. [CrossRef]

134. Chen, J.; Yoo, W.J.; Tan, Z.Y.; Wang, Y.; Chan, D.S. Investigation of etching properties of HfO based high-k dielectrics using inductively coupled plasma. *J. Vac. Sci. Technol. A Vac. Surf. Films* **2004**, *22*, 1552–1558. [CrossRef]

135. Kim, M.; Efremov, A.; Lee, H.W.; Park, H.H.; Hong, M.; Min, N.K.; Kwon, K.H. HfO_2 etching mechanism in inductively-coupled Cl_2/Ar plasma. *Thin Solid Films* **2011**, *519*, 6708–6711. [CrossRef]

136. Hélot, M.; Chevolleau, T.; Vallier, L.; Joubert, O.; Blanquet, E.; Pisch, A.; Mangiagalli, P.; Lill, T. Plasma etching of HfO_2 at elevated temperatures in chlorine-based chemistry. *J. Vac. Sci. Technol. A Vac. Surf. Films* **2006**, *24*, 30–40. [CrossRef]

137. Sha, L.; Puthenkovilakam, R.; Lin, Y.S.; Chang, J.P. Ion-enhanced chemical etching of HfO_2 for integration in metal–oxide–semiconductor field effect transistors. *J. Vac. Sci. Technol. B Microelectron. Nanometer Struct.* **2003**, *21*, 2420. [CrossRef]

138. Sungauer, E.; Pargon, E.; Mellhaoui, X.; Ramos, R.; Cunge, G.; Vallier, L.; Joubert, O.; Lill, T. Etching mechanisms of HfO_2, SiO_2, and poly-Si substrates in BCl_3 plasmas. *J. Vac. Sci. Technol. B Microelectron. Nanometer Struct.* **2007**, *25*, 1640. [CrossRef]

micromachines

MDPI

Review

A Review for Compact Model of Thin-Film Transistors (TFTs)

Nianduan Lu [1,2,3], Wenfeng Jiang [1,2,3], Quantan Wu [1,2,3], Di Geng [1,2,3], Ling Li [1,2,3,*] and Ming Liu [1,2,3]

[1]   Key Laboratory of Microelectronic Devices & Integrated Technology, Institute of Microelectronics, Chinese Academy of Sciences, Beijing 100029, China; lunianduan@ime.ac.cn (N.L.); jiangwenfeng1912@163.com (W.J.); wuquantan@ime.ac.cn (Q.W.); gengdi@ime.ac.cn (D.G.); liuming@ime.ac.cn (M.L.)
[2]   School of Microelectronics, University of Chinese Academy of Sciences, Beijing 100049, China
[3]   Jiangsu National Synergetic Innovation Center for Advanced Materials (SICAM), Nanjing 210009, China
*   Correspondence: lingli@ime.ac.cn

Received: 16 October 2018; Accepted: 9 November 2018; Published: 15 November 2018

Abstract: Thin-film transistors (TFTs) have grown into a huge industry due to their broad applications in display, radio-frequency identification tags (RFID), logical calculation, etc. In order to bridge the gap between the fabrication process and the circuit design, compact model plays an indispensable role in the development and application of TFTs. The purpose of this review is to provide a theoretical description of compact models of TFTs with different active layers, such as polysilicon, amorphous silicon, organic and In-Ga-Zn-O (IGZO) semiconductors. Special attention is paid to the surface-potential-based compact models of silicon-based TFTs. With the understanding of both the charge transport characteristics and the requirement of TFTs in organic and IGZO TFTs, we have proposed the surface-potential-based compact models and the parameter extraction techniques. The proposed models can provide accurate circuit-level performance prediction and RFID circuit design, and pass the Gummel symmetry test (GST). Finally; the outlook on the compact models of TFTs is briefly discussed.

Keywords: thin-film transistors (TFTs); compact model; surface potential

---

1. Introduction

A thin-film transistors (TFTs) is a special kind of field-effect transistor (FET) fabricated by depositing thin films of an active semiconductor layer, as well as the dielectric layer and metallic contacts over a supporting (but non-conducting) substrate [1,2]. In the past 15 years, TFTs has grown into a huge industry based on display, memory, E-paper applications, and so on [3–6]. Generally, a common substrate in TFTs is glass, which differs from the conventional transistor, where the semiconductor material typically is the substrate, such as a silicon wafer. TFTs include three basic elements: (1) a thin semiconductor film; (2) an insulating layer; and (3) three electrodes (gate, source and drain) [7–9]. Three basic elements for configuration of TFT have been illustrated clearly in Figure 1. The source and drain, are in contact with the semiconductor film at a short distance from one another. The gate is separated from the semiconductor film by the insulating layer [10].

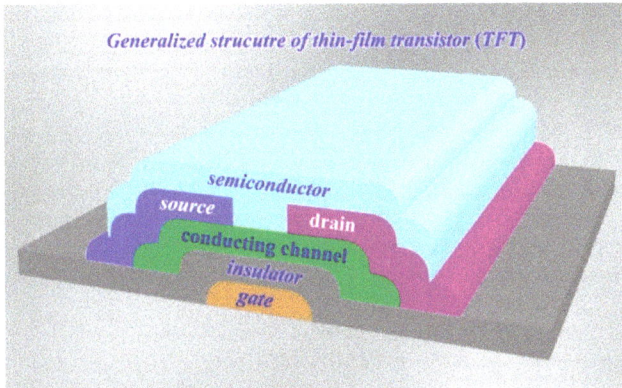

Figure 1. Schematic structure of a generalized thin-film transistor.

The history of TFT really began with the work of P. K. Weimer at Radio Corporation of America (RCA) Laboratories in 1962 [11]. At that time Weimer fabricated the first TFT based on thin films of polycrystalline cadmium sulfide as the semiconductor materials. In the 1970s, the realization of crystalline silicon as the active materials with low cost dramatically changed the prospects of TFTs [12]. In 1979, amorphous silicon as a new active material was introduced by LeComber et al. [13], which had profound implications for TFTs. In 1980, Depp et al. reported polysilicon TFT which achieved good mobility and TFT characteristics [14]. In 1986, the first transistor based on organic semiconductor was reported [15]. As compared with conventional Si TFTs, organic TFT (OTFT) displays much less complex in fabrication processes and can be naturally compatible with plastic substrates for lightweight and foldable products [16]. To develop large-scale TFTs, processing temperatures must be getting lower and lower. In 2004, Nomura et al. used a complex In-Ga-Zn-O (IGZO) semiconductor layer in a TFT, which achieved the room-temperature processing of the semiconductor layer [17]. Looking back into the past half-century, TFTs moved endlessly forward from the initial requirement of performance to today's application of large area and low cost.

During the development of TFTs, the semiconductor device model represents an essential bridge between the semiconductor manufactures and the circuit design. Integrated circuit (IC) designers usually utilize various kinds of software (such as Cadence, SPICE, PHILIPAC) for design circuit [18–20]. The core of the corresponding software is the model of each unit device. Because the IC is consisted of several transistors, if all unit devices would need to run the complicated model of transistor, the system level simulation will beyond computer ability and hence causes non-convergence in calculation. Otherwise, for ensuring the reliability of the simulation, the device model should also be able to accurately describe the physical properties [21]. Compact model is a critical step in the design cycle of modern IC products [22]. It refers to the development of models for integrated semiconductor devices for use in circuit simulations. Compact model is usually used to reproduce device terminal behaviors with accuracy, computational efficiency, ease of parameter extraction, and relative model simplicity for a circuit or system-level simulation, for future technology nodes [23].

Accurate and physical compact models are essential for digital and analog circuits. Generally speaking, an excellent compact model should include the following requirements [21,24]:(i) Representing consistently the behavior; (ii) Being symmetrical to reflect the symmetry of TFT structure; (iii) Being analytical, without differentials or integrals; (iv) Being simple and easily derivable; (v) Parameters that can be characterized easily, or even guessed; (vi) Being upgradable and reducible; (vii) Relations can be physically justified; (viii) Being similar form and correspondence to compact models for other TFTs; (ix) Being tunable to inaccurate (or uncertain) experimental data.

The first compact model could date back to 1983, in which Kacprzak et al. proposed a compact DC model of GaAs FETs for large-signal computer calculation [25]. In 1986, based on one-dimensional (1-D) solution of Poisson's equation, Ahmed et al. reported a compact model for accumulation mode poly-Si devices [26]. Later, plenty of methods, such charge sheet model, effective medium approach (EMA), semi-empirical approach, generation-recombination model, and surface-potential based model, have been introduced for the compact models of the silicon-based TFTs [27–32]. Then, with the emergence of new TFTs, e.g., OTFT and IGZO TFTs, some excellent compact models based on interesting methods have been developed [33–37]. Strictly speaking, all of the proposed compact models can be divided into two categories. One is charge-based and another is surface-potential-based. As compared with the charge-based model, the surface-potential-based compact model is believed to have high accuracy and strong physical property, and be easily simplified into the charge-based and threshold-voltage-based model [21]. It can also describe the operation of transistor more accurately without any smooth functions [38].

Over the past two decades, although some excellent reviews have been published [39–42], a completed review for the compact models of TFTs based on different active materials is still lacking. In this review, we will provide an updated review of surface-potential-based compact model of TFTs with different active materials, such as polysilicon, amorphous silicon, organic and IGZO semiconductors. In Section 2, the charge transport property of different active materials is discussed. In Section 3, we discuss the surface-potential-based compact models for silicon-based TFTs and presented our surface-potential-based compact models for organic and IGZO TFTs, respectively. In Section 4, the comparison of various compact models will be summarized. Finally, the future outlook for this field is briefly discussed in Section 5.

2. Charge Transport Property

In order to achieve an accurate compact model for TFTs, the key is to correctly describe the charge transport characteristics. For TFTs with different active materials, the charge transport has displayed various properties. This section will, in detail, introduce the charge transport properties of TFTs for different active materials.

2.1. Grain-Boundary Trapping Theory

Based on its structure characteristics, the charge transport property of polysilicon has been described in terms of two distinct models: segregation theory and grain-boundary trapping theory [43]. In the segregation theory, impurity atoms tend to segregate at the grain boundary where they are electrically inactive. While the grain boundary trapping theory assumed that the presence of a large amount of trapping states at the grain boundary able to capture, and therefore immobilize, free carriers. The basic limitation of segregation theory is that it does not explain the temperature dependence of the film resistivity which is thermally activated and exhibits a negative temperature coefficient. The grain-boundary trapping theory can explain most of electrical properties in polysilicon.

In the grain-boundary trapping theory, a polysilicon is assumed to be composed of small crystallites joined together by the grain boundaries usually consisted of a few atomic layers of disordered atoms [43]. Inside each crystallite the atoms are arranged in a periodic manner so that it can be considered as a small single crystal. Atoms in the grain boundary represent a transitional region between the different orientations of neighboring crystallites. Although polysilicon is a three-dimensional substance, it is sufficient to treat the problem in one dimension to calculate the transport properties. The traps are assumed to be initially neutral and become charged by trapping a carrier. Figure 2 shows the schematic diagram of crystal structure, charge distribution and energy band structure of polysilicon films.

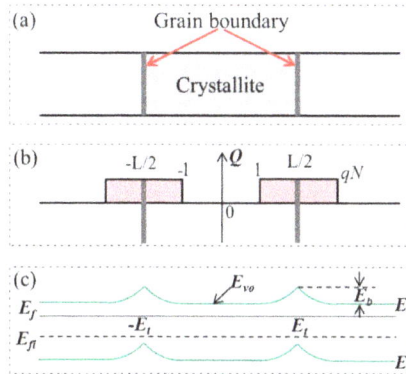

Figure 2. Schematic diagram of (a) crystal structure; (b) charge distribution; and (c) energy band structure of polysilicon films.

The grain-boundary trapping theory considers just the resistance of the grain-boundary region, which includes two important contributions to the current: thermionic emission and tunneling (field emission) [44]. Thermionic emission results from those carriers possessing high enough energy to surmount the potential barrier at the grain boundary. The tunneling current arises from carriers with energy less than the barrier height. When the barrier is narrow and high, the tunneling current can become comparable to or larger than the thermionic emission current. In the polysilicon the potential barrier is the highest when the barrier width is the widest. Because of this, tunneling current may be neglected. Then, for an applied voltage the thermionic emission current density across a grain boundary following Bethe is expressed as the following [44]:

$$J_{th} = q p_a \left(\frac{k_B T}{2m^* \pi} \right)^{1/2} exp\left(-\frac{q V_B}{k_B T} \right) \left[exp\left(-\frac{q V_a}{k_B T} \right) - 1 \right] \tag{1}$$

where q is the elemental charge, p_a is the average carrier concentration, m^* is the effective mass of the carrier, k_B is the Boltzmann constant, V_B is the potential barrier height, and V_a is the applied voltage. Equation (1) neglects collisions within the depletion region and the carrier concentration in the crystallite was assumed to be independent of the current flow, so that it is applicable only if the number of carriers taking part in the current transport is small compared to the total number of carriers in the crystallite. This condition restricts the barrier height to be larger than or comparable to $k_B T$. If V_a is small, $q V_a \ll k_B T$, Equation (1) can be expanded to give the following:

$$J_{th} = q^2 p_a \left(\frac{1}{2\pi m^* k_B T} \right)^{1/2} exp\left(-\frac{q V_B}{k_B T} \right) V_a \tag{2}$$

which is a linear current–voltage relationship. Based on Equation (2), the conductivity of a polysilicon film with a grain size L is written as:

$$\sigma = L q^2 p_a \left(\frac{1}{2\pi m^* k_B T} \right)^{1/2} exp\left(-\frac{q V_B}{k_B T} \right) \tag{3}$$

Then, the effective mobility is expressed as:

$$\mu_{eff} = L q \left(\frac{1}{2\pi m^* k_B T} \right)^{1/2} exp\left(-\frac{E_b}{k_B T} \right) \tag{4}$$

here E_b is the energy barrier.

2.2. Hopping Transport

Differing from crystalline materials, such as polysilicon, the charge transport in amorphous materials exhibits very different properties. Amorphous semiconductor materials, including inorganic and organic, have in common, that their atomic or molecular structure is completely disordered. For inorganic amorphous semiconductors, such as, pure and hydrogenated amorphous silicon (a-Si, a-Si:H), a band structure similar to the one of crystalline materials still exists [45,46]. The electronic states in the conduction and valence bands are therefore delocalized. Thus some of the concepts from crystalline semiconductor physics are still suitable for the inorganic amorphous materials. However, in the band gap between valence and conduction band, some localized states exist in which charge carriers can be trapped. For organic amorphous semiconductors, the intermolecular bonds are due to relatively weak van der Waals interactions, the electronic wave functions usually do not extend over the entire volume of the organic solid, but rather, are localized to a finite number of molecules, or even to individual molecules [47,48]. Due to the spatial and energetic disorder, the charge transport in amorphous semiconductor materials is limited by trapping in the localized states. This means that the charge carrier mobility is expected to be thermally activated, that is, the charge transport always happens to jump from one localized site to another. This type of transport mechanism is called hopping transport. The transition of hopping between two sites depends on the overlap of the electronic wave functions of these two sites [49]. Whenever a charge carrier hops to a site with a higher (lower) site energy than the site that it came from, the difference in energy is accommodated for by the absorption (emission) of a phonon. Figure 3 is a schematic diagram of carrier hopping transport with the density of states [50].

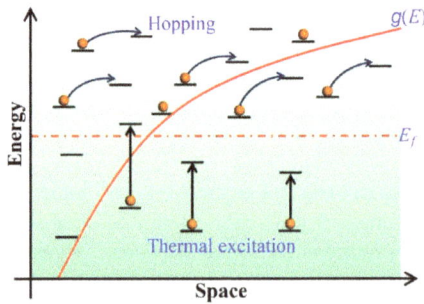

Figure 3. Schematic diagram of hopping transport with the density of states.

The intrinsic transition rate for a carrier hopping from an initial site i to an empty site j is expressed by $\gamma_{ij} = \gamma(R_{ij}, E_i - E_j)$. The average transition rate from site i to site j is then [51]:

$$\nu_{ij} = < m_i(1 - m_j)\,\gamma_{ij} > \tag{5}$$

where m_i and m_j are the occupation numbers for sites i and j, respectively. The energy dependence of γ_{ij} is then a good approximation to take the Miller–Abrahams form to write as [52]:

$$\gamma_{ij} = \nu_0 exp\left(-2\varphi R_{ij} - \frac{\theta(E_j - E_i)}{k_B T}\right) \tag{6}$$

where ν_0 is the attempt-to-jump frequency, φ is the inverse localized length of the inverse wave function, R_{ij} is the distance between site i and site j, E_i and E_j are the energies of sites i and j, respectively, and $\theta(x) = x\varepsilon(x)$ with $\varepsilon(x)$ being the step function.

2.3. Multiple Trapping and Release Theory

For some special materials, such as the small-molecule organic semiconductor and the IGZO semiconductor, which have a strong tendency to form polycrystalline films [50,53,54]. These semiconductors display the regular arrangement, and the delocalized orbitals partially overlap, thereby facilitating more efficient charge transfer and carrier mobility that is much larger than in amorphous films. The charge transport properties of these materials cannot be explained by the grain-boundary trapping theory and hopping transport. In contrast to the grain-boundary trapping theory or hopping theory, the multiple trapping and release (MTR) theory is adapted for the materials [55,56]. MTR theory assumes that the charge transport occurs in extended states, and that most of the charge carriers are trapped in localized states [57]. Energy of localized state is separated from mobility edge energy. When the energy of localized state is slightly lower mobility edge, then the extended states acts as shallow trap, from which the charge carrier can be released (emitted) by the thermal excitations. But, if that energy is far below mobility edge energy, then charge carriers cannot be thermally excited (emitted). The number of carriers available for transport depends on the difference in energy between the trap level and the extended-state band. Figure 4 is a transport diagram of MTR theory.

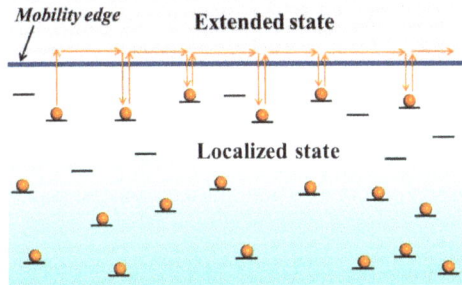

Figure 4. Transport diagram of multiple trapping and release (MTR) theory. The charge carrier (orange balls) is trapped and released into and from localized states (black lines). Conduction happens above the mobility-edge (gray area).

In the MTR theory, total charge carriers' densities, n_{total}, is equal to sum of density in extended states, n_e, and in localized states, as in Ref. [57]:

$$n_{total} = n_e + \int_{-\infty}^{0} g(E)f(E)dE \tag{7}$$

where the upper limit of integral $E = 0$ corresponds to the mobility edge, $g(E)$ is the trap density of states (DOS) energy distribution. $f(E) = \left(1 + exp\left(\frac{E-E_f(x)}{k_B T}\right)\right)^{-1}$ is the Fermi–Dirac distribution, $E_f(x)$ is the quasi-Fermi level. Two methods within the MTR theory usually describe the effect of trapping [58]. One is that, all carrier fields induced can contribute to the current flow at any moment of time, but the effective mobility is reduced in comparison with its intrinsic, trap-free value:

$$\mu_{eff} = \mu_0(T)\frac{\tau(T)}{\tau(T) + \tau_{tr}(T)} \tag{8}$$

here, μ_0 is the carrier mobility in extended state, $\tau_{tr}(T)$ is the average trapping time on shallow traps, and $\tau(T)$ is the average time that a polaron spends diffusively traveling between the consecutive trapping events. Another is that only a fraction of the carrier field induced is moving at any given moment of time:

$$n_{eff} = n_{total} \frac{\tau(T)}{\tau(T) + \tau_{tr}(T)} \tag{9}$$

3. Surface-Potential-Based Compact Models

In spite of the fact that the transport characteristics in TFTs is very different for different active materials, the current–voltage characteristics can, to first order, be described with the same formalism as [53]:

$$I_{ds} = \begin{cases} \frac{\mu C_i W}{L}\left((V_g - V_{th})V_{ds} - \frac{V_{ds}^2}{2} \right) & for \ |V_g - V_{th}| > |V_{ds}| \ (linear \ regime) \\ \frac{\mu C_i W}{2L}(V_g - V_{th})^2 & for \ |V_{ds}| > |V_g - V_{th}| > 0 \ (saturation \ regime) \end{cases} \tag{10}$$

where Equation (10) describes the relationship between the drain current I_{ds}, the gate-source voltage V_g and the drain-source voltage V_{ds} in linear and saturation regimes, respectively. C_i is the gate dielectric capacitance per unit area, μ is the carrier mobility in the semiconductor, W and L is the channel width and length of the transistor, respectively. For silicon TFTs, the threshold voltage V_{th} is defined as the minimum gate-source voltage required to induce strong inversion [59]. However OTFTs and IGZO TFTs usually operate in accumulation region, thus strictly speaking the threshold voltage cannot be defined for OTFTs and IGZO TFTs. Since the threshold voltage concept is nonetheless useful, the compact models will show very different for TFTs with different active materials. Otherwise, the central aim of compact models is to accurately and physically describe the current–voltage characteristics of TFTs in Equation (10). As mentioned above, the surface-potential-based compact model is believed to have high accuracy and strong physical properties. The following will review the surface-potential-based compact models for polysilicon and amorphous silicon TFTs, and then present our compact models for OTFTs and IGZO TFTs based on surface-potential-based.

3.1. Polysilicon TFT Compact Models

Polysilicon TFTs have gotten considerable applications, especially in active matrix liquid crystal displays (AMLCDs), printers, scanners, Static Random-Access Memories (SRAMs) and three-dimensional large scale integration (LSI) circuits [60]. In early time, researchers usually built the polysilicon TFT models based on the one-dimensional solution of Poisson's equation and the effects of grain-boundary traps [26,61]. However, these earlier models were unclear for inversion mode devices due to the "reverse" charge shielding concept defined in its derivation [62]. Later, some authors adopted the EMA method to well address the question of non-uniform polysilicon sample with the grain boundaries [29,63]. In 1999, Benjamín et al. also adopted EMA to develop a unified model for long and short-channel polysilicon TFTs [28]. This method is attractive because it accounts for field effect mobility enhancement in the moderate inversion regime and for mobility degradation at high gate voltages, for drain-induced barrier lowering (DIBL) effect, kink effect, off-state current and channel-length modulation. A few years later, Wu et al. proposed a compact model by approximating the generation rate for poly Si TFTs in the leakage region [50]. Although several models for poly-Si TFTs have been proposed so far, based on different equations for the subthreshold, linear, and saturation regions [64,65], these methods always lead to a significant error in evaluating derivatives such as transconductance [66].

To capture more accurate features of poly-Si TFTs, Shimizu et al. developed a compact model based on a new surface-potential-based [67]. Firstly, in the model the states are approximated by the sum of exponential distributions for the deep and tail states as:

$$g(E) = g_{de}exp\left(\frac{E - E_c}{E_{de}} \right) + g_{ta}exp\left(\frac{E - E_c}{E_{ta}} \right) \tag{11}$$

where E_{de} and E_{ta} are the inverse slope of deep states and tail states, respectively, g_{de} and g_{ta} are the density of deep state and tail state at bottom of conduction band E_c, respectively.

By integrating the 1-D Poisson equation, the surface potentials at the source side as a function of gate voltage can be calculated numerically as the following [68]:

$$C_i\left(V_g - V_{fb} - \varphi_{s0}\right) = \sqrt{\frac{2q\varepsilon_s N_{sub}}{\beta}}\left[exp(-\beta\varphi_{s0}) - exp(-\beta\varphi_{b0}) + \beta(\varphi_{s0} - \varphi_{b0})\right.$$
$$+ \left(\frac{n_i}{N_{sub}}\right)^2[exp(\beta\varphi_{s0}) - exp(\beta\varphi_{b0})] + \left(\frac{\beta N_{deep}}{\gamma N_{sub}}\right)[exp(\gamma\varphi_{s0}) - exp(\gamma\varphi_{b0})]$$
$$\left. + \left(\frac{N_{tail}}{N_{sub}}\right)[exp(\beta\varphi_{s0}) - exp(\beta\varphi_{b0})]\right]^{\frac{1}{2}} \tag{12}$$

where $\gamma = q/E_{de}$, β is the inverse of thermal voltage, ε_s is the dielectric constant, N_{sub} is the dopant concentration, n_i is the intrinsic carrier concentration, φ_{s0} and φ_{b0} are the front and back surface potentials at the source side, respectively, N_{deep} and N_{tail} are the densities of trapped electrons in deep states and tail states under a flat band condition, respectively.

Then, the inversion layer charge density at the source ($x = 0$) or drain ($x = L$) side can be written as [69] the following:

$$Q_i(x) = -C_i\left(V_g - V_{fb} - \varphi_{s0}\right) + \sqrt{\frac{2q\varepsilon_s N_{sub}}{\beta}}\left[exp(-\beta\varphi_{sx}) - exp(-\beta\varphi_{bx}) + \beta(\varphi_{sx} - \varphi_{bx})\right.$$
$$\left. + \left(\frac{\beta N_{deep}}{\gamma N_{sub}}\right)[exp(\gamma\varphi_{sx}) - exp(\gamma\varphi_{bx})] + \left(\frac{N_{tail}}{N_{sub}}\right)[exp(\beta\varphi_{sx}) - exp(\beta\varphi_{bx})]\right]^{\frac{1}{2}} \tag{13}$$

Obviously, Equations (12) and (13) could be only solved by iteration. To determine the surface potentials at the source side or at the drain side, the authors used a method from the literature [70]. In Equation (13), the charge densities of inversion layer are derived based on the charge-sheet approximation. Figure 5 shows a comparison of the front surface potentials obtained from Equation (12) and the exact numerical calculations.

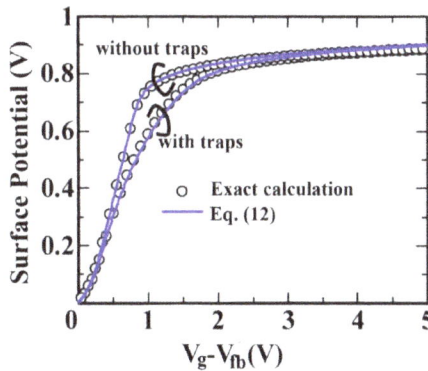

Figure 5. Comparison of calculated front surface potential obtained using Equation (12) (lines) and the exact numerical calculations (circles) with and without traps as a function of gate voltage.

After obtaining the surface potentials, based on the drift-diffusion approximation [71], the authors calculated the drain current as:

$$I_{ds} = \frac{W\mu}{L\beta}\left[C_i\left(\beta\left(V_g - V_{fb}\right) + 1\right)(\varphi_{sL} - \varphi_{s0}) - \frac{\beta}{2}C_i\left(\varphi_{sL}^2 - \varphi_{s0}^2\right) - \frac{\beta}{2}(q_i(0) + (L))\right.$$
$$\left. (\varphi_{sL} - \varphi_{s0}) - (q_i(0) - q_i(L))\right] \tag{14}$$

where φ_{sL} and φ_{bL} are the front and back surface potentials at the drain side, respectively, $q_i(x) = Q_i(x) + C_i\left(V_g - V_{fb} - \varphi_{sx}\right)$. Note, Equation (14) can describe the drain current in all the regions of operation using the unified equation. At the same time, the model did not include the threshold voltage.

Figure 6 shows a comparison of simulated and measured drain current characteristics as a function of gate voltage for an n-channel poly-Si TFT in the subthreshold and above-threshold regions. In the linear and saturation regions, a comparison of simulated and measured drain current characteristics is shown in Figure 7.

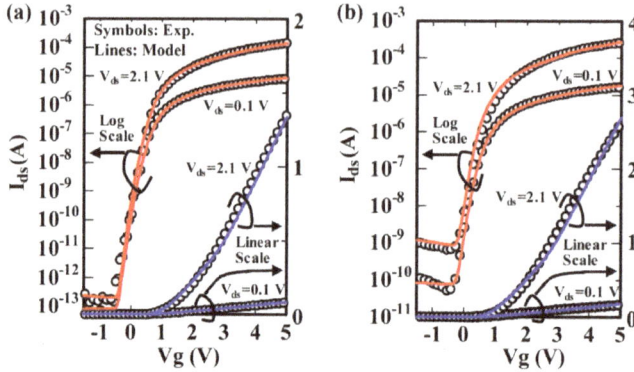

Figure 6. Comparison of measured (circles) and simulated (lines) drain current characteristics as a function of gate voltage on logarithmic (left axis) and linear (right axis) scales for an n-channel poly-Si thin-film transistors (TFT) with (a) W/L = 2 μm/2 μm and (b) W/L = 2 μm/1 μm.

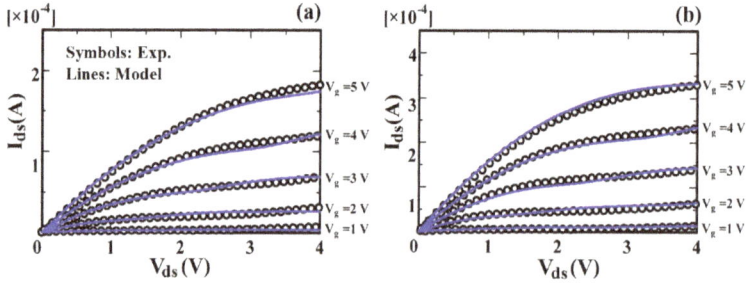

Figure 7. Comparison of measured (circles) and simulated (lines) drain current characteristics as a function of drain voltage for an n channel poly-Si TFT with (a) W/L = 2 μm/2 μm and (b) W/L = 2 μm/1 μm. The parameters used in the simulation are the same as those used in Figure 6.

Differing from iterative solution of the surface potential in Shimizu et al.'s model, Chen et al. have developed an analytical solution to the surface potential of poly-Si TFTs by using the Lambert W function [72]. In Chen et al.'s model, the surface potential of poly-Si TFTs can be expressed as,

$$\left(V_g - V_{fb} - \varphi_s\right)^2 = \gamma^2\left[\left(1 + \frac{N_T}{L_g N_A}\right)\varphi_s + \varphi_t exp\left(\frac{\varphi_s - \varphi_n - 2\varphi_f}{\varphi_t}\right)\right] - \frac{N_T}{L_g N_A}\varphi_t \ln(1 + K_m) \quad (15)$$

where φ_s is the surface potential, γ denotes a body factor: $\gamma = \sqrt{2\varepsilon_s q N_A}/C_i$, L_g is the grain size, N_T and N_A are located traps and acceptor density, respectively, φ_t is the thermal voltage, φ_f is the Fermi potential, φ_n is the channel voltage, $K_m = 0.5exp\left(\frac{E_t + q\varphi_t}{k_B T}\right)$. To derive an analytical and non-iterative evaluation, the normalized form of Equation (15) can be written as the following:

$$\left(v_g - x_W\right)^2 = G_{TFT}^2[x_W + \Delta_{TFT} \exp(x_W) + A] \quad (16)$$

where $x_W = \varphi_s/\varphi_t$ is the normalized surface potential, $v_g = \left(V_g - V_{fb}\right)/\varphi_t$ the normalized effective gate voltage, $G_{TFT} = \gamma\sqrt{\frac{1+N_T/L_gN_A}{\varphi_t}}$, $\Delta_{TFT} = \exp\left(\frac{-\varphi_t - 2\varphi_f}{\varphi_t}\right)/\left(1 + \frac{N_T}{L_gN_A}\right)$, and $A = -\frac{N_T\ln(1+K_m)}{N_T+L_gN_A}$.

Then, with a simple mathematical procedure and using the principal branch of the Lambert W function [73], the authors obtained the physics-based analytical solution of the normalized surface potential as follows:

$$x_W = -W_0[f \times \Delta_{TFT}exp(v_G - f \times A)] + v_G - f \times A \tag{17}$$

where $v_G = \left(v_g + \frac{G_{TFT}^2}{2}\right) - G_{FET}\sqrt{v_g + G_{TFT}^2/4}$ and $f = G_{FET}/2\sqrt{v_g + G_{TFT}^2/4}$.

In order to improve the accuracy, some corrections by using the Schroder series in the surface potential expression have been provided for Equation (17). Finally, the complete solution to the physics-based surface potential of poly-Si TFTs with absolute error only in nanovolt range can be expressed as:

$$\varphi_s = \left[x_W + \omega\left(y_W, y_W', y_W''\right) + \varepsilon\right]\varphi_t \tag{18}$$

Based on Equation (18), the surface potential derivative with respect to the gate voltage has been calculated [72], as shown in Figure 8. Figure 8 shows that no splits and peaks exist near the flatband regions, which suggests that the analytical solution to the surface potential is better than the algorithm in the Penn State Philips (PSP) model [74].

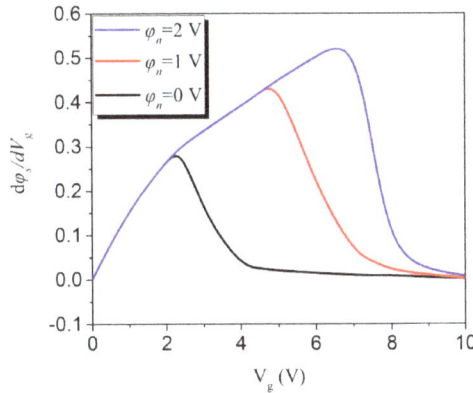

Figure 8. The characteristics of the surface potential derivative with respect to the gate voltage for different channel potential φ_n.

Based on the formulas of surface potential from Chen et al., subsequently some researchers presented a complete modeling for surface potential in partially depleted poly-Si TFTs with undoped or lightly doped body by including both monoenergetic and exponential trap distributions [32]. The proposed closed-form algorithm is able to accurately calculate the surface potential and has the advantage of both accuracy and computational efficiency, which is useful for compact modeling and CAD applications.

3.2. Amorphous Silicon TFTs

Amorphous Silicon, especially hydrogenated amorphous-silicon (a-Si:H), has been considered as the most well-studied materials for TFTs. Generally speaking, the most important features of amorphous silicon TFT characteristics can be described by analyzing the device behavior in two regimes: below-threshold, when the electron quasi-Fermi level is in the deep states; and above-threshold, when the Fermi level enters the tail states [75]. A current model for the below- and above-threshold regimes had been proposed by considering the sheet carrier density as a function of Fermi lever position by Shur et al. [76]. In 1997, Shur et al. again developed a physically based analytical model for n-channel amorphous silicon thin film transistors and for n- and p-channel polysilicon thin film transistors, which covered all regimes of transistor operation: leakage, subthreshold, above-threshold conduction, and the kink regime in polysilicon thin film transistors [63]. Only in the last few years several models have been built, based on the description of below-threshold and above-threshold, respectively [77–79]. However, with gradual accumulation of the requirements imposed on the compact models and simultaneous realization of the limitations associated with the traditional modeling techniques, new physical phenomena become essential for the accurate reproduction of the device characteristics. On the other hand, due to these drawbacks in the regional approach, analytical models based on surface potential have been paid more attentions in the development of device models [40].

In terms of the consideration above, in 2008, Liu et al. presented an analytical a-Si:H TFTs model based on the surface potential [80]. In the model, when TFT is biased, the majority of the induced charges in the channel are trapped in the acceptor-like states, which divided into two groups: deep states and tail states. The distribution of localized acceptor states can be expressed as Equation (11). The localized trapped charge density is expressed as:

$$n_{trapped} = \int_{-\infty}^{E_c} \frac{g(E)}{1 + \frac{1}{g}exp\left(\frac{E-E_f}{k_B T}\right)} \qquad (19)$$

here g is the degenerescence factor of localized states. When the density of trapped charges in the tail states are considered, the integral (Equation (19)) can be rewritten as [68] the following:

$$n_{tail} = g_t g^{T/T_0} \frac{k_B T}{q} f(T, T_t) exp \frac{q\varphi - qV_{ch}(y) - E_{f0}}{k_B T_0} \qquad (20)$$

where g_t is the tail states density at E_c, T_t is the tail state characteristic temperature, V_{ch} stands for the channel quasi-Fermi level which is the channel voltage equal to 0 at the source and V_{ds} at the drain.

To obtain the potential, the authors then solved the Poisson's equation:

$$\frac{\partial^2 \varphi}{\varphi x^2} = -\frac{dF}{dx} = \frac{q}{\varepsilon_s}\left(n_{deep} + n_{tail} + n_{free}\right) \qquad (21)$$

where n_{deep}, n_{tail} and n_{free} are the densities for deep trap, free, tail trap charges, respectively, $n_{free} = N_c exp\left(\frac{q\varphi_s - qV_{ch}(y) - E_{f0}}{k_B T}\right)$. According to Gauss' law, and introducing electrical field effect, the relationship between the gate-source voltage and the surface potential can be found as follows

$$C_i\left(V_g - V_{fb} - \varphi_s\right) = r_t\varepsilon_s exp\left(\frac{q\varphi_s - qV_{ch}(y) - E_{f0}}{k_B T_0}\right) + r_d\varepsilon_s exp\left(\frac{q\varphi_s - qV_{ch}(y) - E_{f0}}{k_B T_d}\right)$$
$$+ r_f\varepsilon_s exp\left(\frac{q\varphi_s - qV_{ch}(y) - E_{f0}}{k_B T}\right) \qquad (22)$$

here $r_t = \sqrt{\frac{2k_B T_0 g_t g^{T/T_0} f(T,T_t) k_B T}{q\varepsilon_s}}$, $r_d = \sqrt{\frac{2k_B T_d g_d g^{T/T_d} \pi k_B T}{q\varepsilon_s sin\left(\frac{\pi T}{T_d}\right)}}$ and $r_f = \sqrt{\frac{2k_B T N_c}{\varepsilon_s}}$.

To derive analytical and noniterative evaluation from Equation (22), the normalized form of Equation (22) can be written as follows:

$$(xg - x) = G_t exp(x - xn) + G_d[exp(x - xn)]^{T_0/T_d} + G_f[exp(x - xn)]^{T_0/T} \qquad (23)$$

where $xg = -\frac{V_g - V_{fb}}{2V_{to}}$, $x = \frac{\varphi_s}{2V_{to}}$, $xn = \frac{\frac{E_{f0}}{q} + V_{ch}(y)}{2V_{to}}$, $G_t = \frac{r_t\varepsilon_s}{2C_i V_{to}}$, $G_d = \frac{r_d\varepsilon_s}{2C_i V_{to}}$, $G_f = \frac{r_f\varepsilon_s}{2C_i V_{to}}$, and $V_{to} = \frac{k_B T_0}{q}$.

Then, by using the two-order Taylor expansion, the solution for the surface potential of amorphous silicon TFTs is expressed by:

$$\varphi_s = x \cdot 2V_{to}. \tag{24}$$

Based on the solution for the surface potential of amorphous silicon TFTs, the authors compared the analytical results with the numerical results, as shown in Figure 9a. And the absolute errors of the new analytical approximation were shown in Figure 9b. The absolute errors introduced by analytical approximation are less than 0.02 V in all cases.

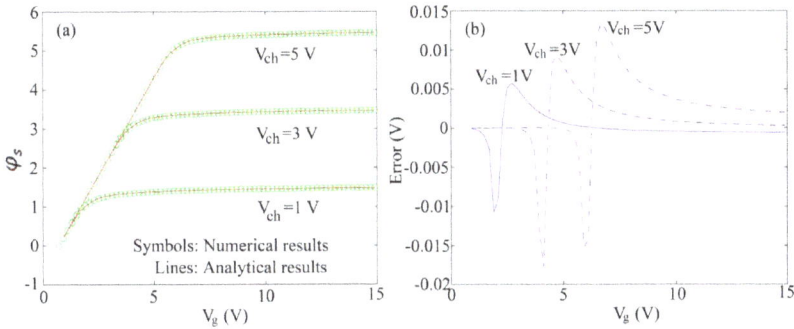

Figure 9. (**a**) Comparison of analytical results with the numerical results; and (**b**) absolute error of the new analytical approximation for the surface potential.

After the surface potential is solved precisely, the authors then discussed the drain current by dividing the new derivation of the DC model into the below threshold region and the above threshold region.

Below threshold region, the static current of amorphous TFTs is written as:

$$
\begin{aligned}
I_{dsd} = {} & \mu_n \frac{W}{L} N_c \varepsilon_s \frac{2k_B T_d k_B T}{2k_B T_d - k_B T} \left(\frac{1}{r_d \varepsilon_s} \right)^{\frac{2T_d}{T}} C_i^{\frac{2T_d}{T} - 1} \left[\frac{T}{2T_d} \left(\Delta\varphi_{ss}^{\frac{2T_d}{T}} - \Delta\varphi_{sd}^{\frac{2T_d}{T}} \right) \right. \\
& \left. + 2V_{td} \frac{T}{2T_d - T} \left(\Delta\varphi_{ss}^{\frac{2T_d}{T} - 1} - \Delta\varphi_{sd}^{\frac{2T_d}{T} - 1} \right) \right]
\end{aligned} \tag{25}
$$

Above threshold region, similarly, the expression of drain current in the above threshold regime can be obtained as:

$$
\begin{aligned}
I_{dst} = {} & \mu_n \frac{W}{L} N_c \varepsilon_s \frac{2k_B T_0 k_B T}{2k_B T_0 - k_B T} \left(\frac{1}{r_t \varepsilon_s} \right)^{\frac{2T_0}{T}} C_i^{\frac{2T_0}{T} - 1} \left[\frac{T}{2T_0} \left(\Delta\varphi_{ss}^{\frac{2T_0}{T}} - \Delta\varphi_{sd}^{\frac{2T_0}{T}} \right) \right. \\
& \left. + 2V_{t0} \frac{T}{2T_0 - T} \left(\Delta\varphi_{ss}^{\frac{2T_0}{T} - 1} - \Delta\varphi_{sd}^{\frac{2T_0}{T} - 1} \right) \right]
\end{aligned} \tag{26}
$$

According to the expression of drain current, the calculated transfer characteristics for a-Si:H TFT is shown in Figure 10a. It is noted that a smooth transition is achieved in the below- and above-threshold regions without any use of smooth functions. Furthermore, the threshold voltage is not required in the whole calculations. Figure 10b displays the measured characteristics and the calculated current–voltage characteristics of an a-Si:H TFT. It is demonstrated that the model exhibits a reasonable agreement in both the linear region and the saturation region.

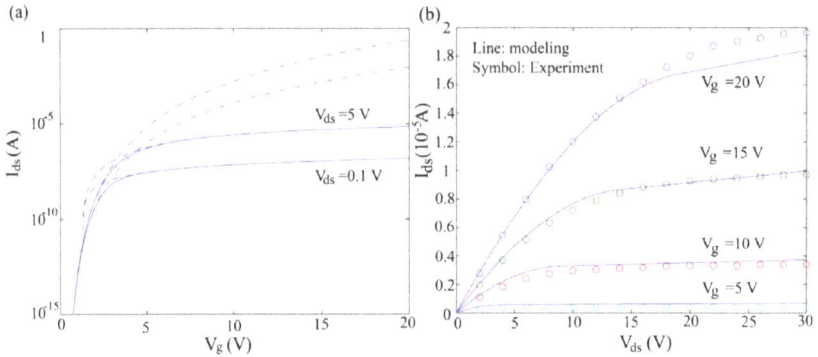

Figure 10. (**a**) Calculated transfer characteristics and (**b**) calculated output characteristics for a-Si:H TFT, with the measured data for comparison.

To calculate the surface potential, other methods have been used. For example, very recently, Qin et al. developed a novel scheme for surface potential of amorphous silicon TFTs by taking deep Gaussian and tail exponential distribution of the density of states into account [81]. In Qin et al.'s model, the authors adopted Taylor expansion below threshold regime, and the principle of Lamber W function and Schroder series above threshold regime, as well as Chen et al.'s model in Section 3.1.

3.3. OTFT Compact Models

In OTFTs, the energy disorder is usually described by Gaussian DOS as [82]:

$$g(E) = \frac{N_t}{\sqrt{2\pi}\sigma} exp\left(-\frac{E^2}{2\sigma^2}\right) \tag{27}$$

where N_t is the total localized states, and σ indicates the width of the DOS. By connecting Gauss law $C_i\left(V_g - V_{fb} - \varphi_s\right) = \varepsilon_s F(0)$, one can obtain the following:

$$C_i\left(V_g - V_{fb} - \varphi_s\right) = \sqrt{\frac{2q\varepsilon_s N_t}{\sqrt{2\pi}\sigma} \int_0^{\varphi_s} \int_{-\infty}^{\infty} \frac{exp(-E^2/2\sigma^2)}{1 + exp\left(\frac{E - E_{f0} - q(\varphi - V)}{k_B T}\right)} dE d\varphi} \tag{28}$$

where $F(0)$ is the electric field perpendicular to the interface at the interface, V is the channel voltage, and E_{f0} is the Fermi level far from the semiconductor-insulator interface.

By approximating the Fermi–Dirac distribution with the Boltzmann distribution, Equation (28) can be rewritten as:

$$C_i\left(V_g - V_{fb} - \varphi_s\right) = \sqrt{\frac{2q\varepsilon_s N_t}{\sqrt{2\pi}\sigma} \int_0^{\varphi_s} \int_{-\infty}^{\infty} exp\left(-\frac{E^2}{2\sigma^2} - \frac{E - E_{f0} - q(\varphi - V)}{k_B T}\right) dE d\varphi} \tag{29}$$

Since the localized states mainly lie in the higher energy of Gaussian DOS, $E - E_{f0} > 2k_B T$ is usually achieved. As the carrier density varies over a narrow range, then the Fermi–Dirac distribution can be approximated by the Boltzmann distribution. According to Equation (29), the surface potential can be calculated as:

$$\left(V_g - V_{fb} - \varphi_s\right)^2 = k exp\left(-\frac{V}{\varphi_t}\right)\left(exp\left(\frac{\varphi_s}{\varphi_t}\right) - 1\right) \tag{30}$$

where $k = \frac{\varepsilon_0 \varepsilon_s k_B N_t}{C_i^2} exp(-0.5\sigma^2)$. The solution of Equation (30) actually is numerical. However, under low gate voltage OTFTs operate in weak accumulation mode, that is, $\varphi_s \ll \varphi_t$. In this situation, the surface potential φ_{sw} is small and can be obtained as

$$\varphi_{sw} = V_g - V_{fb} + \frac{kexp\left(-\frac{V}{\varphi_t}\right)}{2\varphi_t} - \sqrt{\left(V_g - V_{fb} + \frac{kexp(-v/\varphi_t)}{2\varphi_t}\right)^2 - \left(V_g - V_{fb}\right)} \quad (31)$$

Under high gate voltage, OTFTs operate in a strong accumulation mode, that is, $V_g - V_{fb} \gg \varphi_s \gg \varphi_t$. In this case, the surface potential φ_{ss} reads as

$$\varphi_{ss} = 2\varphi_t ln\left(\frac{V_g - V_{fb}}{\sqrt{k}}\right) + V \quad (32)$$

Connecting Equations (31) and (32), the unified surface potential of OTFTs is expressed as

$$\varphi_s = \sqrt{\frac{\varphi_{sw}{}^\gamma \cdot \varphi_{ss}{}^\gamma}{\varphi_{sw}{}^\gamma + \varphi_{ss}{}^\gamma}} \quad (33)$$

Figure 11a shows the comparison between the surface potential calculated using the Boltzmann distribution and Fermi–Dirac distribution functions under different channel voltages, respectively [83]. One can see that a good agreement is observed. Figure 11b shows the absolute and relative error of the Boltzmann function approximation from Figure 11a, revealing that the maximum of relative error is less than 0.6%, as shown by the maximum peak in Figure 11b. This approximation displays good accuracy for weak, moderate and strong accumulation at various channel voltages. Otherwise, the absolute error of the surface potential introduced by the Boltzmann function approximation decreases with channel voltage and is always lower than 0.035 V.

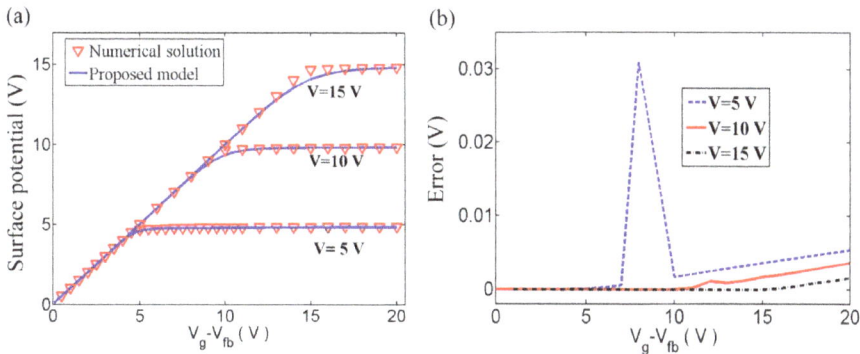

Figure 11. (**a**) Comparison between the surface potential calculated using Boltzmann distribution and Fermi–Dirac distribution functions for different channel voltages and (**b**) absolute error of the Boltzmann approximation from (**a**).

For OTFTs, the field-effect mobility can be written as [84]:

$$\mu = \mu_0 exp\left(C_1(2n/N_t)^{C_2}\right) \quad (34)$$

here C_1 and C_2 are given as $C_1 = 0.5(S^2 - S)$ and $C_2 = 2\frac{ln(S^2 - S) - ln(ln(4))}{S^2}$, which only depend on the disorder, n is the carrier concentration, $\mu_0 = \mu_{00} exp(-aS - bS^2)$, $S = \sigma/k_B T$, and μ_{00} is the mobility in the limit $n \to 0$.

Using the same method in the literature [85], the field-effect mobility μ_{eff} is calculated with the following:

$$\mu_{eff} = \frac{L}{C_i W V_{ds}} \frac{\partial I_{ds}}{\partial V_g} = \frac{\sqrt{\varepsilon_s \varepsilon_0 / 2\hat{q}}}{C_i} \frac{n(\varphi_s)\mu(n(\varphi_s))}{\sqrt{\int_0^\varphi \int_{-\infty}^\infty g(E)dEd\varphi}} \times \frac{2C_i\sqrt{\int_0^{\varphi_s} \int_{-\infty}^\infty g(E)dEd\varphi}}{\sqrt{\varepsilon_s \hat{q} n(\varphi_s)}}$$
$$\approx \mu_0 exp\left(C_1 \frac{(2C_i)^{2C_2}}{(2\varepsilon_s k_B T N_t)^{C_2}}\left(V_g - V_{fb} - \gamma V_{ds}\right)^{2C_2}\right) \tag{35}$$

where $n(\varphi_s) = \int_{-\infty}^\infty g(E)\left(1 + exp\left(\frac{E - E_{f0} - q(\varphi - V)}{k_B T}\right)\right)^{-1} dE$, γ is a parameter that accounts for channel-length modulation.

Then, according to Gauss's law, the sheet density of total induced charges in the channel is given by:

$$Q_i = C_i\left(V_g - V_{fb} - \phi_s\right) \approx C_i\sqrt{K}exp\left(\frac{(\varphi_s - V)}{2\varphi_t} - 1\right) \tag{36}$$

By differentiating Equation (36) with respect to φ_s, we then obtain:

$$\frac{dV}{d\varphi_s} = \frac{2\varphi_t}{\sqrt{K}}exp\left(-\frac{\varphi_s - V}{2\varphi_t} + 1\right) + 1 = 2\varphi_t\frac{C_i}{Q_i} + 1 \tag{37}$$

Using the gradual channel approximation, I_{ds} is given by:

$$I_{ds} = -\mu_{eff} W Q_i \frac{dV}{dy} = -\mu_{eff} W Q_i\left(2\varphi_t\frac{C_i}{Q_i} + 1\right)\frac{d\varphi}{dy} \tag{38}$$

By integrating Equation (38) from $\varphi_s = \varphi_{ss}$ to $\varphi_s = \varphi_{sd}$, the static current of OTFTs becomes:

$$I_{ds0} = \frac{\mu_{eff} W}{L}\left(2\varphi_t C_i(\varphi_{sd} - \varphi_{ss}) - \frac{C_i}{2}\left(\left(V_g - V_{fb} - \varphi_{sd}\right)^2 - \left(V_g - V_{fb} - \varphi_{ss}\right)^2\right)\right) \tag{39}$$

where φ_{ss} and φ_{sd} are the surface potentials at the source and drain side, respectively. Both of them can be analytically calculated by Equation (33). When OTFTs are biased to the saturation region, channel-length modulation becomes significant in short channel devices. In this case, the expression of I_{ds} can be rewritten as:

$$I_{ds} = I_{ds0}(1 + \lambda V_{ds}) \tag{40}$$

Based on Equation (40), the OTFT characteristics can be described by a new formula that does not contain the threshold voltage.

Figure 12 shows the measured characteristics from pentacene transistors and the calculated current–voltage characteristics of OTFT. The model agrees well with the experimental results in both the linear and saturation regions [83].

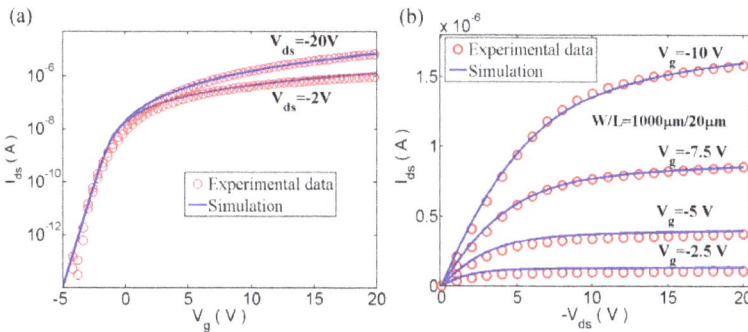

Figure 12. (**a**) Simulated and experimental results for transfer characteristics of organic Thin-film transistors (OTFT); and (**b**) comparison between the simulated and experimental results for output characteristics of OTFT for different gate voltages.

We also verified our proposed model by comparing it to measurements of OTFTs with channel lengths from 25 μm to 5 μm (*W* = 1000 μm), as shown in Figure 13 [83]. The extracted λ values are 0.55 and 0.27 for *L* = 5 μm and *L* = 10 μm, respectively.

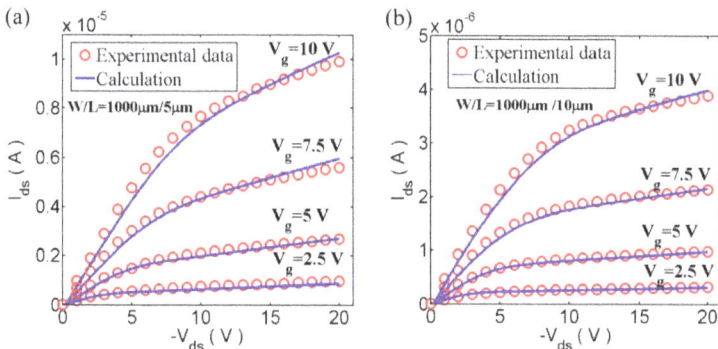

Figure 13. Simulated and experimental results of output characteristics of OTFT: (**a**) for W/L = 1000 μm /5 μm and (**b**) for W/L = 1000 μm/10 μm.

3.4. a-IGZO TFTs

As mentioned in Section 2.3, the MTR theory is responsible for the charge transport of a-IGZO TFTs. We have combined the MTR theory with the surface potential to develop the compact model of IGZO TFTs [86,87]. Generally speaking, in TFTs, due to the accumulated carriers in semiconductor-insulator interface under the gate voltage, the gate-induced potential $\varphi(x)$ shifts the difference between the mobility edge and the Fermi level. The quasi-Fermi level $E_f(x)$ is

$$E_f(x) = E_{f0} + q\varphi(x) \tag{41}$$

The variation of $\varphi(x)$ with respect to the distance x is determined by the Poisson equation as [57]:

$$F(x)^2 = \frac{2q}{\varepsilon_s}n_{total} = \frac{2q}{\varepsilon_s}\left[N_t v_0 \tau_0 exp\left(\frac{E_f(x)}{k_B T}\right) + \int_0^{\varphi(x)}\int_{-\infty}^{0}\frac{g(E)}{1+exp\left(\frac{E-E_f(x)}{k_B T}\right)}dE d\varphi(x)\right] \tag{42}$$

151

where $F(x)$ is the electric field perpendicular to the interface. At the interface, the electric field $F(0)$ can be expressed through Gauss's law as:

$$\varepsilon_s F(0) = C_i\left(V_g - V_{fb} - \varphi_s\right) = \sqrt{2q\varepsilon_s\left[N_t v_0 \tau_0 exp\left(\frac{E_f(x)}{k_B T}\right) + \int_0^{\varphi(x)}\int_{-\infty}^0 \frac{g(E)}{1+exp\left(\frac{E-E_f(x)}{k_B T}\right)}dEd\varphi(x)\right]} \tag{43}$$

where T_{TA} is the characteristic temperature of the exponential DOS, τ_0 is the lifetime of carriers, and v_0 is the attempt-to-escape frequency. Then, the field effect mobility could be written as [88]:

$$\mu_{eff} = \frac{\mu_e}{1+\left(\frac{1}{v_0\tau_0}\Gamma(1+T/T_{TA})\Gamma(1-T/T_{TA})exp\left(\frac{E_{f0}+q\varphi_s}{k_B T_{TA}}\left(\frac{T}{T_{TA}}-1\right)\right)\right)^{-1}} \tag{44}$$

where μ_e is the band mobility and $\Gamma(z) = \pi z/\sin\pi z$. Under the low gate voltage, Fermi level lies in the deep states and hence free carriers above the mobility edge can be neglected, and carriers of localized states will dominate the transport of IGZO TFTs (corresponding to the sub-threshold regime of transistor). Thus, the total carrier concentration is reasonably written as

$$n(x) \approx \int_{-\infty}^0 \frac{g(E,x)}{1+exp\left(\frac{E-E_f(x)}{k_B T}\right)}dE = N_t\Gamma\left(1+\frac{T}{T_{TA}}\right)\Gamma\left(1-\frac{T}{T_{TA}}\right)exp\left(\frac{E_{f0}+q\varphi_s}{k_B T_{TA}}\right) \tag{45}$$

Substituting Equations (45) into (43), one can get the following expression:

$$C_i\left(V_g - V_{fb} - \varphi_s\right) = \sqrt{q\varepsilon_s\left[N_t v_0 \tau_0 exp\left(\frac{E_f(x)}{k_B T}\right) + N_t\Gamma\left(1+\frac{T}{T_{TA}}\right)\Gamma\left(1-\frac{T}{T_{TA}}\right)exp\left(\frac{E_{f0}+q\varphi_s}{k_B T_{TA}}\right)\right]} \tag{46}$$

To achieve the analytic solution of the surface potential, we transformed Equation (46) as:

$$V_g - V_{fb} - \varphi_s = G_T exp\left(\frac{q\varphi_s - qV_{ch}}{2k_B T}\right) + G_{TA}exp\left(\frac{q\varphi_s - qV_{ch}}{2k_B T_{TA}}\right) \tag{47}$$

G_T and G_{TA} can be expressed as:

$$\begin{cases} G_T = \frac{1}{C_i}\sqrt{q\varepsilon_s v_0 \tau_0 N_t exp\left(\frac{E_{f0}}{k_B T}\right)} \\ G_{TA} = \frac{1}{C_i}\sqrt{q\varepsilon_s N_t\Gamma\left(1+\frac{T}{T_{TA}}\right)\Gamma\left(1-\frac{T}{T_{TA}}\right)exp\left(\frac{E_{f0}}{k_B T_{TA}}\right)} \end{cases} \tag{48}$$

Through estimating the order of magnitudes, in Equation (47) the first term is much smaller than the second term. Thus, we only consider the second term and ignore the first term. By using two-order Taylor expansion, one can get:

$$x_i = xg\left\{\left[(xg+1)^2 + 2xn + 2log\left(\frac{xg}{G_T}\right)\right]^{1/2} - xg - 1\right\} \tag{49}$$

However, if one considers only the second term in Equation (47), some errors in the surface potential calculation maybe occur. In order to improve the accuracy, we add some corrections by using the Schroder series method to cover the influence of the first term in Equation (47). Finally, the analytical solution of the surface potential can be written as:

$$\begin{cases} \varphi_s = 2\frac{k_B T_{TA}}{q}\left[x_i - \frac{f}{\partial f}\left(1+\frac{\partial^2 f}{2\partial f}\frac{f}{\partial f}\right)\right] \\ f = (xg-x) - G_{TA}exp(x-xn) - G_T(exp(x-xn))^{\frac{T_{TA}}{T}} \end{cases} \tag{50}$$

Figure 14 shows a comparison of calculated surface potential between analytic solution and numerical result [86,87]. The percentage error between the numerical and analytical solutions is always below 0.2%. The parameters are $T = 300$ K, $T_{TA} = 405$ K, $\nu_0 \tau_0 = 1$, $V_{fb} = 0.5$ V, $C_i = 8.85 \times 10^{-8}$ F/cm^2, and $\mu_e = 19.7$ cm^2/Vs.

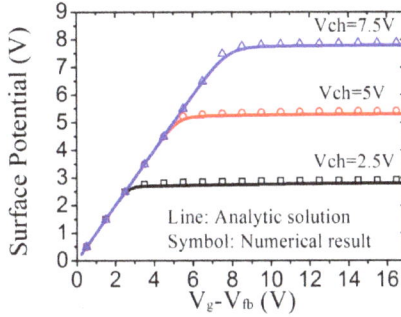

Figure 14. Comparison of the calculated surface potential between analytic solution and the numerical results for different channel voltages.

Using the gradual channel approximation, the current equation is given as:

$$I_{ds} = -\mu_{eff} W Q_i \frac{dV}{dy} = -\mu_{eff} W Q_i \left(2\varphi_t \frac{C_i}{Q_i} + 1 \right) \frac{d\varphi}{dy} \tag{51}$$

By integrating Equation (51) from $\varphi_s = \varphi_{ss}$ to $\varphi_s = \varphi_{sd}$, the static current of a-IGZO TFTs is expressed as:

$$I_{ds0} = \mu_{eff} \frac{W}{L} \left[2\varphi_t C_i (\varphi_{sd} - \varphi_{ss}) - \frac{1}{2} \left(\left(V_g - V_{fb} - \varphi_{sd} \right)^2 - \left(V_g - V_{fb} - \varphi_{ss} \right)^2 \right) \right] \tag{52}$$

where φ_{ss} and φ_{sd} are the surface potential at source and drain side, respectively. Both of them can be analytically calculated from Equation (50).

Figure 15 shows the output and transfer characteristics curve. The good agreement between our modeling results and the experimental data has been observed [86,87]. Figure 16 shows the drain conductance and trans-conductance curves [86,87]. Our model well agrees with the measured results.

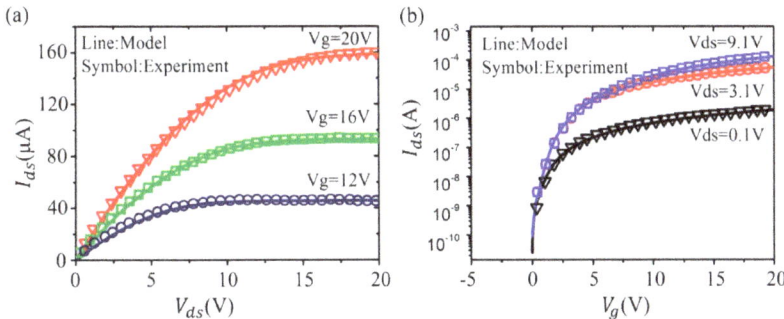

Figure 15. Comparison between the calculation and experimental data; (**a**) for output characteristics of In-Ga-Zn-O (IGZO) under different gate voltages; and (**b**) transfer characteristics of IGZO under different drain-source voltages.

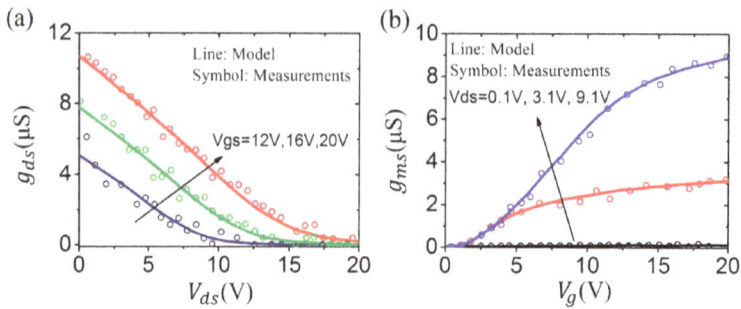

Figure 16. Model $g_{ds} - V_{ds}$ curves (**a**) and trans-conductance curves (**b**).

4. Comparison of Various Compact Models

As mentioned above, the most difference between silicon-based TFTs and TFTs with new active material (e.g., OTFTs and IGZO TFTs) derived from the fact that whether the threshold voltage can be defined in TFT device. Since OTFTs and IGZO TFTs usually operate in accumulation region, the formulation of compact model should discard the influence of the threshold voltage. The following will give a comparison for different compact models.

4.1. Comparison of Model Accuracy

For the compact models, the central aim is to accurately and physically describe the current–voltage characteristics of TFTs. Here, we will discuss various compact models and their accuracies verified for TFTs. For polysilicon TFTs, we firstly compare $V_{ds} - I_{ds}$ characteristics based on the surface-potential-based by Chen et al. [72] and the EMA method by Iñiguez et al. [28], respectively, as shown in Figure 17. It is obvious that the surface-potential-based model agrees well with experimental data. However, the simulated results from Iñiguez et al. show a well consistent between model and experiment under low drain voltage, with increasing the drain voltage, the model seriously deviated from the experiment. Similar errors of model accuracy have also been found in the OTFT compact models. Figure 18 shows a comparison of $V_{ds} - I_{ds}$ characteristics based on the surface potential and the generic model, respectively [24,83]. For the surface-potential-based model in Figure 18a, the accuracy is good in all regions, but for the generic model in Figure 18b the errors increase with the gate voltage increasing.

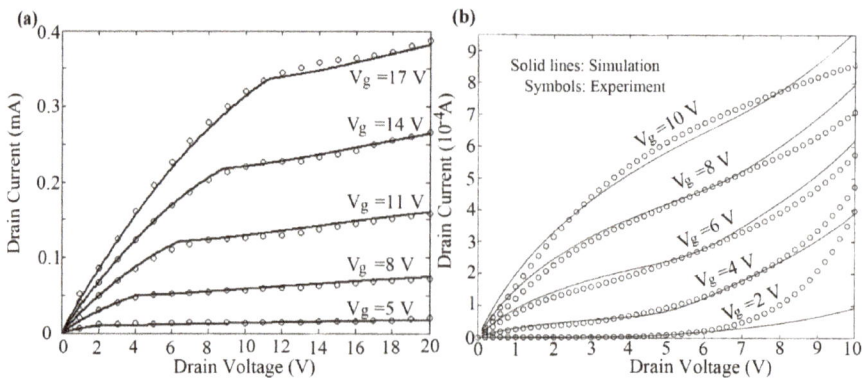

Figure 17. Comparison of $V_{ds} - I_{ds}$ of polysilicon TFTs based on the surface-potential from Chen et al. (**a**), and the effective medium approximation from Iñiguez et al. (**b**).

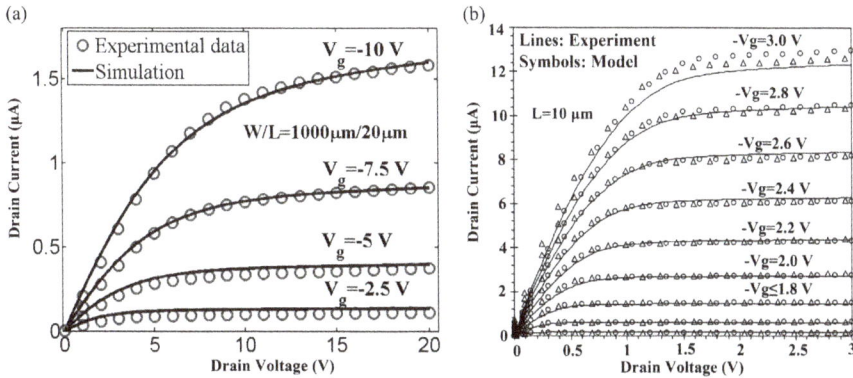

Figure 18. Comparison of $V_{ds} - I_{ds}$ of OTFTs based on the surface-potential (**a**), and the generic model (**b**), respectively.

Strictly speaking, the errors derive from the transformation from the numerical equation to the analytical solution. To obtain analytical solution, the authors usually transferred the numerical model to analytical expression by using some reasonable assumption. Intuitively, various assumptions will generate different error values, which finally affect the model accuracy. Thus, the analysis of the error is essential to transfer the numerical equation to the analytical expression. For the surface-potential-based compact models, reducing the errors of the calculated surface potential has become an important criterion. However, as compared with surface-potential-based models, the errors for the charge-based models are always ignored, which thus results in the lower accuracy.

4.2. Parameter Comparison and Extraction

Apart from the accuracy and comprehensive nature, an excellent compact model should include as few parameters as possible fitting the TFT characteristics. Tables 1 and 2 give a summary of the parameter comparison for the surface-potential-based compact models and OTFT compact models based on different approaches, respectively. It is found that the researchers always aspired for as few parameter numbers as possible during developing compact models. Actually, for the compact model, the fewer the non-physical parameters (fitting parameters), the better the model is considered. From Tables 1 and 2, one can see that the parameter numbers in our model is just 12, which is superior, compared with other models.

Table 1. Comparison of fitting parameter numbers for the surface-potential-based compact models.

Types	Parameter Numbers	Authors	Years
Polysilicon TFTs	15 [67]	Y. Shimizu, et al.	2006
	14 [72]	R. S. Chen, et al.	2007
	11 [32]	W. L. Deng, et al.	2011
Amorphous TFTs	16 [80]	Y. Liu, et al.	2008
	22 [89]	Y. Liu, et al.	2009
	14 [81]	J. Qin, et al.	2014
Organic TFTs	12 [83]	Our work	2015
IGZO TFTs	14 [90]	A. Tsormpatzoglou, et al.	2013
	12 [86]	Our work	2014

Table 2. Comparison of fitting parameter numbers for OTFT compact model based different approaches.

Years	Parameter Numbers	Authors	Method
1995	14 [39]	M. S. Shur, et al.	Effective medium approach
1999	28 [27]	B. Iñiguez, et al.	Effective medium approximation
1999	14 [29]	M. D. Jacunski, et al.	Semi-empirical approach
2006	15 [67]	Y. Shimizu, et al.	Surface potential
2007	14 [72]	R. S. Chen, et al.	Surface potential
2007	12 [30]	W. J. Wu, et al.	Generation-recombination model
2011	11 [32]	W. L. Deng, et al.	Surface potential
2015	12 [83]	Our work	Surface potential

In addition to using as few parameters as possible, parameter extraction also plays an important role in understanding TFT characteristics. Generally speaking, parameter extraction aims at being physical. To achieve higher level, the parameter sequence should introduce physical effect [21]. However, considering the continuity and accuracy of compact model, the fitting parameters will be used for smoothing the output curves and reducing the error. It is anticipated that the compact models of TFTs with the parameter setting will be suitable to circuit design and can provide accurate insight into the performance. The main criterion for a good set of parameters is the balance of error, efficiency and continuity. For IGZO TFTs, we have developed an extraction flow of the key physical parameters of the surface-potential-based compact model, as shown in Figure 19 [86,87]. Based on the corresponding equations shown in Figure 19, four key parameters can be extracted, that is, the maximum mobility μ_0, the characteristic temperature T_{TA}, the product of the escape frequency v_0 and carrier lifetime τ_0.

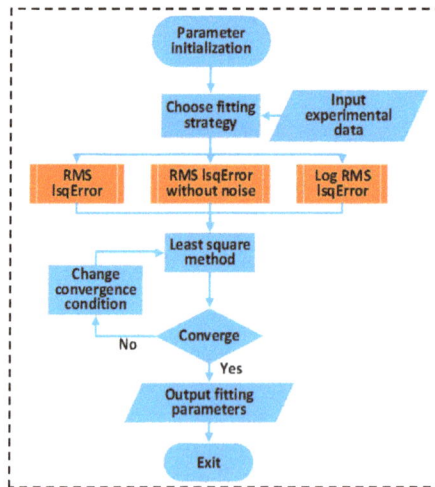

Figure 19. Extraction flow of key physical parameters of the model.

4.3. Criterion and Continuous Test of Compact Models

A compact model must satisfy several rather restrictive requirements imposed by their use in advanced circuit simulators. From the mathematical point of view, the equations of the models should meet three classes at least [40], that is, "class 1" in order to be compatible with Newton–Raphson-based circuit simulators, with "class 2" or better preferred in order to achieve faster convergence, and "class 3" required for circuit simulation of active-matrix organic light-emitting diode (AMOLED) displays or distortion modeling in RF circuits. Currently, the most compact models are satisfied to the "class 1". A small numbers of compact models can meet the requirements of "class 2" and "class 3" together. For the "class 3" requirement, the application is completely based on the active layers of TFTs.

For example, silicon-based TFTs (poly-Si and a-Si:H) are mainly used in AMOLED displays. OTFTs can be applied to logic circuit design and flat-panel display. IGZO TFTs can be used in constructing RFID tags or inverter. Thus, the compact model of TFTs should be established according to their application.

In addition, it would be specially mentioned that, in order to meet the requirement of "class 2", the compact model must fulfill one of the benchmark tests, i.e., Gummel symmetry test (GST) [21,91,92]. Based on our surface-potential-based compact model for IGZO TFTs, the GST has been provided [86,87], as shown in Figure 20. Figure 20a shows a GST circuit for IGZO TFTs. Generally, the higher-order derivatives in TFT compact models are obtained as a function of V_x, which is symmetry for $V_x = 0$. This symmetry roots in the symmetry device structure and channel. Figure 20b shows the GST for the 1, 2, 3-order derivative of the drain current of IGZO TFTs, which display a good continuity and symmetry. Thus, our compact model in IGZO TFTs can pass the GST.

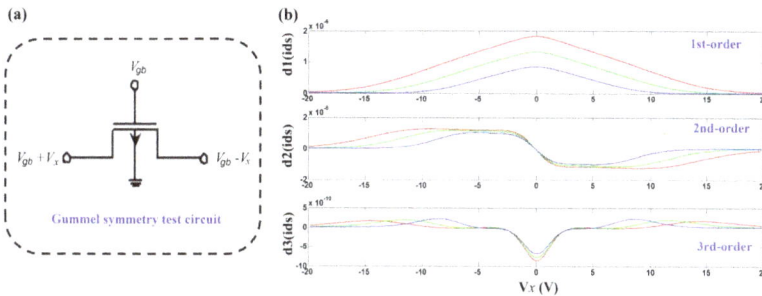

Figure 20. (**a**) Gummel symmetry test circuit for IGZO TFTs; and (**b**) Gummel symmetry test for the 1, 2, 3-order derivative of the drain current under different gate voltages.

5. Conclusions and Outlook

Compact models form a critical link between the manufacturing teams and the chip design teams by mathematically capturing the properties of devices. We have reviewed the concept, development and application of compact model of TFTs. Based on different active materials in TFTs, the charge transport characteristics has also been discussed in detail. Based on the different approaches, especially the surface-potential-based, the merits and shortcomings for current compact models have discussed. We also proposed our surface-potential-based compact models for organic and IGZO TFTs and parameter extraction technology. The comparison of various compact models has been summarized.

Currently, the compact model is still open and evolving. To achieve the excellent compact model, the following should be considered: accurate in all regions of operation and types, suitable for all simulation modes, excellent convergence, and intuitive and easy to extract parameters. In addition, to keep pace with the increase of circuit operating frequencies and device tolerances scale down, the compact model of TFTs should account for the bias dependent contact resistances, gate tunneling, interface effect and scaling effect. The dynamic behavior, aging and hysteresis of TFTs also should be considered in developing the compact model to pursue the future circuit design.

Author Contributions: N.L. and L.L. conceived the idea and designed this work; N.L. drafted the paper; N.L., and L.L. discussed the results and commented on the manuscript; all authors have given approval to the final version of the manuscript.

Acknowledgments: This work was supported in part by National key research and development program (Grant Nos. 2016YFA0201802, 2017YFB0701703, 2018YFA0208503), by the Opening Project of Key Laboratory of Microelectronic Devices and Integrated Technology, Institute of MicroElectronics Chinese Academy of Sciences, by the Beijing Training Project for the Leading Talents in S&T under Grant No. Z151100000315008, and by the National Natural Science Foundation of China (Grant Nos. 61725404, 61574166, 61874134, 61221004, 61376112, and 61404164), by International cooperation project of CAS under Grant 172511KYSB20150006, and by the Strategic Priority Research Program of Chinese Academy of Sciences (Grant No. XDB30000000, XDB12030400).

Conflicts of Interest: The authors declare no conflict of interest.

References

1. Kanatzidis, M.G. Quick-set Thin Films. *Nature* **2004**, *428*, 269–270. [CrossRef] [PubMed]
2. Lu, N.D.; Wei, W.; Chuai, X.C.; Li, L.; Liu, M. Carrier thermoelectric transport model for black phosphorus field-effect transistors. *Chem. Phys. Lett.* **2017**, *678*, 271–274. [CrossRef]
3. Wang, L.; Wang, W.; Xu, G.; Ji, Z.; Lu, N.; Li, L.; Liu, M. Analytical Carrier Density and Quantum Capacitance for Graphene. *Appl. Phys. Lett.* **2016**, *108*, 013503. [CrossRef]
4. Saremi, M. Carrier Mobility Extraction Method in ChGs in the UV Light Exposure. *Micro Nano Lett.* **2016**, *11*, 762–764. [CrossRef]
5. Saremi, M.; Rajabi, S.; Barnaby, H.J.; Kozicki, M.N. The Effects of Process Variation on the Parametric Model of the Static Impedance Behavior of Programmable Metallization Cell (PMC). *MRS Proc.* **2014**, *1692*, 9–39. [CrossRef]
6. Saremi, M. A Physical-based Simulation for the Dynamic Behavior of Photodoping Mechanism in the Chalcogenide Materials Used in the Lateral Programmable Metallization Cells. *Solid State Ion.* **2016**, *290*, 1–5. [CrossRef]
7. Horowitz, G. Organic Thin Film Transistors: From Theory to Real Devices. *J. Mater. Res.* **2004**, *19*, 1946–1962. [CrossRef]
8. Liu, W.Y.; Wu, Q.S.; Ren, Y.K.; Cui, P.; Yao, B.B.; Li, Y.B.; Hui, M.; Jiang, T.Y.; Bai, L. On the bipolar DC flow field-effect-transistor for multifunctional sample handing in microfluidics: A theoretical analysis under the Debye-Huckel limit. *Micromachines* **2018**, *9*, 82. [CrossRef] [PubMed]
9. RSporea, A.; Trainor, M.J.; Young, N.D.; Shannon, J.M.; Silva, S.R.P. Source-gated transistors for order-of-magnitude performance improvements in thin-film digital circuits. *Sci. Rep.* **2014**, *4*, 4295. [CrossRef] [PubMed]
10. Itoh, T.; Kobayashi, A.; Ueno, K.; Ohta, J.; Fujioka, H. Fabrication of InGaN thin-film transistors using pulsed sputtering deposition. *Sci. Rep.* **2016**, *6*, 29500. [CrossRef] [PubMed]
11. Weimer, P.K. The TFT-A New Thin-Film Transistor. *Proc. IRE* **1962**, *50*, 1462–1469. [CrossRef]
12. Fortunato, E.; Barquinha, P.; Pimented, A.; Goncalves, A.; Marques, A.; Pereira, L.; Martins, R. Recent Advances in ZnO Transparent Thin Film Transistors. *Thin Solid Films* **2005**, *487*, 205–211. [CrossRef]
13. LeComber, P.G.; Spear, W.E.; Gaith, A. Amorphous-Silicon Field-Effect Device and Possible Application. *Electron. Lett.* **1979**, *15*, 179–181. [CrossRef]
14. Depp, S.W.; Juliana, A.; Huth, B.G. Polysilicon FET Devices for Large Area Input/Output Application. In Proceedings of the 1980 International Electron Devices Meeting, Washington, DC, USA, 8–10 December 1980; p. 703.
15. Tsumura, A.; Koezuka, H.; Ando, T. Macromolecular Electronic Device: Field-Effect Transistor with a Polythiophene Thin Film. *Appl. Phys. Lett.* **1986**, *49*, 1210. [CrossRef]
16. Xu, W.; Hu, Z.H.; Liu, H.M.; Lan, L.F.; Peng, J.B.; Wangn, J.; Cao, Y. Flexible all-organic, all-solution processed thin film transistor array with ultrashort channel. *Sci. Rep.* **2016**, *6*, 29055. [CrossRef] [PubMed]
17. Nomura, K.; Ohta, H.; Takagi, A.; Kamiya, T.; Hirano, M.; Hosono, H. Room-Temperature Fabrication of Transparent Flexible Thin-Film Transistors Using Amorphous Oxide Semiconductors. *Nature* **2004**, *432*, 488–492. [CrossRef] [PubMed]
18. Han, K.; Qiao, G.H.; Deng, Z.L.; Zhang, Y.N. Asymmetric drain extension dual-kk trigate underlap FinFET based on RF/Analog circuit. *Micromachines* **2017**, *8*, 330. [CrossRef] [PubMed]
19. Li, J.F.; Mao, S.M.; Xu, Y.H.; Zhao, X.D.; Wang, W.B.; Guo, F.J.; Zhang, Q.F.; Wu, Y.Q.; Zhang, B.; Chen, T.S. An Improved Large Signal Model for 0.1 μm AlGaN/GaN High Electron Mobility Transistors (HEMTs) Process and Its Applications in Practical Monolithic Microwave Integrated Circuit (MMIC) Design in W band. *Micromachines* **2018**, *9*, 396. [CrossRef] [PubMed]
20. Buccella, P.; Stefanucci, C.; Zou, H.; Moursy, Y.; Iskander, R.; Sallese, J.M.; Kayal, M. Methodology for 3-D Substrate Network Extraction for SPICE Simulation of Parasitic Currents in Smart Power ICs. *IEEE Trans. Comput.-Aided Des. Integr. Circuits Syst.* **2016**, *35*, 1489–1502. [CrossRef]
21. Lu, N.D.; Wang, L.F.; Li, L.; Liu, M. A Review for Compact Model of Graphene Field-Effect Transistors. *Chin. Phys. B* **2017**, *26*, 036804. [CrossRef]
22. Krishnamoorthy, S.; Chowdhury, M.H. Investigation and a practical compact network model of thermal stress in integrated circuits. *Integr. Comput.-Aided Eng.* **2009**, *16*, 131–140. [CrossRef]

23. Available online: https://www.src.org/program/grc/ds/research-needs/2003/compact-modeling.pdf (accessed on 14 April 2003).

24. Marinov, O.; Deen, M.J.; Zschieschang, U.; Klauk, H. Organic Thin-Film Transistors: Part I-Compact DC Modeling. *IEEE Trans. Electron Dev.* **2009**, *56*, 2952–2961. [CrossRef]

25. Kacprzak, T.; Materka, A. Compact DC Model of GaAs FETs for Llarge-Signal Computer Calculation. *IEEE J. Solid-State Circuits* **1983**, *18*, 211–213. [CrossRef]

26. Ahmed, S.; Kim, D.; Shichijo, H. A Comprehensive Analytic Model Accumulation-Mode MOSFETs in Polysilicon Thin Films. *IEEE Trans. Electron Devices* **1986**, *33*, 973–985. [CrossRef]

27. Siddiqui, M.J.; Qureshi, S. Surface-Potential-Based Charge Sheet Model for The Polysilicon Thin Film Transistors without Considering Kink Effect. *Microelectron. J.* **2001**, *32*, 235–240. [CrossRef]

28. Iñiguez, B.; Xu, A.; Fjeldly, T.A.; Shur, M.S. Unifed Model for Short-Channel Poly-Si TFTs. *Solid-State Electron.* **1999**, *43*, 1821–1831. [CrossRef]

29. Jacunski, M.D.; Shur, M.S.; Owusu, A.A. A Short-Channel DC SPICE Model for Polysilicon Thin-Film Transistors Including Temperature Effects. *IEEE Trans. Electron Devices* **1999**, *46*, 1146–1159. [CrossRef]

30. Wu, W.J.; Yao, R.H.; Li, S.H.; Hu, Y.F.; Deng, W.L.; Zheng, X.R. A Compact Model for Polysilicon TFTs Leakage Current Including the Poole-Frenkel Effect. *IEEE Trans. Electron Devices* **2007**, *54*, 2975–2984. [CrossRef]

31. Lee, M.; Tai, C.W.; Huang, J.J. Correlation between Gap State Density and Bias Stress Reliability of Nanocrystalline TFTs Comparing with Hydrogenated Amorphous Silicon TFTs. *Solid-State Electron.* **2013**, *80*, 72–75. [CrossRef]

32. Deng, W.L.; Huang, J.K. A Physics-Based Approximation for the Polysilicon Thin-Film Transistor Surface Potential. *IEEE Electron. Dev. Lett.* **2011**, *32*, 647–649. [CrossRef]

33. Li, L.; Debucquoy, M.; Genoe, J.; Heremans, P. A Compact Model for Polycrystalline Pentacene Thin-Film Transistor. *J. Appl. Phys.* **2010**, *107*, 024519. [CrossRef]

34. Hao, Z.W.S.; Duval, J.; Ravelosona, D.; Klein, J.O.; Kim, J.V.; Chappert, C. A Compact Model of Domain Wall Propagation for Logic and Memory Design. *J. Appl. Phys.* **2011**, *109*, 07D501. [CrossRef]

35. Valletta, A.; Demirkol, A.S.; Maira, G.; Frasca, M.; Vinciguerra, V.; Occhipinti, L.G.; Fortuna, L.; Mariucci, L. A Compact SPICE Model for Organic TFTs and Applications to Logic Circuit Design. *IEEE Trans. Nanotech.* **2016**, *15*, 754–761. [CrossRef]

36. Hsieh, H.H.; Kamiya, T.; Nomura, K.; Hosono, H.; Wu, C.C. Modeling of Amorphous InGaZnO4 Thin Film Transistors and Their Subgap, Density of States. *Appl. Phys. Lett.* **2008**, *92*, 133503. [CrossRef]

37. Ghittorelli1, M.; Torricelli1, F.; Steen, J.J.V.D.; Garripoli, C.; Tripathi, A.; Gelinck, G.H.; Cantatore, E.; Kovács-Vajna, Z.M. Physical-Based Analytical Model of Flexible a-IGZO TFTs Accounting for Both Charge Injection and Transport. In Proceedings of the 2015 IEEE International Electron Devices Meeting (IEDM), Washington, DC, USA, 7–9 December 2015; Volume 28, pp. 723–726.

38. Duavll, S.G. Statical Circuit Modeling and Optimization. In Proceedings of the IEEE 2000 5th International Workshop on Statistical Metrology, Honolulu, HI, USA, 10 June 2000.

39. Ortiz-Conde, A.; García-Sánchez, F.J.; Muci, J.; Malobabic, S.; Liou, J.J. A Review of Core Compact Models for Undoped Double-Gate SOI MOSFETs. *IEEE Trans. Electron Devices* **2007**, *54*, 131–140. [CrossRef]

40. Gildenblat, G.; Li, X.; Wu, W.M.; Wang, H.L.; Jha, A.; Langevelde, R.V.; Smit, G.D.J.; Scholten, A.J.; Klassen, D.B.M. PSP: An Advanced Surface-Potential-Based MOSFET Model for Circuit Simulation. *IEEE Trans. Electron Devices* **2006**, *53*, 1979–1994. [CrossRef]

41. Song, J.; Yu, B.; Yuan, Y.; Taur, Y. A Review on Compact Modeling of Multiple-Gate MOSFETS. *IEEE Trans. Electron Devices* **2009**, *56*, 1858–1869.

42. Cheng, X.; Lee, S.; Yao, G.Y.; Nathan, A. TFT Compact Modeling. *J. Disp. Technol.* **2016**, *12*, 898–906. [CrossRef]

43. Baccarani, G.; Riccò, B. Transport Properties of Polycrystalline Silicon Films. *J. Appl. Phys.* **1978**, *49*, 5565–5571. [CrossRef]

44. Seto, J.Y.W. The Electrical Properties of Polycrystalline Silicon Films. *J. Appl. Phys.* **1975**, *46*, 5247–5254. [CrossRef]

45. Rech, B.; Wagner, H. Potential of Amorphous Silicon for Solar Cells. *Appl. Phys. A* **1999**, *69*, 155–157. [CrossRef]

46. Bagolini, L.; Mattoni, A.; Collins, R.T.; Lusk, M.T. Carrier Localization in Nanocrystalline Silicon. *J. Phys. Chem. C* **2014**, *118*, 13417–13423. [CrossRef]

47. Lu, N.D.; Li, L.; Liu, M. Universal Carrier Thermoelectric-Transport Model Based on Percolation Theory in Organic Semiconductors. *Phys. Rev. B* **2015**, *91*, 195205. [CrossRef]

48. Li, L.; Lu, N.D.; Liu, M.; Bässler, H. General Einstein Relation Model in Disordered Organic Semiconductors under Quasiequilibrium. *Phys. Rev. B* **2014**, *90*, 214107. [CrossRef]

49. Lu, N.D.; Li, L.; Banerjee, W.; Liu, M. Physical Model of Seebeck Coefficient under Surface Dipole Effect in Organic Thin-Film Transistors. *Org. Electron.* **2016**, *29*, 27–32. [CrossRef]

50. Lu, N.D.; Li, L.; Liu, M. A Review of Carrier Thermoelectric-Transport Theory in Organic Semiconductors. *Phys. Chem. Chem. Phys.* **2016**, *18*, 19503–19525. [CrossRef] [PubMed]

51. Ambegaokar, V.; Halperin, B.I.; Langer, J.S. Hopping Conductivity in Disordered Systems. *Phys. Rev. B* **1971**, *4*, 2612–2620. [CrossRef]

52. Miller, A.; Abrahams, E. Impurity Conduction at Low Concentrations. *Phys. Rev.* **1960**, *120*, 745–755. [CrossRef]

53. Klauk, H. Organic Thin-Film Transistors. *Chem. Soc. Rev.* **2010**, *39*, 2643–2666. [CrossRef] [PubMed]

54. Nomura, K.; Kamiya, T.; Ohta, H.; Ueda, K.; Hirano, M.; Hosono, H. Carrier Transport in Transparent Oxide Semiconductor with Intrinsic Structural Randomness Probed Using Single-crystalline $InGaO_3(ZnO)_5$ Films. *Appl. Phys. Lett.* **2004**, *85*, 1993–1995. [CrossRef]

55. Germs, W.C.; Adriaans, W.H.; Tripathi, A.K.; Roelofs, W.S.C.; Cobb, B.; Janssen, R.A.J.; Gelinck, G.H.; Kemerink, M. Charge Transport in Amorphous InGaZnO Thin-Film Transistors. *Phys. Rev. B* **2012**, *86*, 155319. [CrossRef]

56. Lu, N.; Li, L.; Sun, P.; Banerjee, W.; Liu, M. A Unified Physical Model of Seebeck Coefficient in Amorphous Oxide Semiconductor Thin-Film Transistors. *J. Appl. Phys.* **2014**, *116*, 104502. [CrossRef]

57. Li, L.; Lu, N.D.; Liu, M. Field Effect Mobility Model in Oxide Semiconductor Thin Film Transistors With Arbitrary Energy Distribution of Traps. *IEEE Electron Device Lett.* **2014**, *35*, 226–228. [CrossRef]

58. Podzorov, V. Charge Carrier Transport in Single-crystal Organic Field-Effect Transistors. In *Organic Field Effect Transistor*; CRC Press: Boca Raton, FL, USA, 2007; pp. 27–72.

59. Sze, S.M.; Ng, K.K. *Physics of Semiconductor Devices*, 3rd ed.; Wiley: New York, NY, USA, 2007.

60. Jung, L.; Damiano, J.; Zaman, J.R.; Batra, S.; Manning, M.; Banerjee, S.K. A Leakage Current Model for Sub-Micron Lightly-Doped Drain-Offset Polysilicon TFTs. *Solid-State Electron.* **1995**, *38*, 2069–2073. [CrossRef]

61. Khakzar, K.; Lueder, E.H. Modeling of Amorphous-Silicon Thin-Film Transistors for Circuit Simulations with SPICE. *IEEE Trans. Electron Devices* **1992**, *39*, 1428–1435. [CrossRef]

62. Faughnan, B. Subthreshold Model of A Polycrystalline-Silicon Thin Film Field-Effect Transistor. *Appl. Phys. Lett.* **1987**, *50*, 290–292. [CrossRef]

63. Shur, M.S.; Slade, H.C.; Jacunski, M.D.; Owusu, A.A.; Ytterdal, T. SPICE Models for Amorphous Silicon and Polysilicon Thin Film Transistors. *J. Electrochem. Soc.* **1997**, *144*, 2833–2839. [CrossRef]

64. Siddiqui, M.J.; Qureshi, S. An Empirical Model for Leakage Current in Poly-Silicon Thin Film Transistor. *Solid State Electron.* **2000**, *44*, 2015–2019. [CrossRef]

65. Iñiguez, B.; Fjeldly, T.A.; Shur, M.S. *Thin Film Transistor Modeling, in Silicon and Beyond: Advanced Circuit Simulators and Device Models*; Shur, M.S., Fjeldly, T.A., Eds.; World Scientific Publishers: Singapore, 2000; pp. 703–723.

66. Tsividis, Y.P.; Suyama, K. MOSFET Modeling for Analog Circuit CAD: Problems and Prospects. *IEEE J. Solid-State Circuits* **1994**, *29*, 210–216. [CrossRef]

67. Tsuji, H.; Kuzuoka, T.; Kishida, Y.; Kirihara, M.; Kamakura, Y.; Morifuji, M.; Shimizu, Y.; Miyano, S.; Taniguchi, K. A New Surface Potential Based Poly-Si TFT Model for Circuit Simulation. In Proceedings of the 2006 International Electron Devices Meeting, San Francisco, CA, USA, 11–13 December 2006; pp. 179–186.

68. Leroux, T. Static and Dynamic Analysis of Amorphous-Silicon Field-Effect Transistors. *Solid-State Electron.* **1986**, *29*, 47–58. [CrossRef]

69. Brews, J.R. A Charge-Sheet Model of the MOSFET. *Solid-State Electron.* **1978**, *21*, 345–355. [CrossRef]

70. Sleight, J.W.; Rios, R. A Continuous Compact MOSFET Model for Fully- and Partially-Depleted SOI Devices. *IEEE Trans. Electron Devices* **1998**, *45*, 821–825. [CrossRef]

71. Pao, H.C.; Sah, C.T. Effects of Diffusion Current on Characteristics of Metal-Oxide (Insulator)-Semiconductor Transistors. *Solid-State Electron.* **1966**, *9*, 927–937. [CrossRef]

72. Chen, R.S.; Zheng, X.R.; Deng, W.L.; Wu, Z.H. A Physics-Based Analytical Solution to the Surface Potential of Polysilicon Thin Film Transistors Using the Lambert W Function. *Solid-State Electron.* **2007**, *51*, 975–981. [CrossRef]

73. Corless, R.M.; Gonnet, G.H.; Hare, D.E.G.; Jeffrey, D.J.; Knuth, D.E. On Lambert's W function. *Adv. Comput. Math.* **1996**, *5*, 329–359. [CrossRef]

74. G Gildenblat, T.L.C.; Gu, H.W.; Cai, X. SP: An Advanced Surface Potential-Based Compact MOSFET Model. *IEEE J. Solid-State Circuits* **2004**, *39*, 1394–1406. [CrossRef]

75. Hack, M.; Shaw, J. Transient Simulations of Amorphous-Silicon Devices, Materials Research Soc. *Symp. Proc.* **1991**, *219*, 315–320. [CrossRef]

76. Shur, M.S.; Jacunski, M.D.; Slade, H.C.; Hack, M. Analytical Models for Amorphous-Silicon and Polysilicon Thin-Film Transistor for High-Definition-Display Technology. *J. Soc. Inf. Disp.* **1995**, *4*, 223–236. [CrossRef]

77. Servati, P.; Nathan, A. Modeling of the Static and Dynamic Behavior of Hydrogenated Amorphous Silicon Thin-Film Transistors. *J. Vac. Sci. Technol. A* **2002**, *20*, 1038–1042. [CrossRef]

78. Colalongo, L. A New Analytical Model for Amorphous-Silicon Thin-Film Transistors Including Tail and Deep States. *Solid-State Electron.* **2001**, *45*, 1525–1530. [CrossRef]

79. Servati, P.; Striakhilev, D.; Nathan, A. Above-Threshold Parameter Extraction and Modeling for Amorphous Silicon Thin-Film Transistors. *IEEE Trans. Electron Devices* **2003**, *50*, 2227–2235. [CrossRef]

80. Liu, Y.; Yao, R.H.; Li, B.; Deng, W.L. An Analytical Model Based on Surface Potential for a-Si:H Thin-Film Transistors. *J. Disp. Technol.* **2008**, *4*, 180–187.

81. Qin, J.; Yao, R.H. A Physics-Based Scheme for Potentials of a-Si:H TFT with Symmetric Dual Gate Considering Deep Gaussian DOS Distribution. *Solid-State Electron.* **2014**, *95*, 46–51. [CrossRef]

82. Bässler, H. Charge Transport in Disordered Organic Photoconductors, a Monte Carlo Simulation Study. *Phys. Status Solidi B* **1993**, *175*, 15–56. [CrossRef]

83. Wang, L.; Lu, N.D.; Li, L.; Ji, Z.Y.; Banerjee, W.; Liu, M. Compact Model for Organic Thin-Film Transistor with Gaussian Density of States. *AIP Adv.* **2015**, *5*, 047123. [CrossRef]

84. Coehoorn, R.; Pasveer, W.F.; Bobbert, P.A.; Michels, M.A.J. Charge-Carrier Concentration Dependence of The Hopping Mobility in Organic Materials with Gaussian Disorder. *Phys. Rev. B* **2005**, *72*, 155206. [CrossRef]

85. Li, L.; Chung, K.-S.; Jang, J. Field Effect Mobility Model in Organic Thin Film Transistor. *Appl. Phys. Lett.* **2011**, *98*, 023305. [CrossRef]

86. Zong, W.W.; Li, L.; Jang, J.; Lu, N.D.; Liu, M. Analytical Surface-Potential Compact Model for Amorphous-IGZO Thin-Film Transistors. *J. Appl. Phys.* **2015**, *117*, 215705. [CrossRef]

87. Zong, Z.W.; Li, L.; Jang, J.; Li, Z.G.; Lu, N.D.; Shang, L.W.; Ji, Z.Y.; Liu, M. A New Surface Potential-Based Compact Model for a-IGZO TFTs in RFID Applications. In Proceedings of the 2014 IEEE International Electron Devices Meeting, San Francisco, CA, USA, 15–17 December 2014; Volume 35, pp. 860–863.

88. Ghittorelli, M.; Torricelli, F.; Colalongo, L.; Vajna, Z.M.K. Accurate analytical physical modeling of amorphous InGaZnO thin-film transistors accounting for trapped and free charges. *IEEE Trans. Electron Devices* **2014**, *61*, 4105–4112. [CrossRef]

89. Liu, Y.; Yao, R.H.; Li, B.; Xie, W.N. A Physical Model Based on Surface Potential for Double-Gate a-Si:H TFTs. In Proceedings of the 2009 IEEE International Conference of Electron Devices and Solid-State Circuits (EDSSC), Xi'an, China, 25–27 December 2009.

90. Tsormpatzoglou, A.; Hastas, N.A.; Choi, N.; Mahmoudabadi, F.; Hatalis, M.K.; Dimitriadis, C.A. Analytical Surface-Potential-Based Drain Current Model for Amorphous InGaZnO Thin Film Transistors. *J. Appl. Phys.* **2013**, *114*, 184502. [CrossRef]

91. Bendix, P.; Rakers, P.; Wagh, P.; Lemaitre, L.; Grabinski, W.; McAndrew, C.; Gu, X.; Gildenblat, G. RF distortion analysis with compact MOSFET models. In Proceedings of the IEEE 2004 Custom Integrated Circuits Conference (IEEE Cat. No.04CH37571), Orlando, FL, USA, 6 October 2004; pp. 9–12.

92. McAndrew, C.C. Validation of MOSFET Model Source-Drain Symmetry. *IEEE Trans. Electron Devices* **2006**, *53*, 2202–2206. [CrossRef]

micromachines

MDPI

Article

The Balancing Act in Ferroelectric Transistors: How Hard Can It Be?

Raymond J. E. Hueting

MESA+ Institute for Nanotechnology, University of Twente, P.O. Box 217,
7500AE Enschede, The Netherlands; r.j.e.hueting@utwente.nl

Received: 19 October 2018; Accepted: 5 November 2018; Published: 7 November 2018

Abstract: For some years now, the ever continuing dimensional scaling has no longer been considered to be sufficient for the realization of advanced CMOS devices. Alternative approaches, such as employing new materials and introducing new device architectures, appear to be the way to go forward. A currently hot approach is to employ ferroelectric materials for obtaining a positive feedback in the gate control of a switch. This work elaborates on two device architectures based on this approach: the negative-capacitance and the piezoelectric field-effect transistor, i.e., the NC-FET (negative-capacitance field-effect transistor), respectively π-FET. It briefly describes their operation principle and compares those based on earlier reports. For optimal performance, the adopted ferroelectric material in the NC-FET should have a relatively wide polarization-field loop (i.e., "hard" ferroelectric material). Its optimal remnant polarization depends on the NC-FET architecture, although there is some consensus in having a low value for that (e.g., HZO (Hafnium-Zirconate)). π-FET is the piezoelectric coefficient, hence its polarization-field loop should be as high as possible (e.g., PZT (lead-zirconate-titanate)). In summary, literature reports indicate that the NC-FET shows better performance in terms of subthreshold swing and on-current. However, since its operation principle is based on a relatively large change in polarization the maximum speed, unlike in a π-FET, forms a big issue. Therefore, for future low-power CMOS, a hybrid solution is proposed comprising both device architectures on a chip where hard ferroelectric materials with a high piezocoefficient are used.

Keywords: CMOS; field-effect transistor; ferroelectrics; MOS devices; negative-capacitance; piezoelectrics; power consumption

---

1. Introduction

As is commonly known, the key component of the microprocessor, the conventional metal-oxide-semiconductor field-effect transistor (MOSFET), needs some refurbishment. The traditional dimensional scaling of this device as proposed earlier [1] no longer suffices to cope with present day requirements. Other device architectures, such as FinFETs [2,3] and ultrathin-body (UTB) devices [4,5], and integration of other materials, such as silicon-germanium (e.g., [6]), are presently in production.

However, despite these adjustments, for several years, the maximum supply voltage of the microprocessor has been around 0.7–0.8 V. The main reason for this is to limit the static (or off-state) power consumption that is governed by the off-state current (I_{OFF}) [7], and that rises exponentially for a reduced threshold voltage. The latter is because the current below the threshold voltage, i.e., the subthreshold current, is a diffusion or thermionic emission current. Its slope against the gate-source voltage (V_{GS}) has a maximum value dictated by Boltzmann's tyranny being $\sim$60 mV/dec at room temperature, or, equivalently, a minimum ideality factor m equal to unity.

To break this tyranny, alternative device architectures have been proposed based on other physical principles, such as tunnel FETs [8–11] and impact-ionization MOSFETs [12]. Although these architectures have a strong potential, the realization of those are relatively difficult.

Later, various device architectures have been proposed in which a ferroelectric (FE) layer has been embedded in a conventional MOSFET (see Figure 1): the negative-capacitance field-effect transistor (NC-FET) [13,14], and the piezoelectric field-effect transistor (π-FET) [15,16]. In particular, the NC-FET is currently a hot topic since, in principle, it only requires a single additional FE layer that is compatible with CMOS technology. Although the charge transport physics in both device architectures is basically that of a conventional MOSFET, the physics involved inside the gate stack is different, as discussed in the following sections.

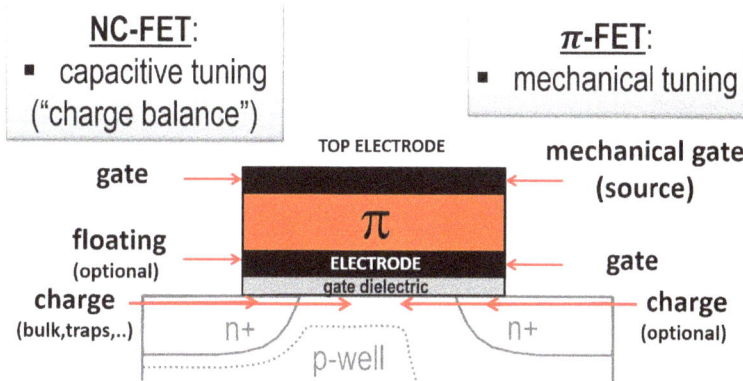

Figure 1. Schematic cross-section of a bulk NC-FET (left) and π-FET. Both device concepts comprise a ferroelectric material for obtaining a positive feedback in the gate control of the current in a conventional MOSFET. In case of the NC-FET, the top electrode represents the gate, and there is an optional floating electrode for technological reasons and to smear out potential fluctuations in the channel. The amount of charge ("charge balance") in or near the body/channel is important here. For the π-FET, the top electrode is a rigid mechanical gate connected to the source, and there is a conventional gate. The electromechanical properties of the semiconductor and mechanical boundary conditions are important here. The figures are not at scale.

Before discussing the device architectures, first the physics and characteristics of FE materials should be briefly explained. For a more thorough overview, refer to [17,18].

FE materials consist of fixed ions in preferably a perfect crystalline lattice. Depending on the asymmetry in the lattice those ions form dipoles, which in turn form domains depending on the quality of the material. Because of those dipoles, a hysteretic polarization-electric field (P-$\mathcal{E}$) curve is obtained (see Figure 2). Important figures-of-merit (FOMs) in FE materials are the coercive electric field ($\mathcal{E}_C$) or coercive voltage, the remnant polarization (P_r), and the saturation (or spontaneous) polarization (P_s). $\mathcal{E}_C$ represents the strength of the applied field for which the polarization direction of (most of) the dipoles flips in the opposite direction. P_r represents the resulting polarization value obtained when the applied varying (increasing or decreasing) electric field becomes zero ($\mathcal{E} = 0$). Furthermore, since FE materials are piezoelectric (π-) materials (not vice versa), the π-coefficient (d_{33}) is also an important FOM in this context. This parameter in turn depends on P_s [19,20]. The reason why for many sensor and transducer applications FE materials are used is because those materials, in particular perovskite ferroics (e.g., lead-zirconate-titanate, PZT), have a relatively high P_s hence d_{33} value. Note that, for most cases, the polarization can be simply be considered to be equal to the areal charge density (Q).

This work is outlined as follows. In Sections 2 and 3, the basic operation principle of the NC-FET respectively π-FET is briefly explained. In Section 4, both devices are compared based on previous reports. Finally, in Section 5, the conclusions are drawn.

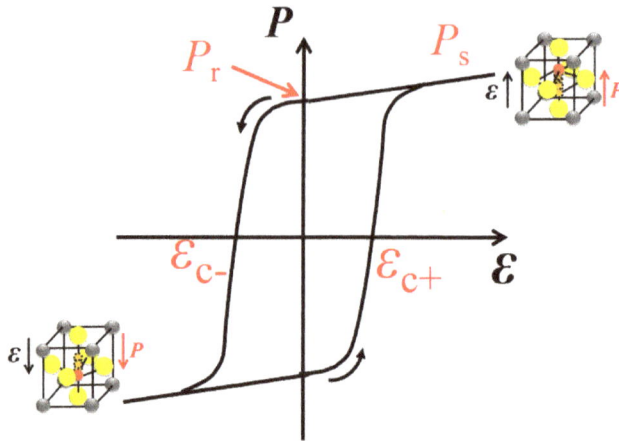

Figure 2. Schematic P-$\mathcal{E}$ curve illustrating the hysteretic effect caused by internal dipoles. The remnant polarization P_r represents the polarization at zero field. The saturation polarization P_s is the maximum polarization at which a spontaneously formed dipole moment has been formed. $\mathcal{E}_{C-}$ and $\mathcal{E}_{C+}$ are the coercive fields at which the hysteresis loop intersects the negative respectively positive field axis.

2. The Negative-Capacitance Field-Effect Transistor

Recently, in the semiconductor community, a wide interest has formed for adopting the negative-capacitance (NC-) effect in reducing the power consumption of the FET, as originally proposed by Salahuddin and Datta [13,14]. Figure 3 shows a schematic cross-section of the NC-FET in which a poled FE layer is integrated, denoted π. The NC-effect originates from the hysteretic P-$\mathcal{E}$ loop that is present in ferroelectrics [17]. According to a proposed theory [21,22], during switching in field polarity, the ferroelectric polarization state does not follow the hysteretic loop. Instead, an internal S-shaped P-$\mathcal{E}$ curve is followed in which a part of its slope becomes negative. Hence, a negative permittivity, and thus a correspondingly negative capacitance are also obtained. This NC-effect is difficult to measure directly; it is energetically an unstable situation. However, by placing the FE layer on top of (or below) a conventional dielectric layer (e.g., SiO_2 or Si) in a metal-insulator-metal (MIM) stack, the total capacitance of the stack increases, as experimentally shown for PZT/STO MIM capacitors [23]. (STO or $SrTiO_3$ stands for strontium-titanate).

The drain current of a long channel MOSFET depends on the voltage division between the oxide or insulator capacitance (C_π) and the silicon capacitance (C_s), or to be more specific by the surface potential [24]:

$$\psi_s = V_{GS} \cdot \frac{C_{tot}}{C_s} = V_{GS} \cdot \frac{C_\pi}{C_s + C_\pi} = V_{GS} \cdot \frac{1}{m}, \tag{1}$$

with C_{tot} being the total capacitance of the gate-stack. If we now consider that C_π is negative such that C_s is just a little bit smaller than C_π, then there is voltage amplification ($\psi_s / V_{GS} \gg 1$) and, therefore, the ideality factor m becomes less than one ($m \ll 1$) [13,23,25]. Consequently, a steep subthreshold swing (SS) can be obtained. In this way, a maximum C_{tot} is obtained excluding any undesired hysteresis effects.

Much research work followed studying the NC-effect (e.g., [26–35]). However, it has been debated that there is a limit to this NC-effect because of the use of multi-domain ferroelectrics [36]. Furthermore, in practice, this so-called "charge-balance" [37] is difficult to obtain partly because the actual NC value is not accurately known. For obtaining this charge-balance, an extensively combined experiment-modeling effort is required for finally obtaining an improved FET. In addition, interface traps in the gate stack could also affect this charge balance [38] which can be understood from Equation (1) in case C_{tot} also includes the interface trap capacitance. Note that in case of multiple interfacial layers in the gate stack C_s in Equation (1) should be replaced by a capacitance representing the stack below the FE material (e.g., C_{MOS} [39]).

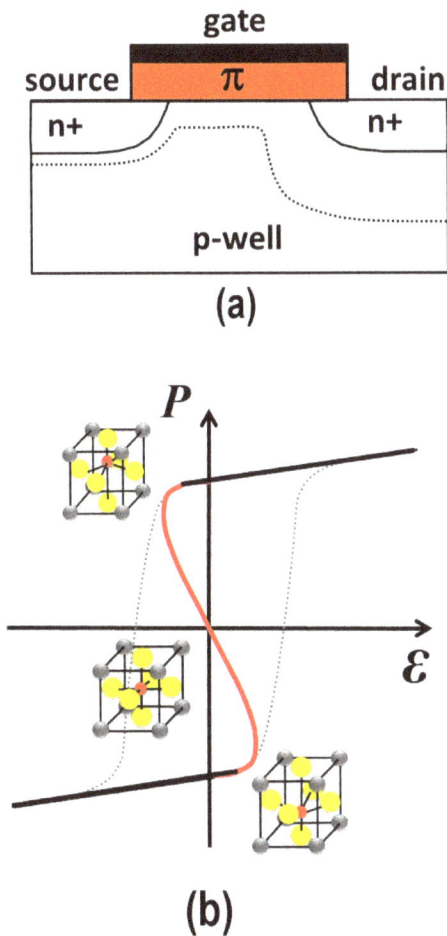

(a)

(b)

Figure 3. (a) schematic cross-section of the originally proposed NC-FET; (b) the "S"-shaped P-$\mathcal{E}$ curve highlighting the negative slope $\partial P/\partial \mathcal{E}$ hence negative capacitance. The NC-FET operation is inside the P-$\mathcal{E}$ loop.

The operation of the NC-FET is inside the hysteretic P-$\mathcal{E}$ loop. This implies that in practice a relatively high $C_\pi = \partial P/\partial \mathcal{E}$ is required and that strain values are relatively low. For an effective use, a relatively wide $P - \mathcal{E}$ loop is needed [40] such that the coercive voltage is higher than the supply voltage of the FET ($V_C > V_{DD}$). However, the optimal P_r depends on the device architecture. For instance, for long channel bulk MOSFETs, it was experimentally shown that a perovskite FE material such as PZT ($P_r \approx 40\ \mu C/cm^2$) results in an SS of ~ 38 mV/dec [34]. Conversely, in fully-depleted (FD) devices, such as a FinFET, a "hard" FE material such as Hafnium-Zirconate (HZO, $P_r \approx 3\ \mu C/cm^2$)) was employed reaching an SS of ~ 55 mV/dec [32].

For illustration purposes, Figure 3 shows a symmetric hysteretic P-$\mathcal{E}$ loop. Such a symmetry strongly depends on the internal charge distribution of the FE layer induced either by the workfunction difference between its incorporated top and bottom electrodes (or semiconductor body) or by its internal parasitic (e.g., fixed) charge distribution. For instance, provided that the workfunction of the top metal is higher than that of the semiconductor body (or bottom electrode), the P-$\mathcal{E}$ curve shifts to the right on the horizontal $\mathcal{E}$-scale. This will cause an asymmetry in the P-$\mathcal{E}$ curve, not desired for the NC-FET since that will reduce the operating bias range. If there was no difference between the electrode workfunctions and no parasitic charge, then the P-$\mathcal{E}$ curve would be symmetric.

3. The Piezoelectric Field-Effect Transistor

As stated before, FE materials are π-materials. In addition, generally, a π-material basically comprises dipoles because of the presence of fixed ions and asymmetry in the lattice. When stressing such material, there is some deformation, hence strain, which also results in a change in polarization, and therefore an electric field has formed. This is called the π-effect. The opposite is also possible, i.e., the converse π-effect [41]. When no mechanical stress (T) is applied to the π-material an external electric field ($\mathcal{E}$) induces a deformational strain S according to (Figure 4):

$$T = c \cdot S + e \cdot \mathcal{E}, \tag{2}$$

with c being Young's modulus and e being the π-charge constant.

In reality, all parameters in Equation (2) are tensors, but, for simplicity, we consider a one-dimensional system. Note that the π-coefficient $d_{33} = -e/c$, i.e., the amount of displacement or deformation in units length per unit applied voltage.

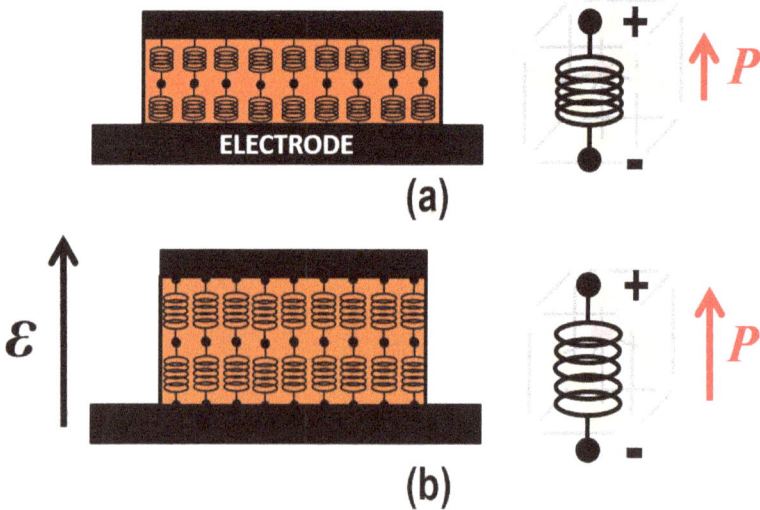

Figure 4. Illustration of the converse piezoelectric effect (**a**) at thermal equilibrium, and (**b**) when an external field has been applied. For illustration purposes, the change in thickness induced by this field has been exaggerated. In conventional piezoelectric materials, the polarization direction is fixed, while, in ferroelectric materials, the polarization depends on the electric field direction.

More recently, we have proposed the so-called piezoelectric field-effect transistor (π-FET) [15,16] in which an FE layer is implemented (see Figure 5). The basic idea is that the strain in the semiconductor body can be tuned by the converse π-effect. As a result, during device operation, the semiconductor body is relaxed in the off-state, resulting in a low off-current (I_{OFF}), and it is strained in the on-state, resulting in a high I_{ON}. The strain influences the semiconductor transport properties: it reduces the band gap and depending on the device/crystal orientation it also increases the charge carrier mobility. This reduces the SS [16].

In our theoretical and numerical work [16,42], we have predicted that there are experimental challenges to reduce the SS effectively: (1) ultrathin and relatively stiff interfacial layers in between the π-layer and semiconductor body are required, all with proper (rigid) mechanical boundary conditions, (2) the semiconductor body should preferably have a low stiffness and a high "effective" deformation potential, and (3) the π-material should have a high π-response d_{33} and a high electrical breakdown field.

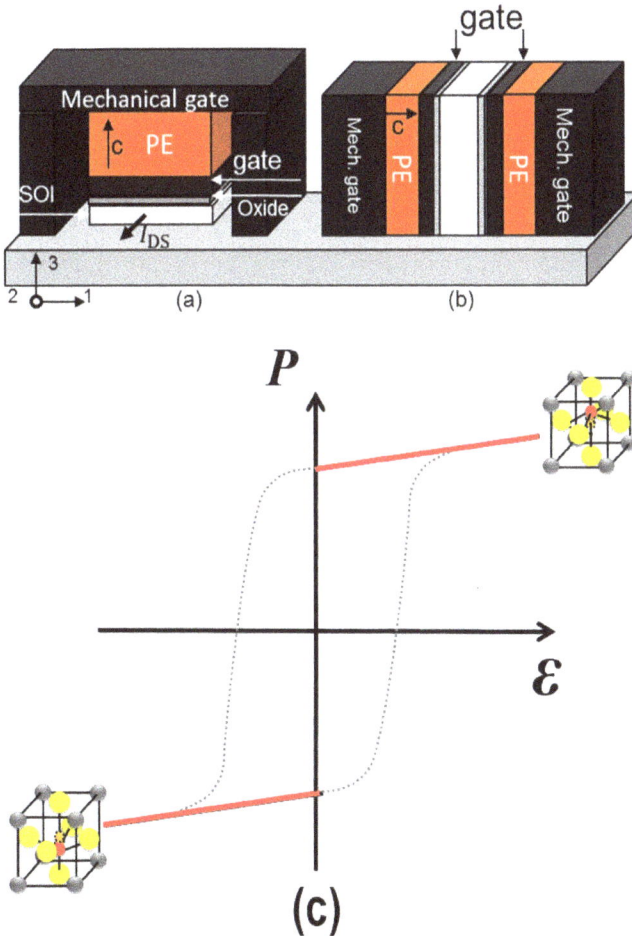

Figure 5. (**a,b**) schematic cross-section of the originally proposed π-FET (after [15]); (**c**) the highlighted operational regime of the π-FET which is outside the P-$\mathcal{E}$ loop.

Opposite to the NC-FET, the operation of the π-FET is outside the hysteretic P-$\mathcal{E}$ loop. This implies that a relatively low $\partial P/\partial \mathcal{E}$ is required and that the strain values are relatively high. As stated before, for improving the converse π-effect in a FET, obviously a relatively high d_{33} is needed. Since this parameter is proportional to P_s [20], this implies that a relatively high P-$\mathcal{E}$ loop is required. In most cases, perovskite FE materials should be used here. PZT, for instance, is such a material, its d_{33} ($\sim$110 pm/V) is much higher than that of conventional π materials such as aluminium-nitride (AlN) or zinc-oxide (ZnO). Therefore, FE materials are commonly used for various sensor and transducer applications [18]. Hard materials such as HZO have a d_{33} of $\sim$10 pm/V ($P_r \approx 8\ \mu C/cm^2$) [43] comparable to that of conventional π-materials such as AlN.

From the view point of the device performance of the π-FET, the symmetry of the hysteretic P-$\mathcal{E}$ loop is less important than it is for the NC-FET, mainly because the overall height of the P-$\mathcal{E}$ curve is essential in this case.

Note that, around the same time, the piezoelectronic transistor (PET) was reported [44] in which the converse π-effect was employed to a highly pressure-sensitive piezoresistive material (e.g., Samarium-selenide, SmSe). Since the transport physics is not based on that of conventional MOSFETs, it has further not been considered in this comparative work. However, there are some interesting features that should be highlighted. First, the PET comprises a perovskite FE material (e.g., PZN-PT, PMN-PT or even PZT) and the mechanics in the PET was treated thoroughly. Second, rigid boundary conditions were used, and also geometry effects were considered. However, very importantly, the breakdown field of the FE material was ignored.

Earlier [16], a combination of technology computer aided design (TCAD) [45] and multiphysics FEM [46] tools was used to estimate the SS in germanium (Ge) FinFETs showing values of around 50 mV/dec. In order to achieve this targeted SS value 5 nm wide, 45 nm high, and 80 nm long fins were adopted. Table 1 summarizes some SS values for both n-type and p-type FETs in various bulk semiconductor materials. Later, other theoretical works [47] reported lower SS values for both Si and Ge FinFETs ($\sim$40 mV/dec) possibly because of the use of a thin π-layer (3 nm) hence operating voltage (0.5 V). Typically, for n-type FETs, lower SS values are obtained which can be explained by the higher deformation potential in the conduction band than that in the valence band [42]. Moreover, Ge is the most attractive material because it has a lower Young's modulus and higher deformation potentials compared to the other bulk materials.

Table 1. Summary of analytically calculated SS values in π-FETs for various bulk semiconductor materials. nMOS and pMOS stands for an n-type and p-type MOSFET, respectively.

	SS (mV/dec), nMOS	SS (mV/dec), pMOS
Si	53.1	56.9
Ge	51.3	56.5
InSb	53.1	56.4

For the first time, in collaboration with the company SolMateS B.V. (Enschede, The Netherlands), we had realized a prototype of the π-FET comprising an Si FinFET wrapped around by (buffered) PZT (see Figure 6) or AlN as a π-material [48]. That work resulted in several new insights. First of all, the buffered PZT layer around the FinFET did not degrade the Si, as confirmed by our electrical measurements and X-ray photoelectron spectroscopy (XPS) analysis. This indicates that a π-material with a high π-response such as PZT can in fact be integrated in Si active devices. Secondly, we observed a lower SS compared to that of the conventional Si FinFET counterpart despite the relatively thick, 11 nm, silicon-dioxide (SiO$_2$) layer. Bulk SiO$_2$ has a relatively low Young's modulus (c_{11} = 57 GPa and c_{12} = 11.4 GPa) and consequently absorbs a big part of the strain [16]. Despite this fact, we observed that the converse π-effect reduced the SS by $\sim$5 mV/dec (but the SS was still higher than 60mV/dec because of interface traps) which can be attributed to the strain-induced reduction of the trap density at the Si/SiO$_2$ interface, as confirmed in the higher electron mobility values obtained in these structures at low vertical electric fields.

We also measured the π-response with the laser Doppler vibrometer [48], for the two π-materials (PZT: d_{33} = 110~110 pm/V, AlN: d_{33} = ~13 pm/V), indicating that the devices function well electromechanically. However, for an improved performance, the device should have been covered with a mechanically rigid layer to prevent mechanical energy losses in addition to the use of less, ultrathin and stiff interfacial layers. Furthermore, as discussed earlier [48], the conformality of the π-layer around the fin should be improved.

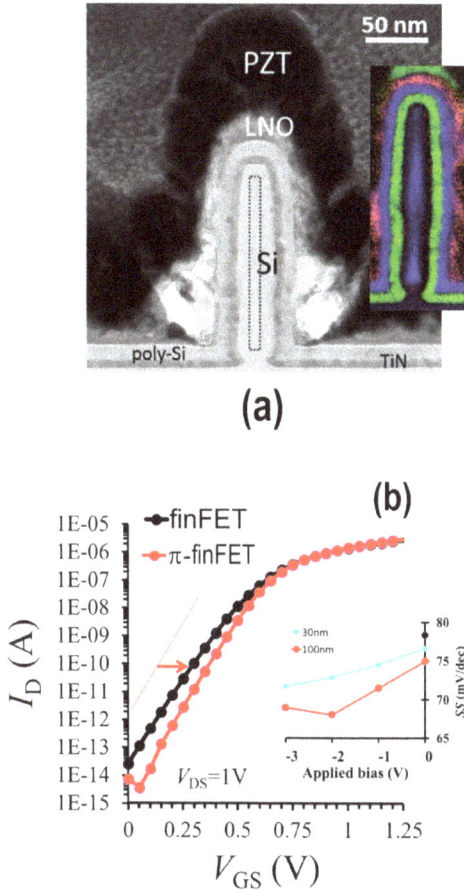

Figure 6. (**a**) High-resolution transmission electron microscopy (HR-TEM) image (incl. an red-green-blue (RGB) map: La is red, Ti is green, and Si is blue) of one fin of a silicon π-FinFET with a 20 nm fin width, and a 10 nm LNO (lanthanum-nickelate) and 100 nm PZT layer stack. The LNO layer is a buffer layer to avoid ferroelectric performance degradation of the PZT layer and atom interdiffusion through interfaces. The fin height is around 150 nm. In addition to the 10 nm LNO layer the gate stack comprises a 12 nm poly-Si, 10 nm SiO_2 and 6 nm TiN (gate metal). (**b**) Current-voltage (I_D-V_{GS}) characteristics of the π-FET compared to that of the conventional counterpart. Note that the π-FET curve has been shifted to match the I_{ON} ($\Delta V_{GS} \approx 0.2$ V). The inset shows the measured SS value against the (separate) bias over the π-layer for two different fin widths (30 and 100 nm).

4. Discussion

In this section, we compare the performance for both device concepts based on previous reports. Then, we comment on some typical issues of those concepts.

As stated before, there have been numerous experimental papers on the NC-FET concept (e.g., [26–35]), all showing impressive results. In particular, the use of PZT in bulk MOSFETs [34] yield a record value in the SS, while, in FinFETs, equally impressive results were shown with HZO [32]. Numerical calculations on short channel NC-FinFETs predicted aggressive SS values [39] as well, though with artificial FE parameters. Unfortunately, an elaborate combined modeling and experimental effort on this device concept has been lacking so far.

For the π-FET, on the other hand, though much less has been reported compared to the NC-FET, less aggressive SS numbers are expected. This means that the NC-FET is superior from the viewpoint of static power consumption (P_{stat}), and also I_{ON}/I_{OFF} ratios since the I_{ON} has dramatically been increased by the elevated gate capacitance in addition to the reduction in threshold voltage. The expected increased channel mobility in the π-FET will result in an increased I_{ON} but less compared to that of the NC-FET.

For dynamic switching of the FE material, other parameters play a role [49–52]. In particular, because of the $\partial P/\partial t$, the resistivity ρ (or damping constant, viscosity) of the FE material is important. This combined, with the fact that the NC-FET operation is based on a relatively high $C_\pi = \partial P/\partial \mathcal{E}$, ultimately show that the delay is determined by the total time constant. In perovskite FE materials, both parameters are relatively high [50,51], but still in hard FE materials, such as HZO, the switching times reach the MHz range [52] instead of the required GHz range for digital logic. Therefore, it was reported [50,51] that hard FE materials are required with lower ρ values. Alternatively, a relatively low P_r could also help but that also depends on the transistor configuration. Note that, even without considering these dynamic FE effects, low switching speeds have been reported for the NC-FET using compact models [40].

For the π-FET, the switching speed is less of an issue since the device operates at relatively low $\partial P/\partial \mathcal{E}$ values. The switching speed is basically determined by the sound velocity and the device dimensions [16,44]. Examples are manifold. There have been several reports on PZT bulk acoustic wave (BAW) resonators reaching GHz speed, for ~1 µm thick layers [53,54]. This even applies to conventional piezoelectric (low polarization) AlN BAW resonators (e.g., [55]).

So far, the dynamic power consumption (P_{dyn}) of both device structures have not been thoroughly investigated. For the NC-FET, some remarks have been given that the damping constant of the FE material governs the switching energy [50]. Moreover, several reports (e.g., [40]) where numerical calculations were performed indicate not a positive outlook (~ tenfold increase in P_{dyn}).

For the π-FET, it was estimated [42] that, depending on the device dimensions, the P_{stat} can drop by a factor of two in the silicon (Si) π-FinFET compared to a conventional FinFET. On the other hand, for the former mechanical power is needed during switching resulting in an increased P_{dyn}, more or less within the same range. Based on prior International Technology Roadmap for Semiconductors (ITRS) roadmap estimations, this implies that, for less than ~8 nm gate length, the total power consumption of the π-FinFET is less than that of the conventional FinFET. However, these are rough estimations and more accurate modeling and experiments are required to have better predictions.

In summary, amongst the discussed device structures, the NC-FET is most attractive for its relatively low P_{stat}. The π-FET, on the other hand, appears to be more attractive from the viewpoint of switching speed and P_{dyn}. Therefore, if no clear solution would have been found on tackling the speed issue of the NC-FET, a potential future low-power CMOS technology could be found in a hybrid solution comprising both device architectures on a chip where hard FE materials with a high d_{33} are used. As indicated in Figure 1, the technology of both device concepts are not that different, which could imply that these are relatively simple to realize within a single production process. Table 2 summarizes the discussion.

For both device concepts, there are also specific requirements regarding the architecture. For the NC-FET, there should be sufficient charge in the close vicinity of the channel or body region. Fully depleted long channel NC-based FinFETs, nanowire (NW-)FETs, or perhaps even some two-dimensional (2D) material FETs, will mainly result in higher I_{ON} rather than lower SS.

Bulk NC-FETs, on the other hand, will result in maximum performance (see, e.g., [34]). The NC-FET is sensitive to dimensional scaling because of this charge balance requirement. For the π-FET, this requirement is not important. For the π-FET, dimensional scaling is also relevant but more from a mechanical point of view and will be less sensitive to process variations. Conversely, the type of semiconductor in the body/channel strongly determines the performance of the π-FET because of its electro-mechanical properties, which is not the case for the NC-FET. For example, a theoretical study in which transition metal dichalcogenides (TMDs) as a channel material in a π-FET configuration was reported, referred to as the 2D-EFET [56], showing promising results.

Table 2. A qualitative summary of the device performance of the NC-FET vs. π-FET.

	NC-FET	π-FET
SS	++	+
I_{ON}	+/++	+
P_{static}	++	+
P_{dyn}	- -	-
τ	- -	+

Furthermore, the choice of FE material is also important. As stated before, for a good functioning of the π-FET, a high P_r is required. For the NC-FET, on the other hand, this strongly depends on the device architecture, though a high $\mathcal{E}_C$ is essential. Of course, materials such as HZO are more compatible in state-of-the-art CMOS technologies, but, for example, PZT can also be integrated in silicon technology, as adopted in Fe-RAM [18], or providing a proper buffer layer is used [48]. This discussion has been summarized in Table 3.

Table 3. A summary of the device architecture requirements for the NC-FET and π-FET. * Charge required in the close vicinity of the body, § May be less effective.

	NC-FET	π-FET
Device type	Bulk/Fin*/NW*/2D*	Bulk§/Fin/NW/2D
Charge body	required	not important
Scaling	very important	important
Body/channel type (e.g., Si, Ge, TMD)	important	very important
Perovskite ferroics (e.g., PZT)	effective (record)	required
"Hard" ferroics (e.g., HfZrO$_2$)	effective (concensus)	insufficient

Obviously, there are many technological challenges, not only from the viewpoint of material science—especially when integrating FE layers in FinFET or NW-based FET technologies. First, the FE layer including the interfacial layers (e.g., gate metal, floating metal, buffer layer) should have a good conformality. In this way, a better gate control can be obtained throughout the device. Second, the conformal layers should be preferably of a minimal layer thickness so that edge or corner effects become less important. Third, for ultrathin FE layers (in particular for PZT [57]), a so-called dead layer is formed which causes a deterioration of the P-$\mathcal{E}$ loop, yielding lower P_r values. A solution for some of those points may be to adopt atomic-layer deposition (ALD) in the process. ALD is an advanced deposition technique that is based on self-limiting surface reactions, yielding practically uniform, conformal layers. Furthermore, the whole gate stack can be formed without a vacuum break, which is important for interface control. Moreover, ALD is suited for an industrial setting. Note that the P-$\mathcal{E}$ loop can be further tuned for instance by adjusting the doping concentration (acceptor doping [Fe,La] for wider loops [19]) or by clamping (e.g., [20]).

Finally, some discussion about the temperature dependence for both types of devices is needed. Despite its importance, surprisingly, however, not much literature can be found regarding this issue. For instance, the work of Jo and Shin [58] report on the temperature dependence of the NC-effect. By connecting a ferroelectric capacitor to a conventional MOSFET, the authors experimentally showed an *SS* increase for elevated temperatures. On the other hand, in the milestone paper of Khan et al. [23], for instance, elevated temperatures were used (up to 500 °C, just near the Curie temperature of PZT) to show experimentally the NC-effect in a composite PZT/STO MIM capacitor. For the π-FET, so far there have been no reports in that direction.

Generally, it can be stated that the temperature has a strong effect on the properties of ferroelectric materials and consequently on the device performance for both types of devices. Therefore, more research is needed in that direction. Firstly, the so-called Curie temperature T_C, which is the minimum temperature at which a ferroelectric material becomes paraelectric (i.e., there is no hysteretic P-$\mathcal{E}$ curve), should be above the desired operating temperature range; otherwise, in principle, the devices won't function properly. For both PZT (e.g., [23]) and HfO_2 (e.g., [59]), it has been reported that this in the range of 500 °C, well above the operating temperature, but this will depend on the material composition and layer thicknesses. Even for high T_C, however, the polarization is sensitive to temperature variation. Second, ferroelectric materials are also so-called pyroelectric materials [17], which means that a surface charge in the ferroelectric material can be formed induced by a change in temperature.

5. Conclusions

The negative-capacitance field-effect transistor (NC-FET) requires capacitive tuning and is quite sensitive to any variation. For proper operation, a relatively wide polarization-field loop is required. The piezoelectric field-effect transistor (π-FET) requires a relatively high piezoelectric coefficient and hence the polarization-field loop, and special attention must be given to the mechanical boundary conditions that are important in this case. The NC-FET is by far superior in subthreshold-swing, on-current and static power consumption. However, literature data show that its dynamic properties (delay and power consumption) are inferior to those of the π-FET. Therefore, if the speed issue was not solved for the NC-FET, a potential future CMOS technology would be a hybrid solution comprising both NC-FET and π-FET in which hard ferroelectric materials with a relatively high piezoelectric coefficient have been incorporated.

Funding: This work was partly supported by the Netherlands Organisation for Scientific Research, Domain Applied and Engineering Sciences (NWO-TTW), the Netherlands.

Acknowledgments: The author would like to thank colleagues from the Integrated Device and Systems (IDeaS) and Inorganic Materials Science (IMS) group at the MESA+ Institute of the University of Twente. In particular, Tom van Hemert and Buket Kaleli, who are presently both with ASML, and Rob Wolters are kindly acknowledged for their contributions.

Conflicts of Interest: The author declares no conflict of interest.

References

1. Moore, G.E. No Exponential is Forever: But "Forever" Can Be Delayed! In Proceedings of the International Solid-State Circuits Conference (ISSCC), San Francisco, CA, USA, 13 February 2003; pp. 20–23.
2. Hisamoto, D.; Lee, W.; Keziekski, J.; Anderson, E.; Takeuchi, H.; Asano, K.; King, T.; Bokor, J.; Hu, C. A Folded-channel MOSFET for Deep-sub-tenth Micron Era. In Proceedings of the International Electron Device Meeting (IEDM), San Francisco, CA, USA, 6–9 December 1998; pp. 1032–1034.
3. Saremi, M.; Afzali-Kusha, A.; Mohamadi, S. Ground plane fin-shaped field effect transistor (GP-FinFET): A FinFET for low leakage power circuits. *Microelectr. Eng.* **2012**, *95*, 74–82. [CrossRef]
4. Colinge, J.-P. *FinFETs and Other Multi-Gate Transistors*; Springer Science+Business Media: New York, NY, USA, 2008.

5. Faynot, O.; Crisoloveanu, S.; Auberton-Hervé, A.J.; Raynaud, C. Performance and Potential of Ultrathin Accumulation-Mode SIMOX MOSFET's. *EEE Trans. Electr. Dev.* **1995**, *42*, 713–719. [CrossRef]
6. Welser, J.; Hoyt, J.L.; Gibbons, J.F. Electron mobility enhancement in strained-Si n-type metal-oxide-semiconductor field-effect transistors. *IEEE Electr. Dev. Lett.* **1994**, *15*, 100–102. [CrossRef]
7. Chandrakasan, A.; Brodersen, R. Minimizing power consumption in digital CMOS circuits. *Proc. IEEE* **1995**, *83*, 498–523. [CrossRef]
8. Seabaugh, A.C.; Zhang, Q. Low-voltage tunnel transistors for beyond CMOS logic. *Proc. IEEE* **2010**, *98*, 2095–2110. [CrossRef]
9. Ionescu, A.M.; Riel, H. Tunnel field-effect transistors as energy-efficient electronic switches. *Nature* **2011**, *479*, 329–337. [CrossRef] [PubMed]
10. Imena Badi, R.M.; Saremi, M.; Vandenberghe, W.G. A Novel PNPN-Like Z-Shaped Tunnel Field-Effect Transistor With Improved Ambipolar Behavior and RF Performance. *IEEE Trans. Electr. Dev.* **2017**, *64*, 4752–4758. [CrossRef]
11. Imena Badi, R.M.; Saremi, M. A Resonant Tunneling Nanowire Field Effect Transistor with Physical Contractions: A Negative Differential Resistance Device for Low Power Very Large Scale Integration Applications. *J. Electron. Mater.* **2018**, *47*, 1091–1098.
12. Gopalakrishnan, K.; Griffin, P.B.; Plummer, J.D. I-MOS: A novel semiconductor device with a subthreshold slope lower than kT/q. In Proceedings of the International Electron Device Meeting (IEDM), San Francisco, CA, USA, 8–11 December 2002; pp. 289–292.
13. Salahuddin, S.; Datta, S. Use of Negative Capacitance to Provide Voltage Amplification for Low Power Nanoscale Devices. *Nano Lett.* **2008**, *8*, 405–410. [CrossRef] [PubMed]
14. Salahuddin, S.; Datta, S. Can the subthreshold swing in a classical FET be lowered below 60 mV/decade? In Proceedings of the International Electron Device Meeting (IEDM), San Francisco, CA, USA, 15–17 December 2008; pp. 1–4.
15. Van Hemert, T.; Hueting, R.J.E. Active Strain Modulation in Field Effect Devices. In Proceedings of the European Solid-State Device Research Conference (ESSDERC), Bordeaux, France, 17–21 September 2012; pp. 125–128.
16. Van Hemert, T.; Hueting, R.J.E. Piezoelectric Strain Modulation in FETs. *IEEE Trans. Electr. Dev.* **2013**, *60*, 3265–3270. [CrossRef]
17. Damjanovic, D. Ferroelectric, dielectric and piezoelectric properties of ferroelectric thin films and ceramics. *Rep. Prog. Phys.* **1998**, *61*, 1267–1324. [CrossRef]
18. Setter, N.; Damjanovic, D.; Eng, L.; Fox, G.; Gevorgian, S.; Hong, S.; Kingon, A.; Kohlstedt, H.; Park, N.Y.; Stephenson, G.B.; et al. Ferroelectric thin films: Review of materials, properties, and applications. *J. Appl. Phys.* **2006**, *100*, 051606. [CrossRef]
19. Kimura, M.; Ando, A.; Sakabe, Y. Lead zirconate titanate-based piezo-ceramics. In *Advanced Piezoelectric Materials. Science and Technology*; Uchino, K., Ed.; Woodhead Publishing Ltd.: Cambridge, UK, 2010; pp. 89–110.
20. Nguyen, D.M.; Dekkers, J.M.; Houwman, E.P.; Steenwelle, R.J.A.; Wang, X.; Wan, X.; Roelofs, A.; Schmitz-Kempen, T.; Rijnders, A.J.H.M. Misfit strain dependence of ferroelectric and piezoelectric properties of clamped (001) epitaxial Pb(Zr$_{0.52}$,Ti$_{0.48}$)O$_3$ thin films. *Appl. Phys. Lett.* **2011**, *99*, 252904. [CrossRef]
21. Landau, L.D.; Khalatnikov, I.M. On the anomalous absorption of sound near a second order phase transition point. *Dokl. Akad. Nauk.* **1954**, *96*, 469–472.
22. Devonshire, A.F. Theory of ferroelectrics. *Adv. Phys.* **1954**, *3*, 85–130. [CrossRef]
23. Khan, A.I.; Bhowmik, D.; Yu, P.; Kim, S.J.; Pan, X.; Ramesh, R.; Salahuddin, S. Experimental evidence of ferroelectric negative capacitance in nanoscale heterostructures. *Appl. Phys. Lett.* **2011**, *99*, 11350.
24. Sze, S.M.; Ng, K.K. *Physics of Semiconductor Devices*, 3rd ed.; John Wiley & Sons, Inc.: Hoboken, NJ, USA, 2007.
25. Catalan, G.; Jiménez, D.; Gruverman, A. Negative capacitance detected. *Nat. Mater.* **2015**, *14*, 137–139. [CrossRef] [PubMed]
26. Salvatore, G.A.; Bouvet, D.; Ionescu, A.M. Demonstration of Subthreshold Swing Smaller Than 60mV/decade in Fe-FET with P(VDF-TrFE)/SiO$_2$ Gate Stack. In Proceedings of the International Electron Device Meeting (IEDM), San Francisco, CA, USA, 14–17 December 2008; pp. 168–170.

27. Rusu, A.; Salvatore, G.A.; Jiménez, D.; Ionescu, A.M. Metal-ferroelectric-metal-oxide-semiconductor field effect transistor with sub-60 mV/decade subthreshold swing and internal voltage amplification. In Proceedings of the International Electron Device Meeting (IEDM), San Francisco, CA, USA, 6–8 December 2010; pp. 395–398.

28. Khan, A.I.; Yeung, C.W.; Hu, C.; Salahuddin, S. Ferroelectric negative capacitance MOSFET: Capacitance tuning & antiferroelectric operation. In Proceedings of the International Electron Device Meeting (IEDM), Washington, DC, USA, 5–7 December 2011; pp. 11.3.1–11.3.4.

29. Lee, M.H.; Lin, J.-C.; Wei, Y.-T.; Chen, C.-W.; Tu, W.-H.; Zhuang, H.-K.; Tang, M. Ferroelectric Negative Capacitance Hetero-Tunnel Field-Effect-Transistors with Internal Voltage Amplification. In Proceedings of the International Electron Device Meeting (IEDM), Washington, DC, USA, 9–11 December 2013; pp. 104–107.

30. Cheng, C.H.; Chin, A. Low-Voltage Steep Turn-On pMOSFET Using Ferroelectric High-k Gate Dielectric. *IEEE Electr. Dev. Lett.* **2014**, *35*, 274–276. [CrossRef]

31. Dasgupta, S.; Rajashekar, A.; Majumdar, K.; Agrawal, N.; Razavieh, A.; Trolier-McKinstry, S.; Datta, S. Sub-kT/q Switching in Strong Inversion in PbZr$_{0.52}$Ti$_{0.48}$O$_3$ Gated Negative Capacitance FETs. *IEEE J. Explor. Solid-State Comput. Devices Circuits* **2015**, *1*, 43–48. [CrossRef]

32. Li, K.-S.; Chen, P.G.; Lai, T.-Y.; Lin, C.-H.; Cheng, C.-C.; Chen, C.-C.; Wei, Y.-J.; Hou, Y.-F.; Liao, M.-H.; Lee, M.-H.; et al. Sub-60 mV-Swing Negative-Capacitance FinFET without Hysteresis. In Proceedings of the International Electron Device Meeting (IEDM), Washington, DC, USA, 7–9 December 2015; pp. 620–623.

33. Khan, A.I.; Chatterjee, K.; Duarte, J.P.; Lu, Z.; Sachid, A.; Khandelwal, S.; Ramesh, R.; Hu, C.; Salahuddin, S. Negative Capacitance in Short-Channel FinFETs Externally Connected to an Epitaxial Ferroelectric Capacitor. *IEEE Electr. Dev. Lett.* **2016**, *37*, 111–114. [CrossRef]

34. Park, J.H.; Joo, S.K. Sub-kT/q subthreshold slope p-metal-oxide-semiconductor field-effect transistors with single-grained Pb(Zr,Ti)O$_3$ featuring a highly reliable negative capacitance. *Appl. Phys. Lett.* **2016**, *108*, 103504. [CrossRef]

35. Jana, R.J.; Snider, G.L.; Jena, D. On the possibility of sub 60 mV/decade subthreshold switching in piezoelectric gate barrier transistors. *Phys. Stat. Sol.* **2013**, *10*, 1469–1472. [CrossRef]

36. Canon, A.; Jiménez, D. Multidomain ferroelectricity as a limiting factor for voltage amplification in ferroelectric field-effect transistors. *Appl. Phys. Lett.* **2010**, *97*, 133509. [CrossRef]

37. Salahuddin, S. Personal Communication, 2016.

38. Rollo, T.; Esseni, D. Influence of Interface Traps on Ferroelectric NC-FETs. *IEEE Electr. Dev. Lett.* **2018**, *39*, 1100–1103. [CrossRef]

39. Hu, C.; Salahuddin, S. 0.2 V Adiabatic NC-FinFET with 0.6 mA/µm I_{ON} and 0.1 nA/µm I_{OFF}. In Proceedings of the Berkeley Symp. Energy Efficient Electronic Systems & Steep Transistors Workshop (E3S), Columbus, OH, USA, 21–24 June 2015; pp. 39–40.

40. Pahwa, G.; Dutta, T.; Agarwal, A.; Chauhan, Y.S. Designing energy efficient and hysteresis free negative capacitance FinFET with negative DIBL and 3.5X I_{ON} using compact modeling approach. In Proceedings of the European Solid-State Device Research Conference (ESSDERC), Lausanne, Switzerland, 12–15 September 2016; pp. 41–46.

41. Auld, B.A. *Acoustic Fields and Waves in Solids*; Wiley: New York, NY, USA, 1973.

42. Hueting, R.J.E.; van Hemert, T.; Kaleli, B.; Wolters, R.A.M.; Schmitz, J. On Device Architectures, Subthreshold Swing, and Power Consumption of the Piezoelectric Field-Effect Transistor (π-FET). *J. Electr. Dev. Soc.* **2015**, *3*, 149–157.

43. Starschich, S.; Schenk, T.; Schroeder, U.; Boettger, U. Ferroelectric and piezoelectric properties of Hf1-xZrxO2 and pure ZrO2 films. *Appl. Phys. Lett.* **2017**, *110*, 182905. [CrossRef]

44. Newns, D.; Elmegreen, B.; Liu, X.H.; Martyna, G. A low-voltage high-speed electronic switch based on piezoelectric transduction. *J. Appl. Phys.* **2012**, *111*, 084509. [CrossRef]

45. Synopsys, Inc. *Sentaurus Device User Guide*; Synopsys, Inc.: Mountain View, CA, USA, 2007.

46. COMSOL, Inc. *Comsol MultiPhysics*, 4th ed.; COMSOL, Inc.: Palo Alto, CA, USA, 2011.

47. Wang, H.; Jiang, X.; Xu, N.; Han, G.; Hao, Y.; Li, S.-S.; Esseni, D. Revised Analysis of Design Options and Minimum Subthreshold Swing in Piezoelectric FinFETs. *IEEE Electr. Dev. Lett.* **2018**, *39*, 444–447. [CrossRef]

48. Kaleli, B.; Hueting, R.J.E.; Nguyen, M.; Wolters, R.A.M. Integration of a Piezoelectric Layer on Si FinFETs for Tunable Strained Device Applications. *IEEE Trans. Electr. Dev.* **2014**, *61*, 1929–1935. [CrossRef]

49. Li, J.; Nagaraj, B.; Liang, H.; Cao, W.; Lee, C.H.; Ramesh, R. Ultrafast polarization switching in thin-film ferroelectrics. *Appl. Phys. Lett.* **2004**, *84*, 1174–1176. [CrossRef]
50. Li, Y.; Yao, K.; Samudra, G.S. Effect of Ferroelectric Damping on Dynamic Characteristics of Negative Capacitance Ferroelectric MOSFET. *IEEE Trans. Electr. Dev.* **2016**, *63*, 3636–3641. [CrossRef]
51. Yuan, Z.C.; Rizwan, S.; Wong, M.; Holland, K.; Anderson, S.; Hook, T.B.; Kienle, D.; Gadelrab, S.; Gudem, P.S.; Vaidyanathan, M. Switching-Speed Limitations of Ferroelectric Negative-Capacitance FETs. *IEEE Trans. Electr. Dev.* **2016**, *63*, 4046–4052. [CrossRef]
52. Kobayashi, M.; Ueyama, N.; Jang, K.; Hiramoto, T. Experimental Study on Polarization-Limited Operation Speed of Negative Capacitance FET with Ferroelectric HfO_2. In Proceedings of the 2016 IEEE International Electron Device Meeting (IEDM), San Francisco, CA, USA, 3–7 December 2016; pp. 314–317.
53. Kirby, P.B.; Su, Q.X.; Komuro, E.; Zhang, Q.; Whatmore, R.W. PZT thin film bulk acoustic wave resonators and filters. In Proceedings of the 2001 IEEE International Frequncy Control Symposium and PDA Exhibition, Seattle, WA, USA, 8 June 2001; pp. 687–694.
54. Conde, J.; Muralt, P. Characterization of Sol-Gel $Pb(Zr_{0.53}Ti_{0.47})O_3$ in Thin Film Bulk Acoustic Resonators. *IEEE Trans. Ultrason. Ferroelectr. Freq. Cntrol.* **2008**, *55*, 1373–1379. [CrossRef] [PubMed]
55. Van Hemert, T.; Reimann, K.; Hueting, R.J.E. Extraction of second order piezoelectric parameters in bulk acoustic wave resonators. *Appl. Phys. Lett.* **2012**, *100*, 232901. [CrossRef]
56. Das, S. Two Dimensional Electrostrictive Field Effect Transistor (2D-EFET): A sub-60mV/decade Steep Slope Device with High ON current. *Sci. Rep.* **2016**, *6*, 34811. [CrossRef] [PubMed]
57. Nguyen, D.M. Ferroelectric and Piezoelectric Properties of Epitaxial PZT Films and Devices on Silicon. Ph.D. Thesis, University of Twente, Enschede, The Netherlands, 2010.
58. Jo, J.; Shin, C. Impact of temperature on negative capacitance field-effect transistor. *Electr. Lett.* **2015**, *51*, 106–108. [CrossRef]
59. Nishimura, T.; Xu, L.; Shibayana, S.; Yajima, T.; Misita, S.; Toriumi, A. Ferroelectricity of nondoped thin HfO_2 films in $TiN/HfO_2/TiN$ stacks. *Jpn. J. Appl. Phys.* **2016**, *55*, 08PB01. [CrossRef]

micromachines

MDPI

Article

Accelerating Flux Calculations Using Sparse Sampling †

Lukas Gnam [1,*, Paul Manstetten [1], Andreas Hössinger [2], Siegfried Selberherr [3] and Josef Weinbub [1]**

1   Christian Doppler Laboratory for High Performance TCAD, Institute for Microelectronics, TU Wien, Vienna 1040, Austria; manstetten@iue.tuwien.ac.at (P.M.); weinbub@iue.tuwien.ac.at (J.W.)
2   Silvaco Europe Ltd., St. Ives PE27 5JL, UK; andreas.hoessinger@silvaco.com
3   Institute for Microelectronics, TU Wien, Vienna 1040, Austria; selberherr@tuwien.ac.at
*   Correspondence: gnam@iue.tuwien.ac.at; Tel.: +43-1-58801-36048
†   This paper is an extended version of our paper published in *Lecture Notes in Computer Science, Volume 10860, Proceedings of the International Conference on Computational Science (ICCS)—Part I, Wuxi, China, 11–13 June 2018*; Springer: Cham, Switzerland, 2018; pp. 694–707.

Received: 20 September 2018; Accepted: 23 October 2018; Published: 26 October 2018

Abstract: The ongoing miniaturization in electronics poses various challenges in the designing of modern devices and also in the development and optimization of the corresponding fabrication processes. Computer simulations offer a cost- and time-saving possibility to investigate and optimize these fabrication processes. However, modern device designs require complex three-dimensional shapes, which significantly increases the computational complexity. For instance, in high-resolution topography simulations of etching and deposition, the evaluation of the particle flux on the substrate surface has to be re-evaluated in each timestep. This re-evaluation dominates the overall runtime of a simulation. To overcome this bottleneck, we introduce a method to enhance the performance of the re-evaluation step by calculating the particle flux only on a subset of the surface elements. This subset is selected using an advanced multi-material iterative partitioning scheme, taking local flux differences as well as geometrical variations into account. We show the applicability of our approach using an etching simulation of a dielectric layer embedded in a multi-material stack. We obtain speedups ranging from 1.8 to 8.0, with surface deviations being below two grid cells (0.6–3% of the size of the etched feature) for all tested configurations, both underlining the feasibility of our approach.

Keywords: flux calculation; etching simulation; process simulation; topography simulation

---

1. Introduction

Semiconductor process simulations can be partitioned into two major types: reactor-scale and feature-scale simulations. The former simulates the full reactor chamber whereas the simulation domain of the latter is a small region of the wafer surface (see Figure 1). Feature-scale simulations (which are the focus of this work) are used when the detailed topographical behavior and the prediction of the surface evolution is of major interest, instead of focusing on the global behavior of the wafer in a reactor-scale simulation [1].

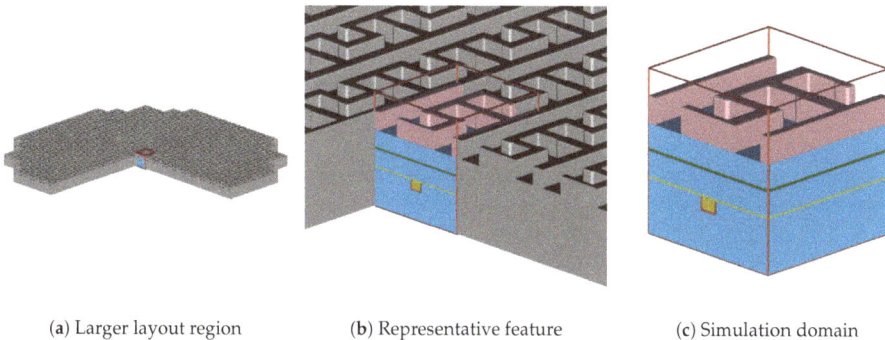

| (a) Larger layout region | (b) Representative feature | (c) Simulation domain |

Figure 1. (a) a larger region of a layout on a wafer; (b) close-up on a representative feature; (c) feature-scale simulation domain with domain boundaries in red.

Particularly challenging is the simulation of etching processes, where the large number of possible process parameters and available materials renders conventional evaluation unreasonable considering costs and time [2], since even the smallest change in an experimental setup can have a major impact on the properties of a device [3]. Additionally, the more and more used three-dimensional (3D) designs (e.g., FinFET [4] or 3D NAND flash memory [5]) prohibits the problem reduction to two dimensions as a simple yet effective technique to minimize computational effort for larger 3D structures. Hence, the development of faster algorithms and methods is a substantial objective to solve the increasingly complex 3D problems and thereby sustaining the high pace of ultra-large scale integrated circuit developments [2].

The common computational tasks in a single timestep of a feature-scale etching simulation are (a) the preparation of a suitable surface representation for the surface flux rate calculations, (b) the calculation of the surface flux rate distributions on the surface, (c) the evaluation of the surface velocity models using the surface flux rate distributions, and (d) the advection of the surface according to the computed surface velocity field. Typically, the majority of the computational time is spent on the calculation of the flux rates [6,7]. In turn, the models for the etch rates depend on the flux rates of the involved etchants where the flux rate calculations constitute the main computational bottleneck.

The common flux calculation methods (which assume ballistic particle transport) rely on a large number of ray-surface intersection tests. Most frameworks are based on an implicit representation of the surface, using the level-set method for advection [8–10]. In [11], the authors study the Bosch process for a micro-electromechanical system (MEMS) application with Silvaco's Victory Process simulator, which uses an implicit representation of the surface during ray casting [12]. However, during ray casting, different (temporary) representations for the surface are also used. Ertl et al. use the ViennaTS [13] simulator to investigate the Bosch process [14]. The process is simulated in three dimensions and the flux calculation is based on a Monte Carlo approach. A representation of the surface as a set of overlapping disks is used to approximate a closed surface. Heitzinger et al. showed a method to accelerate such Monte Carlo based flux calculations by coarsening the explicit surface mesh representing the surface [15]. Recently, Yu et al. presented a 3D topography simulation of a deep reactive ion etching (DRIE) process [16]. Therein, the surface is approximated using a set of spheres during the Monte Carlo ray tracing task.

A common way to increase the computational performance of ray casting is the use of hierarchical spatial acceleration structures, where a bounding volume hierarchy (BVH) is a common choice [17,18]. A similar approach is to apply a bi-level spatial subdivision for reducing the search space for an explicit surface representation in order to accelerate the Monte Carlo based ray tracing as shown by Yu et al. [19]. This method relies on a proper subdivision tree and an optimal choice of the underlying parameters, which need to be determined beforehand. Naeimi et al. accelerate their ray-tracing based heat transfer model by merging empty voxels, explicitly representing their geometry, inside their static simulation domain [20]. Hence, when a ray is traversing through the domain, it can quickly traverse the empty voxels, since they are much bigger than those on the respective surface. Although their algorithm provides a considerable speedup concerning the underlying ray-tracing, the voxel merging process can have a major impact on the performance in transient simulations, when it has to be applied in each timestep. Aguerre et al. proposed a method to reduce the computational effort if the flux calculation is radiosity based [21]. The method relies on a hierarchical strategy and a sorting scheme for acceleration of the necessary calculations based on explicit meshes. Bailey showed that an adaptive sampling of the geometrical elements can improve the computational performance of ray tracing tasks [7]. In [22], it is shown that the overhead introduced by generating a temporary explicit mesh in each timestep is by far compensated using optimized computer graphics libraries for single-precision ray casting.

Another acceleration approach was presented in [6], where the computational performance of the flux calculation is improved by evaluating the surface speed only on a sparse set of cells on an explicit surface mesh. This set is found using an iterative partitioning of the surface, without adapting the surface mesh itself. The approach is evaluated using a simple 3D single-material etching simulation of a cylindric hole. The computational performance of the flux calculation is increased by a factor of 2–8 (compared to the conventional evaluation on the mesh using all cells) while preserving the geometrical features of the surface.

In this paper, we extend the method presented in [6] by including multi-material interfaces into the partitioning scheme and by evaluating practically relevant multi-material problems. We evaluate the new scheme by simulating the etching of a dielectric layer as part of a Dual-Damascene fabrication process sequence, which is an important technique in today's semiconductor industry [23,24]. We show that the novel flux calculation based on the sparse set of surface cells significantly improves the computational performance for practically relevant multi-material simulations, achieving speedups of 1.9 to 8, which is in the range of the speedups reported in [6]. Additionally, we show that the resulting surface deviations for the sparse set are below two grid cells after many timesteps (corresponding to 0.6 to 3% of the via size) for a wide range of surface resolutions. The study is completed with an in-depth analysis of the parallel performance of the sparse and dense flux calculation approaches.

2. Materials and Methods

2.1. Dual-Damascene Process

To reduce the required processing steps in the fabrication pipeline, the Dual-Damascene process was introduced in the 1990s and became an industry standard [25]. In this process, only a single metal deposition step is required, where vias and trenches are filled simultaneously, typically using copper as filling metal [26]. Additionally, the number of chemical-mechanical planarization steps as well as the number of deposition steps for the dielectric are reduced. This yields a shorter overall processing time together with less possibilities for failures.

In general, there are three different major Dual-Damascene process sequences: (1) trench-first, (2) via-first, and (3) self-aligned or buried-via. In the trench-first sequence, the etching of the trenches is conducted before the etching of the vias, whereas, in the via-first approach, the etching is done vice versa. Both methods apply a metal deposition step after each etching step, where the via and the trench are filled with copper. The self-aligned Dual-Damascene process etches both the via and the trench at the same time [25,27] and is the target application of this work.

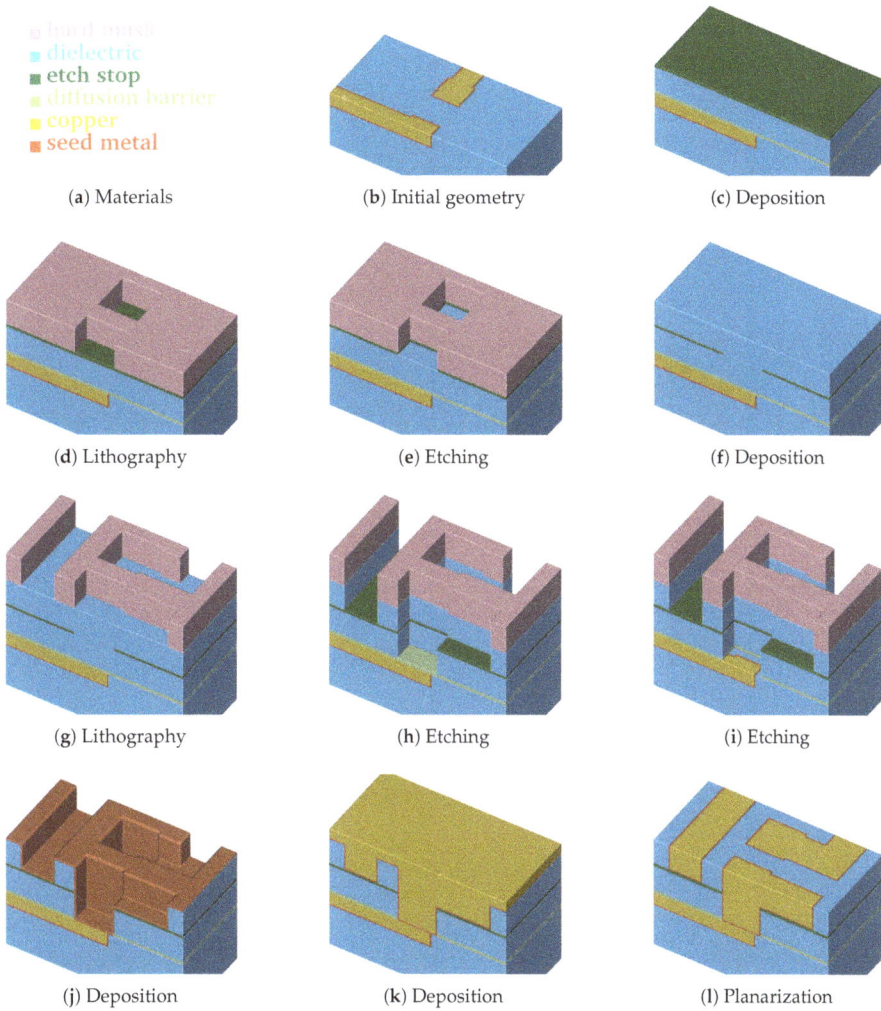

Figure 2. Individual processing steps for the self-aligned Dual-Damascene process. The material legend is shown in (**a**). (**b**–**l**) illustrate all processing steps to create a metalization layer on top of an exiting metalization layer (the square domain is clipped along one axis for better visibility). First row (**b**,**c**): initial patterned planar conductor and deposition of three layers: a diffusion barrier, a dielectric material, and an etch stop material. Second row (**d**–**f**): patterning of the etch stop layer (at the position of the vias) including the deposition of the second dielectric layer. Third row (**g**–**i**): patterning of the dielectric layers (vias and lines) including opening of the diffusion barrier layer around at the position of the via. Last row (**j**–**l**): copper metalization including the deposition of a metal seed layer and chemical-mechanical planarization after the copper deposition.

Figure 2 depicts a typical sequence of processing steps in the self-aligned Dual-Damascene process. The different steps show the initial surface of the wafer (Figure 2b), which is followed by the deposition of the diffusion barrier, a dielectric, and the etch stop (Figure 2c). Afterwards, a hard mask is deposited and subsequently the etch stop layer is patterned (Figure 2d,e). After the removal of the hard mask, a second dielectric is deposited and a hard mask is patterned again (Figure 2f,g). The next step is the combined etching of trenches and vias (Figure 2h). Afterwards, the finalizing steps are the removal of the etch barrier (Figure 2i), the deposition of the metal seed layer (Figure 2j), the copper deposition (Figure 2k), and the chemical-mechanical planarization (Figure 2l).

After switching from Aluminum to Copper as interconnect material in the early 2000s to overcome the problems surfacing in the sub-micron regime [25], today, new challenges arise on the continued path of miniaturization down to a few nanometers. For example, to reduce the interconnect capacitance, new dielectrics such as SiO_2 or porous SiCOH have to be used [28]. As previously discussed, computer simulations provide a highly attractive option to assess the different etching behaviors of various materials to determine process designs and optimizations before proceeding with actual, conventional experimental investigations. Therefore, within this work, we show the applicability of the approach presented in [6] using the etching process of a dielectric layer (Figure 2g–h) as a practically relevant example of a multi-material simulation.

2.2. Multi-Material Simulation Framework

A single-material simulation framework [6] was developed to investigate efficient flux calculation methods for 3D etching and deposition processes. It is based on the sparse volume data structure of OpenVDB [29,30] and uses the accompanying tools to handle surface advection and surface extraction. All advection steps described in the following are performed using the level-set representation of the surface. The explicit representation of the surface, which is only used during the flux calculation, is obtained using OpenVDB's "volumeToMesh" routine producing quads. We subdivide each potentially non-planar quad into two triangles using the distance to the zero-level-set as guidance for the subdivision pattern. The ray-surface intersection tests are performed using Embree [31]. In the following, we provide a short summary of the method to set the stage for the subsequent discussion on the improved multi-material iterative partitioning scheme and the detailed analysis of plasma etching steps for a Dual-Damascene process.

In each timestep of the simulation, the main computational tasks are (a) the calculation of the flux rates R on the surface, (b) the evaluation of the normal surface velocity V_n (which depends on the flux rates), and (c) the advection of the surface. Typically, the normal surface velocity V_n during an etching process depends on the flux rates R of the involved particle species

$$V_n(\mathbf{x}) = f(R_1, R_2, ..., R_k) ,$$ (1)

where k denotes the number of different particle species. The flux rates depend on the incoming flux distribution Γ_{in}. In the simplest case, all incoming directions are equally weighted, which results in

$$R(\mathbf{x}) = \int_\Omega \Gamma_{in}(\mathbf{x}, \omega_{d\Omega}) d\Omega ,$$ (2)

with ω being the incoming direction, and Ω denoting the upper hemisphere facing the source plane.

The surface of this hemisphere is discretized using a subdivided icosahedron. The directions from a surface point towards the centroids of the triangular discretization of the hemisphere are tested for visibility of the source. After obtaining the visibility information for all directions, a numerical integration is performed over the visible solid angles using a centroid rule [22].

Typically, different material regions are present during an etching process simulation, e.g., a dielectric patterned with a hard mask (Figure 2g). A straightforward approach would be to represent each material region with a corresponding level-set function, as shown in Figure 3a. To simultaneously advect all material regions, each region would then be advected separately, leading to potentially mutual penetration. In this case, the parts of a region which are penetrated by another region would be treated inactive, i.e., not subject to advection. One approach would be to perform Boolean operations between material regions to dissolve the penetrations. Hence, a strategy to decide which material fills the former penetrated volume is needed.

We extended the simulation framework presented in [6] using an approach analogous to [32,33], which constructs a total union of all regions and advects the "top layer" (see Figure 3b). To perform the advection with the correct surface speed of the underlying material region, it is necessary to detect the active material for each point in the top layer. This active material for a point **x** is obtained by querying the value of all level-sets at **x**. The material of the level-set with the smallest value is considered active. Additionally, the level-sets have a fixed order and the lower level-set is chosen as active material, if the values are numerically identical. The etching itself is conducted by advecting the top layer, and subsequently transferring the removal of the material to the underlying level-sets representing the material regions using a Boolean operation between the top layer and each material region. One significant advantage of this top layer approach is that material layers can be represented with sub-grid resolution. This is possible if the level-sets representing the materials are chosen to not map directly to the material regions, but are constructed additively.

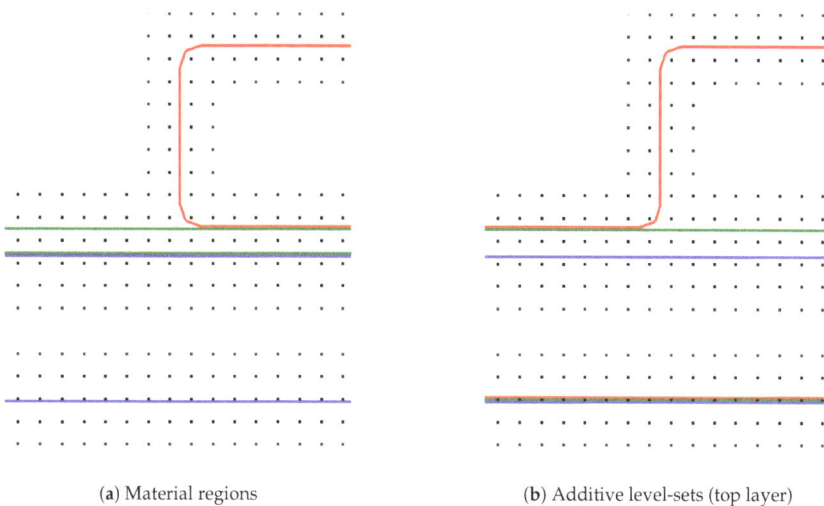

(a) Material regions   (b) Additive level-sets (top layer)

Figure 3. Three materials stack with a thick bottom region (blue), a thin layer (green), and a mask (red). (a) cross section of the material regions showing the level-set grid points and the extracted level-sets; (b) additive level-sets scheme from bottom to top (blue, green, and red).

The chronological development of an exemplary material stack using the top layer advection is shown in Figure 4. Starting from Figure 4a, the green layer is represented with sub-grid resolution (see Figure 4b), until it is fully etched away in Figure 4c. In a time step where a region of a layer is fully etched and the underlying material becomes active, the surface velocities of the involved materials must be averaged. If only the surface velocity for the green layer is considered for the full time step, the surface advection speed is too slow or too fast, if the surface velocity for the blue material is faster or slower, respectively.

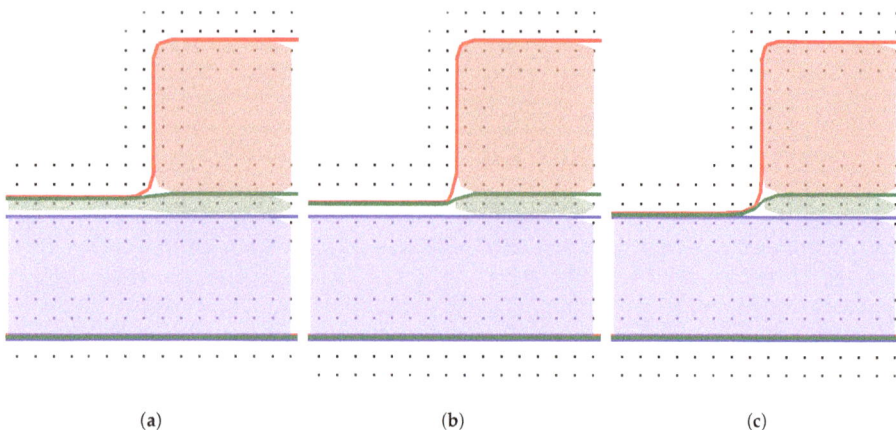

(a)   (b)   (c)

Figure 4. Exemplary development of material stack shown in Figure 3. (a) beginning of etching process; (b) green layer in sub-grid resolution; (c) green layer fully etched.

2.3. Material Interface-Aware Iterative Partitioning

The approach presented in [6] is used to obtain the surface flux rates. The main idea herein is to select a subset (sparse set) of the triangles present in the surface mesh, to minimize the number of integration points used in the flux calculation process. The sparse set is created starting by selecting randomly surface cells with a fixed minimum edge distance of d_{max}, i.e., the minimum number of edges to be crossed to connect two cells on the surface mesh. The iterative partitioning scheme to refine this initial sparse set is guided by two refinement conditions: a threshold for the difference of the normal angles of two neighboring sparse points and a threshold based on the flux difference between two neighboring sparse points. If the normal angle of two neighboring points exceeds the threshold value, the region in between is marked for further refinement. The obtained solution for the sparse set is then extrapolated constantly into the surrounding patch of each sparse point and smoothed using a diffusion approach.

The flux calculation algorithm using the sparse set is comprised of four major parts: (a) the initial partitioning of the surface with a maximum distance d_{max} between sparse points, (b) the calculation of the flux rates at the sparse points, (c) the refinement of the sparse set where new points are added according to the refinement conditions, and (d) the calculation of the flux rates for the recently added points. Steps (c) and (d) are executed iteratively until a minimum distance d_{min} between sparse points is reached. The flux calculations steps (b) and (d) are parallelized with OpenMP, whereas the initial partitioning (a) and the refinement (c) are implemented serially.

The refinement condition presented in [6] is used in the multi-material iterative partitioning scheme for the sparse surface evaluation using the maximal normal deviation v_{max}, the average flux difference u_{avg}, and the maximum flux difference u_{max}. A combination of thresholds for the normal deviation (t_{angle}) and the flux difference (t_{flux}) is used in all of the following results to model the refinement condition RC for a sparse surface location i:

$$RC(i) = \begin{cases} true, & \text{if } v_{max_i} > t_{angle}, \\ true, & \text{if } \frac{u_{avg_i} + u_{max_i}}{2} > t_{flux}, \\ false, & \text{otherwise.} \end{cases} \tag{3}$$

$d_{max} = 32$ is used in all simulations, which gives a total of six iterations, whereas the number of Jacobi iterations (for smoothing of the constant extrapolation) is fixed to $d_{max}/4 = 8$.

The refinement condition aims at capturing geometric features as well as high gradients in flux distribution. However, different simulation scenarios might require a tailored refinement condition to ensure proper refinement of the sparse set in relevant regions.

Since different materials experience potentially high differences in their surface velocities, it is important to ensure a material interface-aware partitioning. Therefore, we extend the scheme presented in [6] by identifying all cells on the top layer embedding a material interface and subsequently set $d_{\max} = 0$ for these surface cells. The effect of this material-interface-aware partitioning is shown in Figure 5, where all cells on the interface are present in the sparse set. Thus, we ensure that the sparse set contains a high amount of cells in the material interface regions.

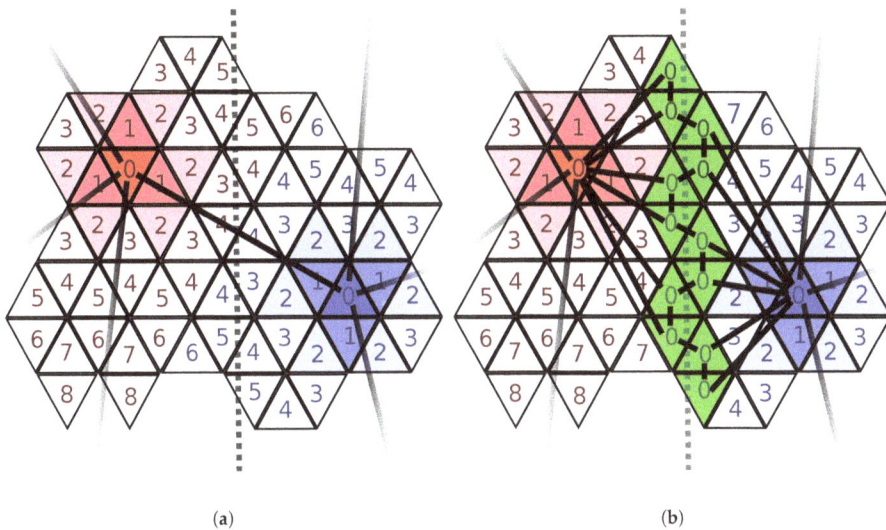

(a)   (b)

Figure 5. Example region of top layer across a material interface (dotted line). Cells of the sparse set are labeled with 0, while the cells forming the surrounding patch are labeled with their edge distance to the sparse set. Solid black lines denote the connection between the cells of the sparse set. (**a**) the region is partitioned into two patches (red and blue) using $d_{\max} = 8$, omitting material information; (**b**) material interface-aware partitioning using $d_{\max} = 0$ for cells embedding the material interface: the green cells, which are part of the sparse set, follow the material interface and thus allow for a proper representation.

2.4. Simulation Setup

From the different process steps in the Dual-Damascene process (see Figure 2), we focus solely on the simulation and evaluation of the etching step indicated in Figure 2g,h. Thus, this section describes the details of the simulation setup of this etching process step.

The used normal surface velocity is defined as

$$V_n(\mathbf{x}) = \alpha(M(\mathbf{x})) \cdot R(\mathbf{x}),$$
(4)

where α is denoting a material dependent scalar weighting factor and R is the direct flux from the ion source. Figure 6 depicts the simulation geometry used in this study and Table 1 lists the applied scalar weighting factors normalized to the weighting factors of the dielectric.

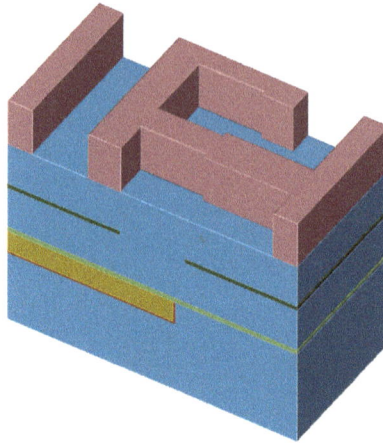

Figure 6. Simulation geometry used in our study (see Figure 2g). Material regions are encoded using colors according to Table 1.

Table 1. Normalized weighting factors corresponding to the etch rates for the different materials in our simulation.

	Material M	Weighting Factor α
	Hard Mask	0.01
	Dielectric	1.0
	Etch Stop	0.02
	Diffusion Barrier	0.02
	Copper	0.1
	Metal Seed	0.1

We set equal weighting factors for the etch stop and the diffusion barrier ($\alpha = 0.02$) as well as for copper and the metal seed layer ($\alpha = 0.1$). The weighting factor of the hard mask is set to $\alpha = 0.01$. These values were arbitrarily chosen to highlight the approach's ability to handle highly different surface reactions in a multi-material simulation.

A power cosine source with exponent $n = 1000$ [22] is used to model a highly vertically focused ion source, where only the direct flux is considered. In order to compute the direct flux rates on the surface elements, we use the method based on an icosahedron presented in [22], with the subdivision factor set to 5.

Each simulation is conducted until the simulation time $T = 2.0$, where the edge stop layer and the diffusion barrier are reached at $T \approx 0.5$ and $T \approx 1.0$, respectively. We investigate level-set resolutions ranging from 32 to 128 cells per unit length resulting in 140 to 560 timesteps (Table 2).

Table 2. Investigated level-set resolutions, corresponding initial domain resolutions, the number of triangles and vertices in the initial surface mesh, and the number of required timesteps until simulation time $T = 2.0$.

Cells per Unit Length	Vertical Cells	Horizontal Cells	Triangles	Vertices	Timesteps
32	80	96 × 96	28,642	56,384	140
64	160	192 × 192	113,986	226,176	280
128	320	384 × 384	455,298	907,008	560

The choice of the threshold values must be chosen according to the simulated process and which information will be extracted from the simulation results. To investigate their influence on the accuracy and performance of the flux calculation, we varied the parameters of the refinement conditions around their default values [6]. The threshold t_{angle} for the difference of the normal angles between two sparse points was set to three different values: 0.05, 0.1, and 0.2. For the flux difference, we set the threshold t_{flux} to 5, 10, and 20% of the global maximum flux calculated in each timestep.

To assess the influence of the sparse surface evaluation, we perform each simulation twice: once using the sparse surface evaluation (sparse approach) and once using a full evaluation (dense approach)—the latter represents the conventional reference approach.

We investigate the scalability for a fixed problem size (i.e., strong scalability) for both the dense and the sparse approach, with 1, 2, 4, 8, and 16 threads. The execution times of the major parts of the sparse flux calculation (a)–(d) are tracked individually.

A single node of the Vienna Scientific Cluster 3 (VSC-3) [34] is used for all simulations. This node has 64 GB main memory and is comprised of two Intel Xeon E5-2660v2 Sandy Bridge EP processors with 10 physical cores on each processor running at 2.20 GHz. Hence, a total of 20 physical and 40 logical cores are available.

3. Results and Discussion

3.1. Performance and Accuracy

The results at different times T during the etching simulation of the dielectric layer when using the sparse surface approach with threshold values $t_{flux} = 10\%$ and $t_{angle} = 0.1$ for the refinement condition (3) are shown in Figure 7. The left and right part of each image in Figure 7 shows the results for level-set resolution of 128 and 32 cells per unit length, respectively. Significant rounding of sharp geometrical features is visible for the lower resolution, which is expected due to the implicit level-set representation. Especially for thin material regions (e.g., the etch stop layer), the influence of the level-set resolution is noticeable.

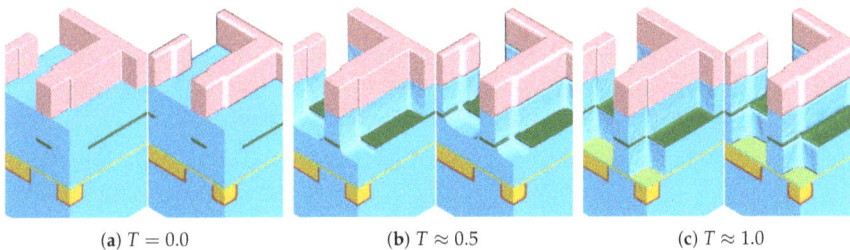

(a) $T = 0.0$ (b) $T \approx 0.5$ (c) $T \approx 1.0$

Figure 7. Resulting material regions of the multi-material etching simulation of the dielectric layer for resolution 128 (left half of each figure) and 32 (right half) at different times T.

The sparse sets (colored in red) for resolution 64 at identical times T as in Figure 7 are shown in Figure 8. Due to the design of the refinement condition (3), the density of the sparse cells increases towards sharp geometrical features as well as in regions of high flux deviation. Additionally, the material interfaces (d_{max} is set to 0 at interfaces) are captured within the sparse set. The ratio of the number of cells used in the dense approach to the number of active cells selected with the sparse approach is defined as

$$x_{rel} = \frac{n_{dense}}{n_{sparse}} \qquad (5)$$

and ranges from 5.5 to 7.4 in Figure 8.

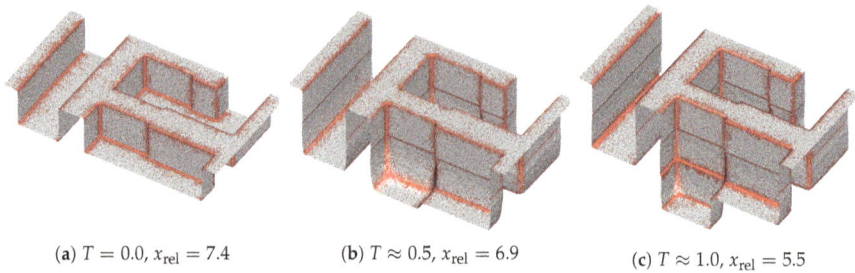

(a) $T = 0.0$, $x_{rel} = 7.4$ (b) $T \approx 0.5$, $x_{rel} = 6.9$ (c) $T \approx 1.0$, $x_{rel} = 5.5$

Figure 8. Active cells marked in red for $d_{max} = 32$ using resolution 64 at different timesteps.

The factor x_{rel} denotes the theoretical limit for the obtainable speedup S (6), which is the ratio of the flux calculation runtime of the dense approach over the flux calculation runtime of the sparse approach:

$$S = \frac{time_{dense}^{flux}}{time_{sparse}^{flux}}. \tag{6}$$

To judge the accuracy of the sparse approach, the differences in the surface positions between the sparse and the dense approach are calculated using a fixed timestep interval. A constant Courant–Friedrichs–Lewy (CFL) number of $\alpha_{CFL} = 0.4$ is used throughout the simulation [35]. For each surface point, the closest distance (surface deviation) between the two surface meshes is computed.

For the three different resolutions (32, 64, 128), the relative cell factors x_{rel}, the obtained speedups, and the maximum surface deviations are depicted in Figure 9. For a level-set resolution of 32, we obtain speedups between 1.9 and 2.5 (see Figure 9a), which is about 46% below the theoretical maximum of 4.6. The obtained speedup increases with increasing level-set resolution, which can be seen in Figure 9b,c: for a resolution of 64, we achieve speedups from 2.8 to 4.4, where x_{rel} reaches at most 7.7. With a resolution of 128, we obtain speedups ranging from 4.1 to 7.0 where x_{rel} peaks at 14.6. The observed gap between the actual speedup S and the theoretical limit (x_{rel}) stems from the computational overhead occurring in the sparse method, which is investigated in detail in Section 3.3. The total runtimes of the dense and the sparse simulations are shown in Table 3 for different resolutions.

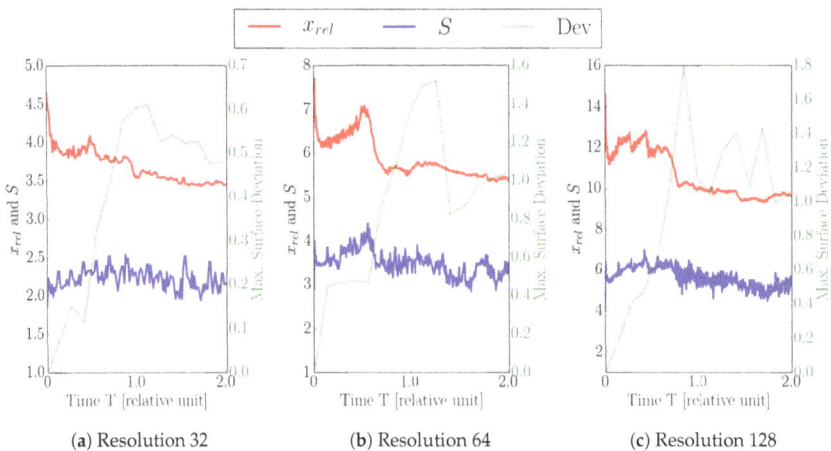

(a) Resolution 32   (b) Resolution 64   (c) Resolution 128

Figure 9. Ratio between dense and sparse number of cells x_{rel}, Speedup S, and maximum surface deviation (Dev) for level-set resolution of (a) 32, (b) 64, and (c) 128 using the default values for both refinement conditions ($t_{flux} = 10\%$ and $t_{angle} = 0.1$). The average values are depicted in Figure 10.

Table 3. Total simulation runtimes for different level-set resolutions using the dense and sparse approach.

Resolution	Runtime Dense [min]	Runtime Sparse [min]
32	22	12
64	159	60
128	1 209	311

Figure 9c shows that the resulting maximum surface deviations are below 1.8 grid cells for the highest resolution (i.e., 128), whereas the maximum deviations for resolution 32 and 64 stay below 0.7 and 1.6 grid cells, respectively. These maximum deviations occur in the vias, where for resolutions 32 and 64 the deviations are about 3% and 1.1% of the size of the via, respectively. For the highest resolution of 128 the maximum deviations correspond to 0.6% of the via size, so the relative maximum deviation decreases with increased resolution. The peak of the maximum surface deviation is for all resolutions around $T = 1.0$, where the etching process reaches the bottom of the via (diffusion barrier), and the high vertical surface velocities occurring in the dielectric are not present anymore. Therefore, in the following timesteps, the maximum surface deviations remain more or less static.

3.2. Impact of Variation of Threshold Values

The sensitivity of the results with respect to the choice of threshold values used in the refinement condition (3) is presented in the following. Figure 10 shows the average speedup and the average x_{rel} obtained for flux thresholds t_{flux} of 5, 10, and 20% and normal angle thresholds t_{angle} of 0.05, 0.1, and 0.2 using three different level-set resolutions (32, 64, and 128).

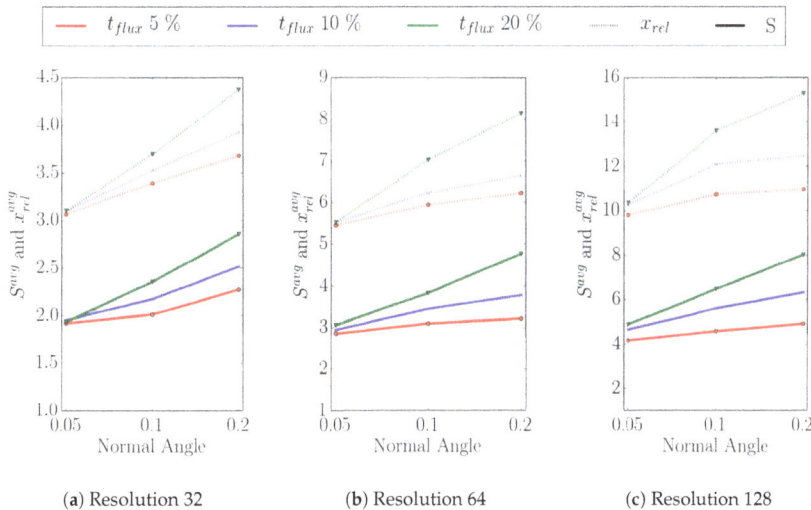

Figure 10. Average speedups S^{avg} (solid lines) and average ratios x_{rel}^{avg} (dotted lines) for three different flux (red, blue, green) and three different normal angle thresholds (x-axis) for level-set resolutions of (a) 32, (b) 64, and (c) 128.

For simulations with a level-set resolution of 32 (Figure 10a), the ratio x_{rel} increases with increasing normal angle as well as with increasing flux rate threshold, most prominently for $t_{flux} = 20\%$. The reason for this effect is that the condition if a point has to be added to the sparse set is based on either a violation of the flux rate threshold *or* the normal angle threshold. Therefore, the higher the flux or the normal angle threshold, the smaller the number of points in the sparse set. This also holds for the other two level-set resolutions depicted in Figure 10b,c. As stated before, the discrepancy between the ratios and the achieved speedup shows the computational overhead introduced with the iterative partitioning scheme (Section 3.3).

The influence of the threshold variations on the surface deviations is summarized in Table 4. We categorize the surface cell deviations in bins of size 0.1 ranging from 0.0 to 0.4 and above 0.4. The table contains the percentage of the respective bin population. For all resolutions, an increase in the thresholds leads to an increase in the deviations. The percentage of the surface with a deviation above 0.4 increases from the smallest ($t_{angle} = 0.05$, $t_{flux} = 5$) to the highest ($t_{angle} = 0.2$, $t_{flux} = 20$) threshold configurations, from 0.02 to 0.68% for resolution 32. For resolutions 64 and 128, the percentage of cells having a deviation above 0.4 increases from 0.28 to 3.8% and 0.27 to 5.62%, respectively.

Table 4. Distribution of surface deviations (normalized to the cell size) at $T \approx 1.0$ for different level-set resolutions, normal angle, and flux thresholds. The population of the bins is given in percent.

		Resolution 32								
	Angle		**0.05**			**0.1**			**0.2**	
	Flux	**5**	**10**	**20**	**5**	**10**	**20**	**5**	**10**	**20**
	0.0–0.1	98.62	98.05	98.45	94.40	93.66	89.67	91.12	86.24	81.33
	0.1–0.2	1.17	1.57	1.26	4.31	4.97	8.65	5.55	7.06	10.82
Deviation	0.2–0.3	0.15	0.22	0.18	1.10	1.27	1.49	2.37	3.37	4.84
	0.3–0.4	0.04	0.10	0.07	0.13	0.09	0.11	0.87	2.74	2.33
	>0.4	0.02	0.06	0.04	0.07	0.01	0.08	0.08	0.58	0.68
		Resolution 64								
	Angle		**0.05**			**0.1**			**0.2**	
	Flux	**5**	**10**	**20**	**5**	**10**	**20**	**5**	**10**	**20**
	0.0–0.1	96.42	92.91	92.55	96.10	89.69	85.26	95.73	88.42	80.42
	0.1–0.2	2.63	5.84	6.05	2.93	7.99	8.52	3.31	9.01	9.88
Deviation	0.2–0.3	0.46	0.71	0.80	0.51	1.67	3.14	0.51	1.88	3.76
	0.3–0.4	0.22	0.25	0.28	0.18	0.36	1.89	0.17	0.37	2.13
	>0.4	0.28	0.29	0.32	0.27	0.30	1.19	0.27	0.31	3.80
		Resolution 128								
	Angle		**0.05**			**0.1**			**0.2**	
	Flux	**5**	**10**	**20**	**5**	**10**	**20**	**5**	**10**	**20**
	0.0–0.1	88.11	84.26	84.26	86.08	80.20	76.90	85.46	78.55	73.65
	0.1–0.2	8.92	9.89	9.59	10.35	11.60	11.69	10.68	12.24	12.11
Deviation	0.2–0.3	2.19	3.66	3.54	2.56	4.56	4.92	2.70	5.16	5.77
	0.3–0.4	0.50	1.38	1.57	0.68	2.13	2.26	0.76	2.26	2.86
	>0.4	0.27	0.81	1.04	0.32	1.51	4.23	0.40	1.79	5.62

3.3. Parallel Scalability

For the scalability investigation of the dense and the sparse approach, we evaluated the flux calculation execution times for both using 1, 2, 4, 8, and 16 threads. The thresholds in the sparse approach were set to ($t_{angle} = 0.1$ and $t_{flux} = 10$).

Figure 11 depicts the average execution times and average speedups of the flux calculation using both, the dense (dotted lines) and the sparse (solid lines) approach. The average execution time using the sparse set for the flux calculation is always below the dense set's time, with the biggest difference occurring for a level-set resolution of 128. Here, the sparse set using one thread and 16 threads is about 7.5 and 3.4 times faster, respectively. The scaling for all resolutions is better with the dense approach where the difference to the sparse set speedup increases for higher resolutions. Using the dense approach with the highest level-set resolution of 128, the resulting speedup with 16 threads is 26% larger than the one from the sparse set. The differences in the execution times, especially for smaller thread counts, originate from the difference in the used number of surface cells for the flux calculation.

However, with increasing thread count, the gap between the execution times of the two approaches shrinks, but the sparse method still outperforms the reference approach for 16 threads. The scaling of the sparse set method is better for the lowest resolution of 32, but the sparse speedup for 16 threads is still 26% below the dense speedup. For resolutions 64 and 128, the sparse speedup is about 40% and 55% below the dense's, respectively.

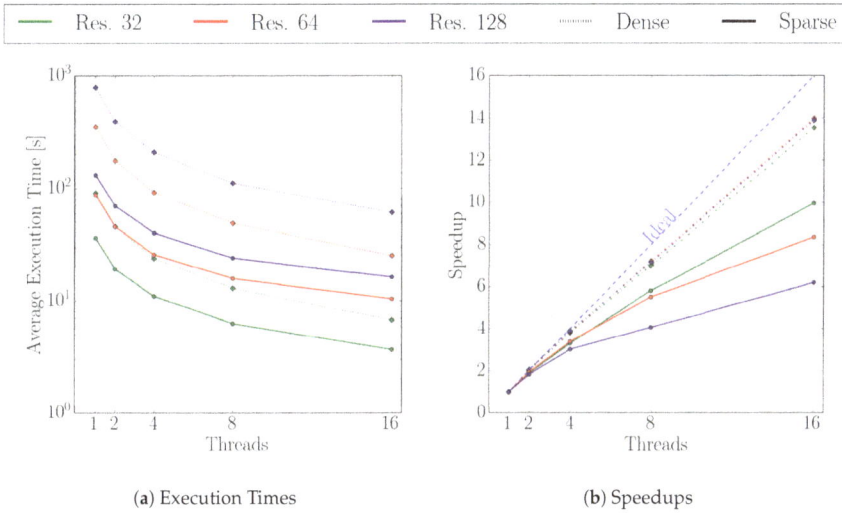

(**a**) Execution Times   (**b**) Speedups

Figure 11. Execution times and speedups at various level-set resolution for the dense and sparse flux calculation approach.

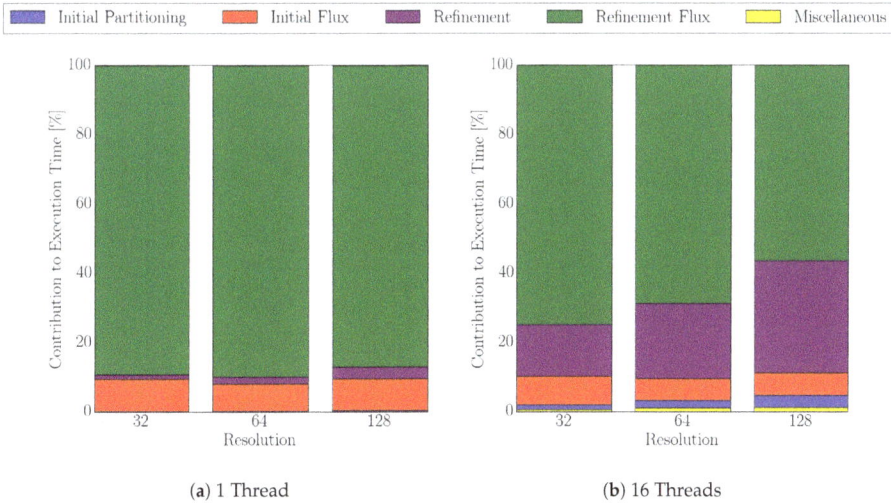

(**a**) 1 Thread   (**b**) 16 Threads

Figure 12. Contribution of the four major steps in the sparse flux calculation algorithm using 1 and 16 threads for different level-set resolutions: initial partitioning of the cells for the sparse set (blue), initial flux calculation (red), iterative refinement (purple), and the corresponding flux calculation (green). Miscellaneous (yellow) accounts for any other occurring overhead within the sparse set algorithm, e.g., computing the local flux deviations from the global maximum flux.

Figure 12 shows the different shares of the overall execution times for 1 to 16 threads for a single timestep. For the serial case (Figure 12a), the time spent for the refinement of the sparse set is only about 3%, whereas the flux calculation for the refined surface cells accounts for about 87%. This relation changes for 16 threads (Figure 12b) where the share for the refinement step increases nearly 10 times to about 32%, and the share of flux calculation for the refined cells decreases to about 56%. The data shown in Figure 12 shows that the serial refinement step for the creation of the sparse set is a bottleneck in our approach. Hence, the limited scalability compared to the dense set approach (cf. Figure 11). Although the scalability within the sparse set is limited, it still significantly outperforms the dense set approach in terms of execution time.

4. Conclusions

We have shown that the flux calculation for a multi-material etching simulation of a dielectric layer in the Dual-Damascene process can be accelerated using our recently developed multi-material interface-aware surface evaluation approach. For different surface resolutions and threshold values for the flux difference and the normal angle deviation in our refinement condition for the sparse set, we obtain speedups from 1.9 to 8.0. Our approach introduces minor surface deviations in the surface positions, where the maxima occur inside the vias and are below two grid cells for the highest resolution, corresponding to about 0.6% of the via size. For the lowest resolution, we obtain deviations below 0.7 grid cells which corresponds to 3% of the via size. We evaluated the scalability for a fixed problem size for up to 16 threads, revealing a nearly linear scaling for up to four threads for both the dense and the sparse set approach. Currently, the serial implementation of the iterative partitioning limits the speedup to roughly 50% of the theoretical limit, i.e., the ratio of the size of the dense and sparse set. However, our sparse approach still massively outperforms the conventional approach in terms of execution time. We plan to overcome this limitation by adopting the partitioning scheme to allow for a parallel implementation.

To increase the general applicability of the presented approach, we also aim to support reemitted flux. Either the already existing sparse set is reused and further refined where necessary or a dedicated sparse set is created in each subsequent reemission step. Additionally, it could be an advantage to adapt the refinement condition for the calculation of the reemitted flux distributions.

Author Contributions: Conceptualization, L.G., P.M., A.H., S.S. and J.W.; Data Curation, L.G. and P.M.; Formal Analysis, L.G. and P.M.; Funding Acquisition, J.W.; Investigation, L.G. and P.M.; Methodology, L.G., P.M., A.H., S.S. and J.W.; Project Administration, A.H. and J.W.; Resources, J.W.; Software, L.G. and P.M.; Supervision, S.S. and J.W.; Validation, L.G. and P.M.; Visualization, L.G. and P.M.; Writing—Original Draft, L.G.; Writing—Review and Editing, P.M., A.H., S.S. and J.W.

Acknowledgments: The financial support by the Austrian Federal Ministry for Digital and Economic Affairs and the National Foundation for Research, Technology and Development is gratefully acknowledged. The computational results presented have been achieved using the Vienna Scientific Cluster (VSC). The authors acknowledge the TU Wien University Library for financial support through its Open Access Funding Program.

Conflicts of Interest: The authors declare no conflict of interest.

References

1. Shaeri, M.R.; Jen, T.C.; Yuan, C.Y.; Behnia, M. Investigating Atomic Layer Deposition Characteristics in Multi-Outlet Viscous Flow Reactors Through Reactor Scale Simulations. *Int. J. Heat Mass Transf.* **2015**, *89*, 468–481. [CrossRef]
2. Ma, T.; Moroz, V.; Borges, R.; Sayed, K.E.; Asenov, P.; Asenov, A. Future Perspectives of TCAD in the Industry. In Proceedings of the International Conference on Simulation of Semiconductor Processes and Devices (SISPAD), Nuremberg, Germany, 6–8 September 2016; pp. 335–339.
3. Willsch, B.; Hauser, J.; Dreiner, S.; Goehlich, A.; Kappert, H.; Vogt, H. Analysis of Semiconductor Process Variations by Means of Hierarchical Median Polish. In Proceedings of the Austrochip Workshop in Microelectronics (Austrochip), Linz, Austria, 12 October 2017; pp. 1–5.

4. Huard, C.M.; Zhang, Y.; Sriraman, S.; Paterson, A.; Kanarik, K.J.; Kushner, M.J. Atomic Layer Etching of 3D Structures in Silicon: Self-Limiting and Nonideal Reactions. *J. Vac. Sci. Technol. A* **2017**, *35*, 031306-1–031306-15. [CrossRef]

5. Oh, Y.T.; Kim, K.B.; Shin, S.H.; Sim, H.; Van Toan, N.; Ono, T.; Song, Y.H. Impact of Etch Angles on Cell Characteristics in 3D NAND Flash Memory. *Microeletron. J.* **2018**, *79*, 1–6. [CrossRef]

6. Manstetten, P.; Gnam, L.; Hössinger, A.; Selberherr, S.; Weinbub, J. Sparse Surface Speed Evaluation on a Dynamic Three-Dimensional Surface Using an Iterative Partitioning Scheme. In *Lecture Notes in Computer Science, Volume 10860, Proceedings of the International Conference on Computational Science (ICCS)—Part I, Wuxi, China, 11–13 June 2018*; Shi, Y., Fu, H., Tian, Y., Krzhizhanovskaya, V., Lees, M., Dongarra, J., Sloot, P., Eds.; Springer: Cham, Switzerland, 2018; pp. 694–707.

7. Bailey, B.N. A Reverse Ray-Tracing Method for Modelling the Net Radiative Flux in Leaf-Resolving Plant Canopy Simulations. *Ecol. Model.* **2018**, *368*, 233–245. [CrossRef]

8. Sethian, J.A. *Level Set Methods and Fast Marching Methods: Evolving Interfaces in Computational Geometry, Fluid Mechanics, Computer Vision, and Materials Science*; Cambridge University Press: Cambridge, UK, 1999; Volume 3.

9. Kokkoris, G.; Brault, P.; Thomann, A.-L.; Caillard, A.; Samelor, D.; Boudouvis, A.; Vahlas, C. Ballistic and Molecular Dynamics Simulations of Aluminum Deposition in Micro-Trenches. *Thin Solid Films* **2013**, *536*, 115–123. [CrossRef]

10. Yanguas-Gil, A. *Growth and Transport in Nanostructured Materials: Reactive Transport in PVD, CVD, and ALD*; Springer: Cham, Switzerland, 2017.

11. Pagazani, J.; Martyl, F.; Babayan, A.; Hössinger, A.; Lissorgues, G.; Nejim, A. DRIE Process Modelling—A MEMS Case Study on a Real Design. In Proceedings of the Symposium on Design, Test, Integration and Packaging of MEMS/MOEMS (DTIP), Barcelona, Spain, 16–14 April 2013; pp. 1–5.

12. Victory Process—3D Process Simulator. Available online: https://www.silvaco.com/products/tcad/process_simulation/victory_process/victory_process.html (accessed on 25 July 2018).

13. Ertl, O.; Selberherr, S. A Fast Level Set Framework for Large Three-Dimensional Topography Simulations. *Comput. Phys. Commun.* **2009**, *180*, 1242–1250. [CrossRef]

14. Ertl, O.; Selberherr, S. Three-Dimensional Level Set Based Bosch Process Simulations Using Ray Tracing for Flux Calculation. *Microelectron. Eng.* **2009**, *87*, 20–29. [CrossRef]

15. Heitzinger, C.; Sheikholeslami, A.; Badrieh, F.; Puchner, H.; Selberherr, S. Feature-Scale Process Simulation and Accurate Capacitance Extraction for the Backend of a 100-nm Aluminum/TEOS Process. *IEEE Trans. Electron. Devices* **2004**, *51*, 1129–1134. [CrossRef]

16. Yu, J.C.; Zhou, Z.F.; Su, J.L.; Xia, C.F.; Zhang, X.W.; Wu, Z.Z.; Huang, Q.A. Three-Dimensional Simulation of DRIE Process Based on the Narrow Band Level Set and Monte Carlo Method. *Micromachines* **2018**, *9*, 74. [CrossRef]

17. Áfra, A.T.; Szirmay-Kalos, L. Stackless Multi-BVH Traversal for CPU, MIC and GPU Ray Tracing. *Comput. Graph. Forum* **2014**, *33*, 129–140. [CrossRef]

18. Vinkler, M.; Havran, V.; Bittner, J. Performance Comparison of Bounding Volume Hierarchies and Kd-Trees for GPU Ray Tracing. *Comput. Graph. Forum* **2016**, *35*, 68–79. [CrossRef]

19. Yu, S.; Wu, B.; Song, J.; Hao, L.; Zheng, X.; Shen, L. Bi-level Spatial Subdivision Based Monte Carlo Ray Tracing Directly Using CAD Models. *Fusion Eng. Des.* **2017**, *122*, 211–217. [CrossRef]

20. Naeimi, H.; Kowsary, F. Macro-Voxel Algorithm for Adaptive Grid Generation to Accelerate Grid Traversal in the Radiative Heat Transfer Analysis via Monte Carlo Method. *Int. Commun. Heat Mass Transf.* **2017**, *87*, 22–29. [CrossRef]

21. Aguerre, J.P.; Fernández, E. A Hierarchical Factorization Method for Efficient Radiosity Calculations. *Comput. Graph.* **2016**, *60*, 46–54. [CrossRef]

22. Manstetten, P.; Weinbub, J.; Hössinger, A.; Selberherr, S. Using Temporary Explicit Meshes for Direct Flux Calculation on Implicit Surfaces. *Procedia Comput. Sci.* **2017**, *108*, 245–254. [CrossRef]

23. Baklanov, M.R.; Adelmann, C.; Zhao, L.; De Gendt, S. Advanced Interconnects: Materials, Processing, and Reliability. *ECS J. Solid State Sci. Technol.* **2014**, *4*, Y1–Y4. [CrossRef]

24. Sharma, A.; Bulaga, J.; Agrawal, S.; Srivastava, R.; Gogna, M.; Singh, S.; Scott, S. Optimization of Wet Clean and its Cost Effectiveness in Dual Damascene 14 nm BEOL. In Proceedings of the IEEE Annual SEMI Advanced Semiconductor Manufacturing Conference (ASMC), Saratoga Spring, NY, USA, 30 April–3 May 2018; pp. 128–130.
25. Wolf, S. *Silicon Processing for the VLSI Era*; Lattice Press: Sunset Beach, CA, USA, 2002; pp. 671–710.
26. Nag, J.; Ray, S.; Kohli, K.K.; Simon, A.H.; Cohen, B.A.; Tijiwa-Birk, F.; Parks, C.J.; Krishnan, S.A. Non-Contact, Sub-Surface Detection of Alloy Segregation in Back-End of Line Copper Dual-Damascene Structures. *IEEE Trans. Semicond. Manuf.* **2015**, *28*, 469–473. [CrossRef]
27. Kriz, J.; Angelkort, C.; Czekalla, M.; Huth, S.; Meinhold, D.; Pohl, A.; Schulte, S.; Thamm, A.; Wallace, S. Overview of Dual Damascene Integration Schemes in Cu BEOL Integration. *Microelectron. Eng.* **2008**, *85*, 2128–2132. [CrossRef]
28. Seshan, K.; Schepis, D. *Handbook of Thin Film Deposition*, 4th ed.; Elsevier: Oxford, UK, 2018.
29. Museth, K. VDB: High-Resolution Sparse Volumes with Dynamic Topology. *ACM Trans. Graph.* **2013**, *32*, 27:1–27:22. [CrossRef]
30. OpenVDB. Available online: http://www.openvdb.org/ (accessed on 2 August 2018).
31. Wald, I.; Woop, S.; Benthin, C.; Johnson, G.S.; Ernst, M. Embree: A Kernel Framework for Efficient CPU Ray Tracing. *ACM Trans. Graph.* **2014**, *33*, 143:1–143:8. [CrossRef]
32. Ertl, O. *Numerical Methods for Topography Simulation. Doctoral Dissertation*; TU Wien: Vienna, Austria, 2010.
33. ViennaTS—The Vienna Topography Simulator. Available online: http://www.iue.tuwien.ac.at/software/viennats/ (accessed on 26 July 2018).
34. VSC—Vienna Scientific Cluster. Available online: http://vsc.ac.at (accessed on 25 October 2018).
35. Courant, R.; Friedrichs, K.; Lewy, H. Über die partiellen Differenzengleichungen der mathematischen Physik. *Math. Annal.* **1928**, *100*, 32–74. [CrossRef]

MDPI

St. Alban-Anlage 66

4052 Basel

Switzerland

Tel. +41 61 683 77 34

Fax +41 61 302 89 18

www.mdpi.com

Micromachines Editorial Office

E-mail: micromachines@mdpi.com

www.mdpi.com/journal/micromachines

www.ingramcontent.com/pod-product-compliance
Lightning Source LLC
Chambersburg PA
CBHW051850210326
41597CB00033B/5846